Telecommunications

J. BROWN

*Professor and Head of the
Department of Electrical Engineering
Imperial College, London*

and

E. V. D. GLAZIER

THIS EDITION
REVISED BY J. BROWN

CHAPMAN AND HALL
AND SCIENCE PAPERBACKS

First published 1964
by Chapman and Hall Ltd
11 New Fetter Lane, London EC4P 4EE
Reprinted 1966, 1969
First published in
Science Paperbacks 1966
Second edition 1974

Printed in Great Britain by
T & A Constable Ltd, Edinburgh

SBN 412 12210 3 (cased edition)
SBN 412 12220 0 (Science Paperback)

Distributed in the U.S.A.
by Halsted Press, a Division
of John Wiley & Sons, Inc, New York
Library of Congress Catalog Card Number 73-13386

Contents

Preface to the Second Edition

Since the first edition was published, many developments have occurred in telecommunications, particularly in relation to digital systems. Much of the fundamental theory remains, of course, unchanged, although once again there have been many significant advances. The changes which have been made in this edition therefore consist mainly of additions, particularly in respect of digital systems for which a new chapter (Chapter 7) has been included. A detailed account of the theoretical advances which have been made would greatly increase the length of the book and would be beyond the target initially set, that of providing an introductory account suitable for undergraduate students. It is hoped that the present level of treatment will also be appropriate for students preparing for the Council of Engineering Institutions Part II examination, the syllabus for Section A of Subject 347 Communication Engineering being fully covered.

A particular difficulty has arisen in relation to the last two chapters which were originally written by Dr. Glazier whose untimely death in 1971 was a severe loss to the engineering profession. His wide knowledge of communication and radio navigational systems provided an excellent basis for the material contained in these chapters. It has therefore been thought desirable to leave them substantially unchanged, apart from the addition of a new section at the end of each.

The original edition contained a list of references at the end of each chapter. These have now been replaced by a Bibliographical Note, which will, it is hoped, provide an introduction to the extensive literature available to those who wish to continue their studies of this fascinating subject.

The task of preparing a new edition has been greatly eased by the receipt of many comments both verbally and in writing. The individuals concerned are too numerous to mention but to all of them the author expresses his appreciation.

J. Brown

Preface to the First Edition

The term " Telecommunications " is generally accepted as including all forms of point-to-point communications by electrical or radio means and also all methods of radiolocation and radio navigation. The subject has expanded at a very rapid rate and has in recent years been thoroughly analysed so that the general principles involved are now clearly understood. An increasing amount of time is now given in University and Technical College Electrical Engineering courses to Telecommunications but there has not been any suitable text-book covering the basic principles of the subject at a level suitable for undergraduate students. This is partly due to the development of Telecommunications as a branch of electronics and the consequent lumping of the two subjects together for teaching purposes. The present book is an attempt to meet the need for a text on the foundations of Telecommunications in which the treatment is firmly based on the principles involved and not coloured by descriptions of the electronic circuitry required for its practical realisation. In such a treatment, the authors feel that the emphasis should be placed on the concepts of modulation, noise and information and on the general properties of four terminal networks and channels which pertain to any form of communication. It is of course clearly necessary to refer from time to time to practical systems, but where this has been done the detail included has been kept to the minimum required to illustrate the general principles.

It is assumed that the reader has a good knowledge of a.c. circuit theory but only an elementary knowledge of electromagnetic wave propagation and electronic circuitry is needed. It is considered that the Fourier Integral method of dealing with nonperiodic waveform is the most instructive approach to a study of telecommunication principles and the main text is largely concerned with providing a physical grasp of the essentials of this method. A more detailed mathematical treatment, including the relation to other methods of network analysis, is given in Appendix A. A certain amount of probability theory is necessary for dealing with noise problems and an introduction to this is given at the appropriate stage. It is hoped that this will suffice to enable the reader to appreciate the significance of the calculations pertaining to noise problems.

The book is designed to meet the needs of University and Technical College Students reading for a first Degree or a Diploma in Technology. It has not been aimed at any particular syllabus but it is hoped that it will provide the essential foundations for a number of examination requirements in the more specialist branches of telecommunications. At the same time, the authors hope that the practising engineer may find the book of value as a reference on basic problems.

The original conception of the book was due to Prof. H.E.M. Barlow of University College, London, and the late Mr. G.H. Parr of Chapman & Hall Ltd., and the authors would like to express appreciation to them, for the advice and encouragement which they have given during the course of preparation.

J. BROWN
E.V.D. GLAZIER

Chapter 1

Introduction

1.1. Communication

The ability to communicate, that is to share information, is one of the characteristics of the human race which has played a major part in its development. The extension of this ability to allow communication over distances arises from the possibilities offered by present-day telecommunication services. The prefix 'tele-' in telecommunications stems from the Greek for 'far off' and emphasizes the attention being given to communication between remote points. It is also a useful reminder to electrical engineers that the technology of telecommunications is but a small part of a very much larger subject, which could range from sciences such as linguistics covering the development of languages as a vehicle of communication, to the branch of civil engineering dealing with communications by transportation.

A complete study of telecommunications in its restricted sense would embrace the sociological considerations which ultimately determine the scale and form of the services required, physiological considerations which determine how communication equipment can best be matched to human observers, and economic considerations which influence the choice of a particular system from the available options. The object of this book is the limited one of presenting the physical principles which underlie the operation of all telecommunication services and to develop these in a way which emphasizes the common features of apparently different techniques. The principal linking factor will be a discussion of the restrictions on the volume of communication attainable by a given physical system.

In all the systems examined here, the information to be communicated is converted to electrical waveforms and our study begins in the next chapter with an examination of the nature of the waveform to be handled. Subsequent manipulation of these waveforms can be examined in general terms, usually by a mathematical representation of the operation being conducted, so that detailed reference to the equipment required is not necessary. However, an appreciation of the nature of these operations is helped by a knowledge of the physical processes which are occurring and reference will therefore sometimes be made to the circuitry which is used. It is assumed that the reader is already familiar with such basic electronic equipment as amplifiers, oscillators and detectors and with the solution of a.c. network problems.

There are some further practical considerations which play a major part in the choice of a communication system for a particular purpose and these will be referred to as necessary, since the aim of this book is to study the design and implementation of practical systems. However some preliminary comments on such considerations follow in this chapter.

1.2. Outline of a Typical Communication System

Well-known examples of communication services include telegraphy and telephony, radio and television. The essential features of such systems can be indicated by reference to the block diagram of Fig. 1.1 and an appreciation of the function of each block in this diagram is the unifying theme which links together the material in this book.

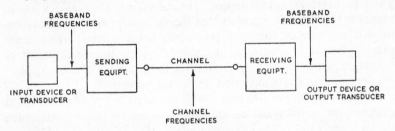

Fig. 1.1. Typical telecommunication system.

We begin by considering the terminals of the system, labelled input and output device or transducer. Almost without exception, the inputs and outputs of telecommunication systems are non-electrical, for example acoustic in telephony and radio, visual in television and alpha-numeric in telegraphy. The conversion of such inputs to electrical form is effected by a transducer, which is the general name given to any device which converts an input of one physical form to an output of a different physical form. A microphone, for example, delivers an electrical output which corresponds to the input sound wave. An ideal transducer is one in which this conversion is perfect in the sense that the output electrical waveform is an exact replica of the input waveform. Practical trans-ducers are not ideal and the performance of communication systems is consequently limited. However, for the purpose of analysis it is possible to represent a practical transducer as the combination of an ideal transducer and an electrical network introducing equivalent imperfec-tions. The departures from ideal behaviour can therefore be studied in exactly the same way as for the other electrical networks in the complete system. It is therefore unnecessary to devote special attention to trans-ducers. A further point of interest in this connection is that the operation of many transducers is often most easily studied by examining an electrical system with analogous behaviour.

The central section of the system in Fig. 1.1 is labelled 'channel'. The term channel covers the path by which communication is established between two distant points and may involve the use of radio waves in free space or guided waves in transmission lines or waveguides. Such paths may be used simultaneously for several communication links, but it is usual to define the term channel in respect of a single link. There is thus a difference between the physical path and the channel in that one path may be shared between several channels. Methods by which sharing can be achieved will be discussed later.

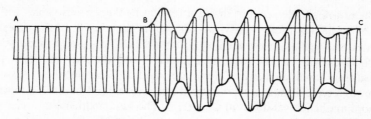

Fig. 1.2. Example of amplitude modulated wave as used in radio telephony.

The waveform at the output of the input transducer is not in general suitable for transmission by the channel. A simple example confirming this arises in radio: the output waveform from the microphone contains frequencies within the audio range, say up to 15 kHz whereas effective radio wave propagation requires the use of much higher frequencies. Part of the function of the block labelled 'sending equipment' is therefore to convert the frequencies associated with the transducer output, referred to as baseband frequencies, to those appropriate for the channel used, called the channel frequencies. In the simplest case, the sending equipment will include a transmitter generating a single frequency carrier wave and a modulator, which modulates (or changes) some property of the carrier wave to correspond to the baseband waveform. For example, in a telephone the amplitude of a carrier wave may be varied in sympathy with the baseband signal delivered by the microphone, thus providing an amplitude modulated wave. The resulting waveform, as a function of time, has the form indicated in Fig. 1.2. From A to B, the sinusoidal carrier is unmodulated, corresponding to zero baseband signal. Between B and C, the carrier amplitude, *i.e.* its peak-to-peak excursion, is proportional to the voltage corresponding to the speech waveform. The simplest forms of modulation are examined in Chapter 3.

A more sophisticated approach is to convert the baseband signal into a waveform of different kind, chosen to provide more efficient use of the available channel. The term coding is applied to such conversions. An example of coding arises in telegraphy where alpha-numeric characters are changed into a series of dots and dashes, as in Morse code. Coding possibilities of a more elaborate kind can be applied in any communication system and it will be shown later that coding is essential if the potential information capacity of a channel is to be fully exploited.

The receiving equipment provides the converse operations to those of the sending equipment. The incoming wave from the channel is demodulated and, if necessary, decoded to restore the baseband frequencies required to operate the output transducer.

1.3. Topics to be Discussed

The above simple account draws attention to the various topics which need to be discussed. Firstly, we will require knowledge of the types of baseband signals to be handled (Chapter 2). The simpler modulation processes will be discussed in Chapter 3 and the way in which the resulting modulated waveforms are modified during their passage through typical channels is discussed qualitatively in Chapter 4 and with more mathematical detail in Chapter 5. The basic mathematics relevant to Chapters 2-5 is Fourier analysis which enables waveforms to be examined either on a time basis or on a frequency basis. An account of the mathematical results required is given in Appendix A and readers unfamiliar with Fourier methods are recommended to study this appendix before embarking on Chapters 2-5.

Amongst the many unwanted waveforms which can arise in communications, by far the most important are those lumped together as 'noise'. This name originates from radio usage in that such waveforms when applied to a loudspeaker cause a noise background which may in extreme cases drown the desired speech or other audio output. Certain forms of noise are fundamental in the sense that they are governed by basic physical laws, so that certain minimum levels of noise are unavoidable in any physical system. Since such minimum levels exist, it follows that electrical waveforms used to transmit information must be of adequate strength to avoid being masked by noise. The properties of noise waveforms and their effect on communication are examined in Chapter 6.

A topic of very great importance in modern communication practice arises from the ability to handle 'continuous' signals such as arise in telephony in exactly the same way as the 'discrete' signals occurring in telegraphy. The term 'digital communications' covers this relatively new field and is the subject of Chapter 7. The discussion in this chapter of a particular form of digital communication—pulse-code modulation —provides a convenient introduction to the first half of Chapter 8. Here the amount of information which may be carried by a communication channel is examined, together with the theoretical limits of the capacity of such a channel in relation to the rate at which information may be transmitted. The remainder of Chapter 8 provides a brief introduction to what may be described as the central problem in communications, viz. the effective detection of a desired signal in the presence of a noise background.

The final two chapters survey the wide range of communication systems currently in use and draw attention to the relevance of the theory presented to the design of such systems.

1.4. Radio Waves

As mentioned in the discussion on Fig. 1.1, the channel may be provided by a transmission line or by radio means. Transmission lines of a variety of kinds are used and their influence on communication can be described in terms of the attenuation and phase shift which they cause. When radio waves are used, a number of other considerations arise and it is convenient to refer to some of these at this stage.

A radio wave is an example of an electromagnetic wave in which propagation occurs as a result of a continuous interchange of electric and magnetic energy. The most remarkable property of such waves is that the velocity of propagation in a vacuum is constant, independent of frequency, with a value of 3×10^8 m/s. The wavelength, λ, associated with a wave of frequency, f, is given by

$$c = f\lambda = 3 \times 10^8 \qquad (1.1)$$

when λ is measured in metres and f in Hertz.

The range of frequencies for which electromagnetic waves exist, is shown in Fig. 1.3 and is subdivided according to generally accepted classifications. The term spectrum is applied to indicate a range of frequencies and the radio spectrum covers frequencies from about 10^4 Hz to 10^{12} Hz. The corresponding range of wavelengths runs from 3×10^4 m to 3×10^{-4} m, *i.e.* from 30 km to 0·3 mm. The best known electromagnetic waves are those in the visible spectrum—light waves—and for most purposes may be regarded as travelling along straight line paths. The extent to which radio waves behave similarly to light waves depends primarily on the wavelength value relative to such dimensions as total path length and those of obstacles in the path. For the highest radio frequencies, the wavelength is much shorter than such dimensions and the similarity of behaviour to that of light is close. Such frequencies are therefore used for line-of-sight applications, *i.e.* when an unobstructed straight-line path exists between the two communication terminals. For longer wavelengths, the restriction to line-of-sight paths is less important since the waves enter those regions which would be in shadow

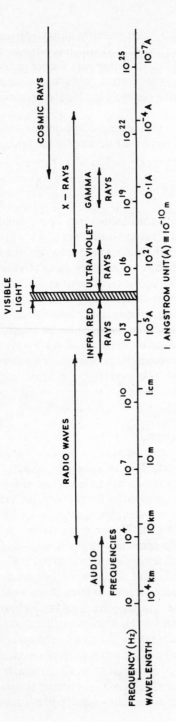

Fig. 1.3. The electromagnetic spectrum.

TABLE 1.1

Classification of Radio Waves

(According to British Standard 204: 1960, 'Glossary of Terms used in Telecommunication and Electronics')

Band No.	Frequency Range	Wavelength Range	Verbal Classifications
4	3 to 30 kHz	100 to 10 km	Very-low frequency (V.L.F.) or myriametric waves*
5	30 to 300 kHz	10 to 1 km	Low-frequency (L.F.) or kilometric waves
6	300 kHz to 3 MHz	1 km to 100 m	Medium frequency (M.F.) or hectometric waves
7	3 to 30 MHz	100 to 10 m	High frequency (H.F.) or decametric waves
8	30 to 300 MHz	10 to 1 m	Very-high frequency (V.H.F.) or metric waves
9	300 kHz to 3 GHz	1 m to 10 cm	Ultra-high frequency (U.H.F.) or decimetric waves
10	3 to 30 GHz	10 to 1 cm	Super-high frequency (S.H.F.) or centimetric waves
11	30 to 300 GHz	1 cm to 1 mm	Extra-high frequency (E.H.F.) or millimetric waves

*These verbal classifications apply to waves with frequencies less than 30 kHz.

if light waves were used. The mechanism which is involved is called diffraction. Diffraction plays a very important part in broadcasting by extending coverage to areas which lie in shadows of hills, buildings, etc.

A number of other factors modify radio wave propagation in ways which depend on the frequencies used. Amongst such factors are the absorption and reflecting properties of buildings, etc., absorption in the atmosphere and the absorption and reflecting properties of the ionosphere, which is the region from 100 to 1000 km in which ionised particles exist in sufficient numbers to interact with radio waves. The choice of frequency must therefore be very dependent on the nature of the service which is required.

A convenient classification of radio waves is shown in Table 1.1, based on a band number N covering the frequency range 0.3×10^N to 3×10^N Hz. The following contractions are used:

10^3 Hz = 1 kHz (kilohertz)
10^6 Hz = 1 MHz (megahertz)
10^9 Hz = 1 GHz (gigahertz)
10^{12} Hz = 1 THz (terahertz)

Table 1.2 provides a general indication of the uses which are made of these bands.

TABLE 1.2

Uses of Radio Bands

VLF	(3-30 kHz)	Very long distance communication
LF	(30-300 kHz)	Broadcasting
		Radio navigation
MF	(0.3-3 MHz)	Broadcasting
HF	(3-30 MHz)	Radio telephony
VHF	(30-300 MHz)	FM Broadcasting
		Television
		Mobile radio
		Radio navigation
UHF	(0.3-3 GHz)	Television
		Mobile radio
		Radio navigation
		Radar
SHF	(3-30 GHz)	Multi-channel telephony links
		Radar
		Satellite Communications
EHF	(30-300 GHz)	

Effective use of radio waves requires that antennas can be constructed to direct the waves in the required directions. An indication of the difficulty involved is given by the antenna size required to limit the radiation to a beam of prescribed angular size. The theory of diffraction provides the desired relation and shows that the angular beam width in the horizontal plane of an antenna of horizontal length, l, operating at wavelength λ is about 60 λ/l degrees. Limits on the practicable size of antennas thus restrict the use of radio waves particularly at lower frequencies. For example, a one degree beam requires an antenna length equal to 60 wavelengths, *i.e.* 18 km at 1 MHz but only 18 m at 1 GHz. It is evident that applications depending on the use of narrow beams must make use of the shorter wavelengths, *i.e.* the higher frequencies.

The attenuation of radio waves by the atmosphere is very dependent on frequency as shown by the curve in Fig. 1.4 from which it may be deduced that there is an upper limit of some 30 GHz at least for applications involving transmission over appreciable distances. The complex nature of the dependence of attenuation on frequency arises from the

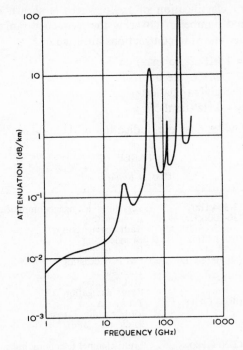

Fig. 1.4. The attenuation of radio waves by the atmosphere under normal conditions.

various resonances associated with the constituent gases of the atmosphere. An extra complication is that the attenuation is increased, particularly at the shortest wavelengths by rain, fog, etc.

The ionosphere, the part of the atmosphere for heights between about 100 and 1000 km, plays a very important part in relation to radio wave propagation. In this region the atmospheric gases are ionised and the free electrons are able to move under the action of the electric field associated with a radio wave. This electron motion causes the propagation of radio waves to be modified. In particular, for frequencies up to about 30 MHz, the ionosphere acts as a reflector and communication over very long distances is possible by using such frequencies. The effective height at which radio waves are reflected depends on frequency and the electron density. The electron density is influenced in a very complicated way by the effect of solar radiation and so radio wave propagation via the ionosphere particularly in the HF band, is dependent on the time of day and the season of the year. To maintain a reliable long-distance link using ionospheric reflection, it is necessary to use predicted information on the state of the ionosphere and to adjust the frequency at intervals of a few hours.

The choice of radio frequency for a particular communication requirement depends on a suitable compromise between the various factors mentioned. Strict international control of the frequency spectrum is needed to minimize interference between different services; frequency allocations between countries and for different services are agreed by international agencies.

1.5. Some Examples of Communication Services

The basic principles of telegraphy, telephony, sound broadcasting and television are sufficiently familiar to require no introduction. One point which deserves emphasis is that the effective development of such services has relied on an understanding of the physiology of hearing and vision. A simple example fundamental to television will serve to demonstrate this. Both television and cinematography are possible in their present forms because of the physiological phenomenon of persistence of vision. The mechanism by which we perceive visual images has a response time of about 1/25 s. A succession of still pictures occurring at a rate of 25 per second is thus interpreted by a human observer as if the scene were changing continuously. We are very far from a complete understanding of the physiology of hearing and vision and even further from comprehending the psychology of perception, *i.e.* in understanding the process by which 'meaning' is extracted from the response of the human nervous system to physical stimuli such as audio waves and light waves. Advances in our understanding of these subjects will inevitably have repercussions on the design of effective communication services.

One or two less well-known forms of communication will be referred to in illustrating the theoretical ideas to be presented in this volume and are introduced in the following sections.

1.5.1. TRANSMISSION OF STILL PICTURES—FACSIMILE

The basic principle employed in the transmission of still pictures (facsimile systems) can be understood by reference to Fig. 1.5. The picture to be transmitted is wrapped around a drum D on a lead screw S, and a small spot is illuminated by a light and lens L. The illuminated

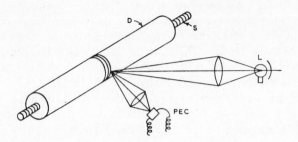

Fig. 1.5. Basic principle of operation of a facsimile transmitter.

spot is focused on to a device (PEC), known as a photo-electric cell.
As the drum rotates the reflected light from the spot fluctuates according
to the picture intensity. The current from the photocell fluctuates in
sympathy. There are many forms of photo-electric devices but for simplicity we shall consider here the selenium photo-conductive cell. As
its name implies it is made from the element selenium which has the
property that light falling upon it causes its resistance to change. If the
cell is connected in the simple circuit of Fig. 1.6 in series with a battery
and a mirror galvanometer, we have a rudimentary method of transmitting light fluctuations. When light is flashed on to the selenium cell,
its resistance falls and the current in the circuit increases. The deflection

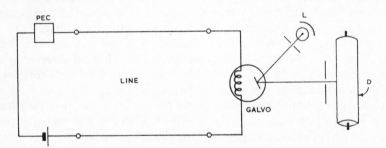

Fig. 1.6. Circuit arrangement for facsimile receiver.

of the mirror galvanometer will increase and the light reflected from its
mirror can be made to pass through a suitable slit and increase the light
falling on a screen. The maximum rate of light fluctuations that can
be transmitted in this way will probably be limited by the inertia of
the moving system of the galvanometer. It would be much more satisfactory if a lamp which was capable of responding rapidly enough could
be used. A metal filament lamp would not be satisfactory because it
could not heat and cool rapidly enough. Suitable gas discharge lamps

are available but they cannot be connected in the simple series arrangement of Fig. 1.6. However, by means of a suitable amplifier of the current fluctuations, the lamp may be flashed and caused to expose a photographic film. At the receiving end of the system a lamp is flashed in this way and is focused on to a spot on a photographic film wrapped round a drum similar to that at the sending end. The strength of the photocell current is dependent upon the intensity of the incident light, so that it is possible to transmit pictures having markings from white through various shades of grey to black.

The transmitting and receiving drums must be rotated in synchronism. In the earliest forms of instrument this was effected by clock mechanisms, but modern systems employ an electronic clock based on an electrically-maintained tuning fork or a vibrating quartz crystal. Synchronizing signals are transmitted to keep the receiving drum in the correct phase relative to the sending drum.

As the sending drum rotates it traverses along the lead screw S, so that the spot of light and the photo-electric cell record the light intensity of the picture along a helical spiral around the drum. If the lead screw is of pitch 50 threads per inch the picture is 'scanned' by 50 lines per inch. If the picture is opened out flat these scanning lines are as shown in Fig. 1.7. The picture transmitted and received is built up from these

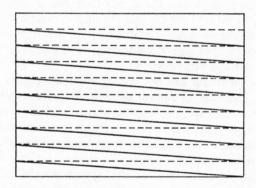

Fig. 1.7. Scanning pattern used in facsimile transmission.

lines which are slightly oblique but not enough so to be noticeable. The dotted lines indicate the path of the scanning spot at the join in the picture when it is on the drum. This method of reading the information in a picture is known as 'scanning'. The latter is a common term in facsimile and television. Ideally the lines should be wide enough so as just to touch one another. When the lines are scanned in sequence, the process is known as sequential scanning; other orders of line scanning may however be used.

We must now return to the receiving end of the simplified facsimile system. The fairly rapid fluctuations of electrical current representing the picture intensity variations along each scanning line are transmitted by line or radio and will probably be received in much weakened form at the receiver. A suitable amplifying device is capable of raising these variations to a level of voltage that will flash the gas discharge lamp. Thus a white spot on the original picture will produce a larger photo-cell current which will produce a larger received voltage and a more intense illumination of the receiving lamp. The photographic film will be more heavily exposed and when developed and printed a correspond-ing white spot will appear on the receiver picture.

Such systems of facsimile usually transmit a photograph of about 10 in. by 9 in. in a time of about 15 minutes. The definition of the received picture obviously improves as the number of scanning lines per inch increases. Modern practice usually employs between 50 and 100 lines per inch.

1.5.2. REMOTE CONTROL: TELECONTROL

This is a form of communication in which a signal has to be passed from a point A to a point B and at the latter point the received signal is amplified so as to control some process. Frequently there are a number of channels between A and B carrying independent signals. Often the point B is some form of vehicle and the signals control its motion. For example, in the remote control of an aircraft or missile from the ground, it is necessary to send signals controlling yaw, pitch and possibly roll as well as speed and other controls.

The requirements on any one telecontrol channel may vary from a simple but reliable on-off control signal to an infinitely variable signal of high accuracy or correspondence between output and input. The latter may call for special treatment of the signal before transmission, *i.e.* the signal or information may need to be transformed before it is suitable for transmission.

At the receiving end of a telecontrol system the electrical signal has often to be converted to some form of mechanical motion. This involves a closely related subject, namely, that of servo-mechanisms.

1.5.3. TELEMETRY

Telemetry is a system of communication in which a number of quan-tities have to be transmitted between two points. The frequency band-width and accuracy requirements on the channels of information may vary considerably. A common requirement is the remote presentation of the readings of a number of electrical meters, for example, those of electric power sub-stations. The need to send several readings over one pair of wires necessitates the use of multichannel techniques, such as those described in Chapter 4. If the readings are those pertaining to the performance of a vehicle such as an aircraft or missile, it is necessary to send the information over a radio link and again it is necessary to employ multichannel techniques.

The actual methods of signal handling and modulation vary very greatly so that it is not feasible to generalize on the techniques employed. Nevertheless, the treatment in the following chapters will be such that a combination of one of the types of signal and one of the types of modulation should cover many of the systems encountered in practice.

1.5.4. RADAR, DIRECTION FINDING AND RADIO NAVIGATION

The location of aircraft, ships and other objects by radio methods hardly falls within our original definition of communications. The problem here is actually to seek out information on the position of objects using radio transmitters and receivers.

In ordinary direction-finding the transmitter is located on a distant aircraft or ship, and a receiver with a special aerial system is employed to determine the bearing of the transmitter. Two such receiving stations are required to determine a fix on the earth's surface or in a geometric plane.

In radar the transmitter sends a signal which is reflected by the object of interest. The reflected signal is used to obtain information about the target object, such as its position and/or velocity. The time delay of the received echo is related to the range of the object. Angular position is determined by using directional aerials, as in direction finding.

In radio navigation an aircraft or ship can locate its position by receiving radio signals from two or more fixed transmitters or 'beacons'.

It is difficult to bring these systems within a generalized treatment of telecommunications, but much of the science of communications may be applied without difficulty.

Chapter 2

The Nature of the Signal

2.1. Introduction

The term 'signal' is often used loosely to cover various types of information sent over a communication channel. For example, we have:

(*a*) the sequence of letters forming a telegram,

(*b*) the sound pressure fluctuations forming the input to a telephone channel,

(*c*) the picture to be sent over a facsimile or television system,

(*d*) the quantity to be transmitted through telemetry or control channels.

It is preferable to use the term 'message' for the information before it has been modified and subjected to the errors and distortion of the channel, and to use the term 'signal' for the message after it has been modified for transmission. It is also more usual to regard the term 'signal' as referring to the electrical waveform which carries the information through the various stages in the channel.

The term baseband has been introduced in Section 1.2 and we shall use 'baseband signal' to refer to the electrical waveform which appears at the output of the transducer and at the input of our communication system. A general distinction can be drawn between continuous and discrete signals. By a continuous signal we mean one which can be

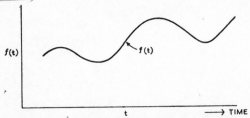

Fig. 2.1. Typical continuous signal waveform.

represented by a continuous function of time, $f(t)$, as shown graphically in Fig. 2.1. Continuous signals usually arise in situations where the electrical waveform is a replica of the input waveform to the transducer, *e.g.* the sound wave incident on a microphone. The term analogue signal determined by the line separation since the maximum detail that can be is therefore also used since the electrical waveform is analogous, or similar to the input waveform. Examples of analogue signals and of their essential characteristics are given in the next section.

Discrete signals arise in situations in which the message is a sequence of letters or numbers, so that each letter or number may be defined by a signal level drawn from a limited number of possibilities. Possible forms of discrete signal are examined in section 2.3.

A result of fundamental importance, namely that an analogue signal can be represented to any desired degree of accuracy by a discrete signal, is discussed in Section 2.5.

2.2. Analogue Signals

The suitability of any piece of communication equipment to handle continuous signals of the kind shown in Fig. 2.1 can be assessed in relation to two relatively simple quantities. The first is the range of values which $f(t)$ covers, *i.e.* the difference between its maximum and minimum values. The range can readily be changed by using an amplifier or an alternator and so does not require further attention at present. The second quantity needed is a measure of how rapidly the signal changes and here for the first time we see the importance of the time-frequency relation for waveforms. The behaviour of most communications equipment is specified in terms of frequency response and we therefore wish to relate rapidity of signal change to frequency.

In some cases, telephony and radio in particular, the relation arises directly since the frequency response of the human ear is well known from experiments and we can specify the desirable frequency content of audio signals from this knowledge. The upper frequency limit of audibility is generally regarded to be 15 kHz and this is taken as the requirement for high quality sound broadcasting such as is used in the VHF band. For many purposes lower quality is acceptable and 9 kHz is taken as the upper frequency limit for MF broadcasting. In telephony, the ultimate criterion is intelligibility of speech and experiments conducted over many years show that 3·4 kHz is an adequate upper frequency limit: an increase beyond this frequency does not significantly improve intelligibility but does naturally increase the cost of telephone services. Similar conclusions apply to the low-frequency limit and the values used range from 100 Hz in high quality broadcasting to 300 Hz in telephony.

A different approach must be used to estimate the bandwidth requirements for a television signal, based on a consideration of the picture definition which is required. Suppose that the number of lines in the picture is N: the picture definition in the vertical direction is therefore determined by the line separation since the maximum detail that can be included corresponds to black and white on alternate lines. Suppose an equivalent definition is desired in the horizontal direction so that a pattern such as that in Fig. 2.2(a) may be reproduced. The number of

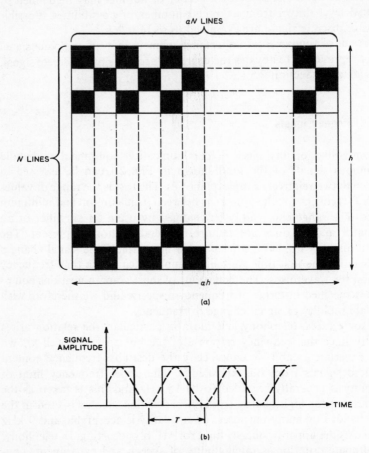

(a)

(b)

Fig. 2.2. (a) Pattern used to estimate television bandwidth requirements. (b) ——— waveform corresponding to (a) ...... acceptable approximation.

possible black and white elements per line is then aN, where a is the aspect ratio, *i.e.* the ratio of picture width to picture height. Ideally, the wave form required is shown in Fig. 2.2(b) by the solid curve but an acceptable approximation can be achieved by transmitting the dotted curve, a sine wave of period T.

The corresponding frequency, f, equal to $1/T$, is given by $N \times (aN/2) \times p$, where p is the number of pictures transmitted per second and f is the smallest possible value for the upper frequency limit. For example, in the UK

$$N = 625, p = 25, a = 4/3$$

hence

$$f = 6 \cdot 5 \text{ MHz}$$

An identical argument is applicable to facsimile, which has been described in Section 1.5.1. In this case, the frequency bandwidth available is usually fixed, e.g. that of a telephone line and the time of transmission is adjusted to provide the required definition. As an example suppose a picture of size 30×30 cm is to be transmitted over a telephone circuit with an upper frequency limit of 3 kHz, the definition to be 40 lines per cm. Comparison with the result for television shows that the time, T, must satisfy

$$\frac{30 \times 30 \times (40)^2}{2T} = 3 \times 10^3$$

i.e.

$$T = 240 \text{ s}$$

A final group of systems, including telemetry, telecontrol etc., may be considered together. In each case, the system performance will be specified in terms of some time constant, e.g. the time constant associated with a measuring instrument in the case of a telemetry system. A rough

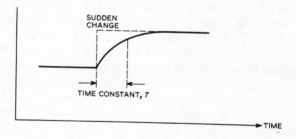

Fig. 2.3. Typical telemetry waveform showing effect of sudden change.

indication of the frequency band required may be obtained by an argument similar to that used for the television case. Suppose a rapid change occurs from a mean position. The resulting waveform will be determined by the time constant, T, and will take the form shown in Fig. 2.3. We therefore require that changes in the waveform arising in the communication link do not cause any significant increase in the time constant. Methods such as those developed in Chapter 5 show that a frequency response up to a value approximately equal to $1/T$ is required.

The results of this section are summarised in Table 2.1.

TABLE 2.1

Bandwidth Requirements for Baseband Signals

	Frequency Range	Comment
Telephony	0·3-3·4 kHz	Adequate for intelligibility of speech
FM Radio	up to 15 kHz	
Hi-fi audio	up to 20 kHz	Extends beyond normal limits of audibility
Monochrome TV		625 line 50 frame (interlaced)
Colour TV		625 line 50 frame (interlaced) (colour provided by modulated subcarrier within same bandwidth as monochrome)
Facsimile	$n^2ab/2T$	Picture size $a \times b$ (cm) Definition n lines/cm Transmission time T s
Data	approximately 0 to s	s is signalling speed in bauds (see equation 2.3)
Radar	approximately $1/T$	T pulse length related to range accuracy
Telemetry	approximately $1/T$	T shortest time constant

2.3. Discrete Signals

Discrete signals arise naturally in telegraph systems designed to transmit a specified number of characters or symbols. The simplest possibility is to allocate one level to each of the symbols to be transmitted, but it is clear that this places a severe requirement on the ability of the receiver to discriminate between adjacent levels. A much more satisfactory arrangement is to reduce the number of levels and to transmit a set of levels for each character. The transformation of the signal into this set of levels is called *coding* and a simple example is the Morse code by means of which the letters of the alphabet are converted into sets of dots and dashes. Coding is an example of a transformation designed to convert a signal into a form more suitable for transmission. The principle involved is a very general one and the many systems of modulation and coding used in radio communication can all be regarded as signal transformations designed to achieve a specific objective, such as a reduction of interference from noise or a decreased probability of errors occurring during transmission.

The discrete type of signal is sometimes called a digital signal, each element of the signal of duration τ seconds in Fig. 2.4 being known as a *digit*. If there are only two levels, as in one form of telegraphy, they are known as binary digits, or, for short, 'bits'. In this case a typical signal will appear as in Fig. 2.5(*a*), this being known as a single current signal and the form in (*b*) as a double current signal. The latter has advantages for transmission over long distances in that the mean power is a maximum for a given peak voltage or power, and the circuits and relays etc., can be designed so as to be symmetrical.

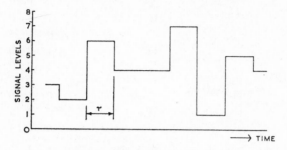

Fig. 2.4. Typical discrete signal waveform.

A three-level digit system that has been used in submarine cables is shown in Fig. 2.5(*c*). This is known as cable code and is used for sending

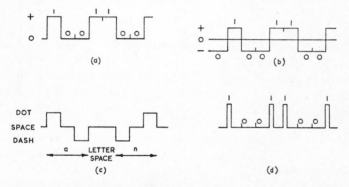

Fig. 2.5. Examples of signal coded in digital form: (a) single current binary; (b) double current binary; (c) three-level system; (d) non-contiguous single current binary.

Morse code over long cables. Positive polarity indicates a dot and negative polarity a dash; a zero level indicates a space as in ordinary Morse code. The binary digits shown in Fig. 2.5(*d*) are non-contiguous. This form is often used in computors because it simplifies synchronizing, but is not suitable for long-distance transmission because of the large bandwidth required, as will be seen in Section 2.4.4.

Figures	Letters	International Telegraph Alphabet no 2				
—	A	1	1	0	0	0
?	B	1	0	0	1	1
:	C	0	1	1	1	0
	D	1	0	0	1	0
3	E	1	0	0	0	0
%	F	1	0	1	1	0
@	G	0	1	0	1	1
£	H	0	0	1	0	1
8	I	0	1	1	0	0
BELL	J	1	1	0	1	0
(	K	1	1	1	1	0
)	L	0	1	0	0	1
.	M	0	0	1	1	1
,	N	0	0	1	1	0
9	O	0	0	0	1	1
0	P	0	1	1	0	1
1	Q	1	1	1	0	1
4	R	0	1	0	1	0
,	S	1	0	1	0	0
5	T	0	0	0	0	1
7	U	1	1	1	0	0
=	V	0	1	1	1	1
2	W	1	1	0	0	1
/	X	1	0	1	1	1
6	Y	1	0	1	0	1
+	Z	1	0	0	0	1
Carriage return		0	0	0	1	0
Line feed		0	1	0	0	0
Figure shift		1	1	0	1	1
Letter shift		1	1	1	1	1
Space		0	0	1	0	0
Unperforated tape		0	0	0	0	0
Order of digits		1	2	3	4	5

Fig. 2.6. The Murray 5-unit code, used in teleprinters.

Although Morse code has many desirable features, it has the disadvantage that different letters are represented by codes of different lengths. This was unimportant in the days of manual telegraphy but presents complications if automatic telegraph senders and receivers are used. A code specifically produced for automatic working, the Murray 5-unit code, shown in Fig. 2.6, is a two-level code of fixed length, each character being represented by five binary digits.

In digital communication systems some form of synchronization is necessary between the two ends of the channel. If accurate time signals were available at each end of the channel this would present no difficulty in principle, but in general some means of synchronization is necessary. One of the earliest forms of synchronization is that known as flywheel synchronization. It was executed in early telegraph systems by mechanical means but in recent times it is usually carried out electrically. In basic principle the digit rate is derived from the signal at the receiving end and pulses, or a sine wave at this rate, are passed through a narrow band filter or tuned circuit. During any cessation of signal the filter output continues to supply the digit rate, thus providing the flywheel effect.

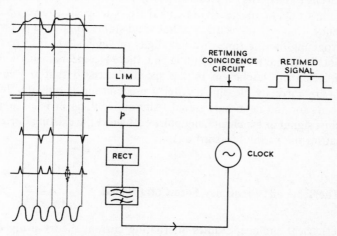

Fig. 2.7. Block diagram of circuit providing fly-wheel synchronization.

The block schematic diagram of the manner in which this is achieved is shown in Fig. 2.7, together with waveforms. The telegraph sequence is limited, differentiated and rectified. The derived sequence of pulses has a strong component at the frequency equal to the number of binary digits per second. (Where a pulse is missing we may imagine there to be equal and opposite pulses, so that we have a full train of pulses with some interfering negative pulses.) This strong component passes through a circuit which rejects all frequencies except this wanted component. This frequency then represents the speed of the device at the sending end which is generating the signals. It is usually called the clock frequency.

The telegraph signals may in fact have some inaccuracies in timing as a result of passing over a communication channel. That is, the changeover from 0 to 1 and from 1 to 0 may be displaced slightly backwards or forwards from the originally timed position. The filtered frequency from the filter can in fact be employed to retime or regenerate the signal. Apart, therefore, from errors which can occur due to noise in transmission (see Chapter 6) the original signal can be accurately reproduced such that it can again be re-transmitted.

The effect of an extremely narrow band filter may in fact be realized by a separate stable clock oscillator which is compared with, and adjusted to, the frequency from the narrow band filter.

The other basic form of synchronization is the so-called start-stop system. This is used in the teleprinter system of automatic telegraphy and it is the basis of the simplest form of television picture synchronizing. It assumes that at the receiving end there is a speed source or clock frequency which is sufficiently accurate and stable to be able to provide synchronizing for the time of a few digits or signals. On the receipt of a distinctive signal from the sending end, there is a mechanism or circuit which effectively latches in to the speed source so that the ensuing signals are correctly identified. After a predetermined time it unlatches and waits for the next start signal. Alternatively, as in television, the next start signal or synchronizing pulse cuts short the existing operation, and causes the process to start again.

2.4. The Time and Frequency Forms of a Signal

The electrical engineer studies alternating current theory using steady state sinusoidal voltages and currents. The whole of this theory can be adapted to communications if signals can be expressed in terms of such voltages and currents. When the radiated signal may be represented by a sequence of pulses at a fixed recurrence rate as in Fig. 2.8(a), the simple Fourier analysis of periodic waveforms can be used immediately to determine the sine waves. An example of such a waveform is the envelope of the signal radiated by a simple radar. Thus, following the terminology of Appendix A, for a periodic signal:

$$f(t) = a_0 + \sum_{n=1}^{\infty} a_n \cos(n\omega t) + \sum_{n=1}^{\infty} b_n \sin(n\omega t)$$

$$= a_0 + \sum_{n=1}^{\infty} c_n \cos(n\omega t + \phi_n)$$

$$= \sum_{n=-\infty}^{\infty} d_n \exp(jn\omega t) \tag{2.1}$$

where $\omega = 2\pi/T$ and T is the period.

The integrals which enable the determination of the various constants are given in Appendix A. We are only concerned with $f(t)$ real, so that the second form shows that the signal can also be represented by the superposition and summation of the d.c. or mean value a_0 and a number of cosine waves, starting with a fundamental frequency ω and containing extra harmonics 2ω, 3ω, etc., of amplitudes depending upon the con-

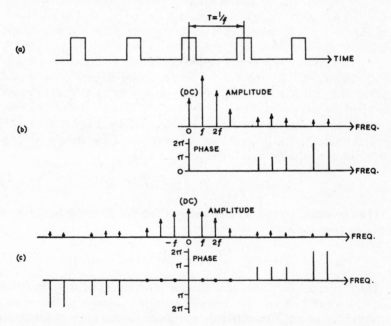

Fig. 2.8. Fourier representation of a continuous pulse train: (a) time waveform; (b) frequency representation in terms of cosine waves; (c) frequency representation in terms of exponentials.

stants. To complete the representation, the phases ϕ_n of the cosine waves at $t = 0$ must also be specified. An alternative way of expressing the signal is therefore to tabulate or display the amplitudes and phases of its components at frequencies 0, ω, 2ω, 3ω, etc. This might be done as in Fig. 2.8 (*b*), where frequency is plotted as abscissa and amplitude and phase as ordinates. This is known as a line spectrum and the two parts are the amplitude spectrum and the phase spectrum.

If the exponential form of Fourier representation is used the line spectrum is as in Fig. 2.8(*c*). It should be noted that there are now lines at negative frequencies, that the amplitude spectrum is even and the phase spectrum odd. The lines of each positive and negative frequency pair are half the amplitude of the corresponding lines in Fig. 2.8(*b*), simply because they must both add up to the amplitude of the actual frequency component. The conception of negative frequency is often useful. If the repetitive waveform is symmetrical about $t = 0$ only cosine terms are present so that the phase spectrum is zero or multiples of π. This happens in the case of the periodic waveform of Fig. 2.8(*a*) which might represent a radar waveform before it has passed through any networks.

Periodic waveforms do not, however, convey any information once they are known. The communications engineer is concerned with functions $f(t)$ which do not follow a regular pattern. One method of extending the theory is to take T as a very long period compared with the period of any low frequency component in the fluctuation of $f(t)$. A typical section of $f(t)$ over this period could then be assumed to repeat itself. This could then be Fourier analysed and the spectrum would consist of a vast number of closely spaced lines in the frequency range occupied by the signal. The extension of the Fourier Series Theorem to the Fourier Integral Theorem is of great usefulness and is dealt with in Appendix A*. The result is that the spectrum of a general waveform $f(t)$ is given by the Fourier Transform:

$$g(\omega) = \int_{-\infty}^{\infty} f(t) \,.\, \exp(-j\omega t) \,.\, \mathrm{d}t \qquad (2.2.\mathrm{A})$$

The waveform $f(t)$ in Fourier Integral form is given by the Inverse Fourier Transform

$$f(t) = \frac{1}{2\pi} \int_{-\infty}^{\infty} g(\omega) \,.\, \exp(j\omega t) \,.\, \mathrm{d}\omega \qquad (2.2.\mathrm{B})$$

The first equation means that the line between ω and $\omega + \mathrm{d}\omega$ has a strength given by $g(\omega) \,.\, \mathrm{d}\omega$. This is a complex quantity so that it also contains the phase. The amplitude and phase spectra are now functions rather than discrete lines and the term continuous spectrum is applied to them. The spectrum extends over positive and negative frequencies, the amplitude portion being even and the phase portion odd as before.

Once the spectrum is found it can, of course, be plotted for positive

* Appendix A contains a self-contained introduction to Fourier transforms and should be studied by the reader who has not previously met this concept. There is inevitably some repetition of material in the present chapter and, where appropriate, detailed proofs are given for the results quoted in this chapter.

frequencies by simply doubling the amplitudes on the positive side. However, there is much to be said for preserving the idea of positive and negative frequencies, as will be seen in Chapter 5. The application of the Fourier Integral Theorem to certain non-repetitive or transient waveforms will now be carried out.

2.4.1. FREQUENCY SPECTRUM OF A DIGITAL PULSE

In all the digital signals considered so far we can always identify the shortest element. In Figs. 2.4 and 2.5 it is the time duration of one digit. In Morse code it is the length of a dot or short space. In binary code of

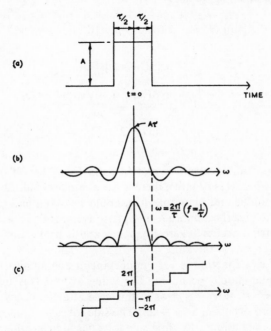

Fig. 2.9. Fourier transform of a single rectangular pulse:
(a) time waveform; (b) transform, $g(\omega)$; (c) magnitude
and phase of $g(\omega)$;

the contiguous type in Fig. 2.5(a) and (b) it is one element, whether it be mark or space. The length of the shortest element is an important quantity when considering the ability of a channel to pass a telegraph signal. An alternative way of specifying this shortest element τ (seconds) is to employ its reciprocal or the number of shortest elements per second. This is often called the signalling speed s. Thus:

$$s = \frac{1}{\tau} \text{ bauds} \qquad (2.3)$$

The unit is known as the baud after the telegraph engineer Baudot. The definition includes time in the same way as a speed is measured in knots.

The importance of this concept follows from the frequency spectrum of such a signal, which will now be determined. In Fig. 2.9(a) the single element $f(t)$ has a voltage or current amplitude A and extends from

$$t = -\frac{\tau}{2} \text{ to } \frac{\tau}{2}.$$

Applying the Fourier Transform, then, because $f(t)$ is zero outside the limits $-\tau/2$ and $\tau/2$, we have

$$g(\omega) = \int_{-\tau/2}^{\tau/2} A \exp(-j\omega t)\,dt$$

$$= \frac{A}{-j\omega} [\exp(-j\omega t)]_{-\tau/2}^{\tau/2}$$

$$= A \times \frac{\exp(j\omega\tau/2) - \exp(-j\omega\tau/2)}{j\omega}$$

$$= A\tau \times \frac{\sin \omega\tau/2}{\omega\tau/2} \qquad (2.4)$$

[handwritten margin notes:] $\cos \omega T_2 + j \sin \omega T_2$ $-\cos \omega T_2 + j \sin \omega \tau$ $T \; 2 j \sin \omega T_2$ $T \; j \omega_2$

Sometimes this spectrum is plotted as in Fig. 2.9(b) it being understood that where it is negative there is a π phase reversal. Strictly speaking, the modulus and phase of $g(\omega)$ should be taken and plotted as in Fig. 2.9(c) to give the amplitude and phase spectra. The maximum value of the spectrum occurs at zero frequency and is proportional to $A\tau$ the area or 'strength' of the pulse.

The $(\sin x)/x$ function is very important in communications and its general shape should be noted. The value at $x = 0$ is unity and the first zero occurs at $x = \pm\pi$ or in this case at a frequency of $f = 1/\tau$. The next zero occurs at $x = \pm 2\pi$. The side lobes are thus one-half the width of the main lobe. The first side lobe has a peak value of approximately $\frac{2}{3}\pi$, the exact value being derived in the Appendix. The spectrum extends to infinity and the maxima only decrease as $1/\omega$, *i.e.* rather slowly.

As a check we may take the inverse transform of equation (2.4). Then

$$f(t) = \frac{1}{2\pi} \int_{-\infty}^{\infty} A\tau \cdot \frac{\sin \omega\tau/2}{\omega\tau/2} (\cos \omega t + j \sin \omega t)\,d\omega$$

$$= \frac{1}{2\pi} \int_{-\infty}^{\infty} A\tau \cdot \frac{\sin \omega\tau/2}{\omega\tau/2} \cos \omega t\,d\omega \qquad (2.5)$$

for $f(t)$ real, showing that the waveform is built up from an infinite number of elementary cosine waveforms. These cancel outside the range $\pm\tau/2$ but within this range they add to produce the pulse.

2.4.2. FREQUENCY SPECTRUM OF A UNIT STEP.

The unit step of voltage or current is a well-known driving function in electrical theory because it enables the solution of problems involving the closing or opening of a perfect switch. The spectrum for the unit step will be given here by way of example. One of the conditions on $f(t)$ before the Fourier integral can be obtained is that

$$\int_{-\infty}^{\infty} f(t)\,dt$$

must converge.

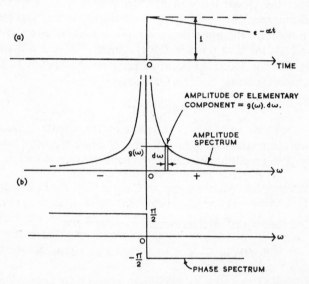

Fig. 2.10. Fourier transform of unit step: (a) unit step waveform; (b) amplitude and phase spectrum.

Consider a function of this type in Fig. 2.10(a) which may be represented in two ranges:

$$f(t) = 0 \quad \text{for} \quad t < 0$$
$$= \exp(-\alpha t) \quad \text{for} \quad t > 0$$

Then the Fourier Transform is obtained by integration between zero and infinity, because the integrand is zero for negative time, thus,

$$g(\omega) = \int_0^{\infty} \exp -(\alpha + j\omega)t\,dt$$

$$= -\left[\frac{\exp -(\alpha + j\omega)t}{\alpha + j\omega}\right]_0^{\infty}$$

$$= \frac{1}{\alpha + j\omega} \tag{2.6}$$

If now $\alpha \to 0$ we obtain the spectrum of a unit step as:

$$g(\omega) = \frac{1}{j\omega}$$

$$= \frac{1}{\omega} \exp(-j\pi/2) \tag{2.7}$$

This amplitude and phase spectrum is illustrated in Fig. 2.10(b).

2.4.3. FREQUENCY SPECTRUM OF A UNIT IMPULSE

Consider the digital pulse in Fig. 2.9(a) in which $A\tau = 1$, i.e. the pulse has unit area. If $\tau \to 0$ then $A \to \infty$, the product always remaining unity. In the limit the pulse is known as the unit impulse.* If $\tau \to 0$ in equation (2.4) the first zeros in the spectrum of Fig. 2.9(b) will move apart, so that the spectrum becomes progressively flatter. In the limit the spectrum becomes:

$$g(\omega) = 1 \tag{2.8}$$

i.e. the amplitude spectrum is constant out to infinite frequency and the phase spectrum is zero, i.e. all frequencies are present with equal intensity. The amplitude of the component between ω and $\omega + d\omega$ is simply:

$$1 \times d\omega$$

as may be seen by taking the inverse transform thus:

$$f(t) = \frac{1}{2\pi} \int_{-\infty}^{\infty} (\cos \omega t + j \sin \omega t) \, d\omega$$

$$= \int_{-\infty}^{\infty} \frac{\cos \omega t}{2\pi} \, d\omega \tag{2.9}$$

because the second term vanishes over positive and negative frequencies.

An alternative method of finding the spectrum of the unit impulse is to consider the differential with respect to time of the unit step. Now, from the inverse transform:

$$f'(t) = \frac{1}{2\pi} \int_{-\infty}^{\infty} g(\omega) j\omega \exp(j\omega t) \, d\omega \tag{2.10}$$

If $f(t)$ and $g(\omega)$ represent a unit step, by substituting from equation (2.7) we obtain:

$$f'(t) = \frac{1}{2\pi} \int_{-\infty}^{\infty} 1 \times \exp(j\omega t) \, d\omega \tag{2.11}$$

which shows that for a unit impulse the spectrum is given by unity.

* The properties of the unit impulse function are discussed in section A.3.4.

2.4.4. COMPARISON OF PULSE SPECTRA

The rectangular pulse shapes of Figs. 2.4, 2.5, 2.8 and 2.9 are useful for descriptive purposes and as signals prior to modification before transmission. However, they are unsuitable for transmission owing to the very wide frequency spectra which they occupy, as this would cause interference with other channels. To be suitable for transmission an information pulse or digit should have the narrowest frequency spectrum commensurate with the need to be able to detect the presence of the pulse at the receiver, or in the case of multi-level digits, the need to detect the pulse amplitude to a given order of accuracy. If pulses of narrower spectrum are desired they may have precursors and tails which will overlap into the region of adjacent pulses. There is no geometric form of pulse which is restricted to a finite time and which has a frequency spectrum restricted to a finite band, but there are pulse shapes restricted to a finite duration which have spectra which fall to values much lower than the side lobes of Fig. 2.9(c).

In approaching this topic the $(\sin x)/x$ pulse will be considered. This pulse is a theoretical concept in that it extends from minus to plus infinity in time, but it has a finite spectrum and it provides an important viewpoint in information theory. To avoid the difficult integration in determining the Fourier Transform of such a pulse we may take the inverse transform of a frequency spectrum which is uniform between frequencies of $\pm B/2$ and zero outside this range. Then, by an integration similar to equation (2.4) we obtain:

$$f(t) = \frac{1}{2\pi} \int_{\omega=-\pi B}^{\pi B} 1 \cdot \exp(i\omega t)\, d\omega$$

$$= B \cdot \frac{\sin \pi Bt}{\pi Bt} \tag{2.12}$$

The spectrum amplitude has been taken as unity between the limits and the phase spectrum as zero. This shows that such a spectrum is consistent with a $(\sin x)/x$ form of pulse. Such a pulse is shown in Fig. 2.11(b), and its spectrum is compared with the spectrum of a rectangular pulse which is repeated again in Fig. 2.11(a). This illustrates the reciprocal nature of pulses and their spectra. The first zero of the $(\sin x)/x$ pulse occurs at $t = 1/B$. If the pulse width between zeros is specified as $\tau = 2/B$ then the spectrum on the positive frequency side extends to a frequency $1/\tau$ as in the case of the rectangular pulse. The maximum amplitude of the pulse is proportional to the bandwidth of the spectrum.

The next pulse to be considered has the Gaussian shape defined by the function $\exp(-x^2)$. It is shown in Appendix A that the Fourier transform of the time waveform defined by $\exp(-2\pi t^2/\tau^2)$ is $(\tau/\sqrt{2})\exp(-\omega^2\tau^2/8\pi)$. The dependence of the transform on ω is the same apart from constants as the dependence of the time waveform on t. The time waveform has been expressed in a form which enables τ to be taken as a suitable measure of the pulse width. When t equals $\pm\tau/2$, the time

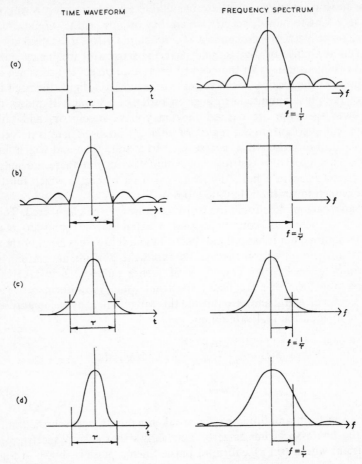

Fig. 2.11. Comparison of pulse amplitude and spectrum shapes: (a) rectangular pulse; (b) $(\sin x)/x$ pulse; (c) gaussian pulse; (d) cosine squared pulse.

function equals $\exp(-\pi/2)$ $[=0\cdot028]$ and this value is 13·6 dB* below the maximum value of unity which occurs for t equal to zero. Similarly, the frequency spectrum falls to 13·6 dB below its maximum if $\omega\tau = \pm2\pi$. The frequency corresponding to the 13·6 dB value is therefore $1/\tau$. Outside these limits both the time waveform and the frequency spectrum fall rapidly to zero as shown in Fig. 2.11(c). The pulse amplitude is 54dB

*The decibel notation is discussed in Section 4.3.

down on the maximum value, if t equals τ, and the spectrum has the same level below its maximum when the frequency is $2/\tau$.

Another pulse shape which has useful properties is the cosine-squared pulse, shown in Fig. 2.11(d). The time waveform consists of a half-cycle of a cosine wave, which is squared, and forms a pulse of length τ. The spectrum is derived in Appendix A and is shown to be the product of two factors, the first being of the form $(\sin x)/x$ with $x = \frac{1}{2}\omega\tau$, and the second being proportional to $1/(1 - f^2\tau^2)$. This second factor has two effects: it considerably reduces the amplitude of the side lobe levels and it removes the first zeros of the spectrum. This removal of zeros arises because the second factor is infinite when $f\tau = \pm 1$ i.e. when $x = \pm\pi$, these being the first zeros of $\sin x$.

The cosine-squared pulse may be expressed as:

$$f(t) = \cos^2 (\pi t/\tau) \text{ for } |t| < \tau/2 \tag{2.13}$$

and use of a standard trigonometrical identity gives the equivalent expression:

$$f(t) = \tfrac{1}{2} + \tfrac{1}{2} \cos (2\pi t/\tau) \text{ for } |t| < \tau/2 \tag{2.14}$$

Equation (2.14) shows that $f(t)$ is formed by raising a cosine wave by an amount which makes the minimum values of $f(t)$ equal to zero. For this reason the cosine-squared pulse is often referred to as a raised-cosine pulse.

The pulses which have been considered indicate a general property in respect of the rate at which the frequency spectrum decays as the frequency tends to infinity. We compare in Table 2.2 features of the time waveforms and the rate of decay (taken as the rate at which the amplitudes of side-lobe peaks decay).

TABLE 2.2

Pulse Waveform	Feature	Decay Rate
Impulse	Infinite discontinuity at $t = 0$ (i.e. discontinuity in integral)	zero (independent of ω)
Rectangular	Finite discontinuities	ω^{-1}
Cosine-squared (Raised-cosine)	Waveform continuous Finite discontinuities in first derivative	ω^{-2}
Gaussian	Waveform and all derivatives continuous	$\exp(-c\omega^2)$ (c-constant)

A study of Table 2.2 suggests that the spectrum decays as $\omega^{-(n-1)}$ for a waveform which is continuous, has continuous first $(n-1)$

derivatives but a discontinuity in the nth derivative. Analysis of the behaviour of the Fourier transform as ω tends to infinity confirms that this suggestion is correct. The Gaussian pulse conforms to this pattern in that $\exp(-c\omega^2)$ tend to zero more rapidly than any power of ω.

2.5. The Sampling Theorem

The possibility mentioned in Section 2.1, of replacing an analogue waveform by a discrete one, depends on the sampling theorem, which specifies the minimum number of samples per second by which the analogue waveform can be represented. An indication of the nature of the result follows by considering the Fourier series representation of an analogue signal of limited frequency bandwidth. Suppose that this signal contains frequency components in the range 0 to W Hz only. A sample of the signal of duration T may be represented by a Fourier series of fundamental frequency $1/T$ and this series contains $(2WT+1)$ real constants, i.e. the amplitudes of WT cosine components, WT sine components and the d.c. component. If T is very large, the number of constants is $2W$ per second duration of the waveform. This argument suggests that these constants may be calculated if the signal value is sampled at $2W$ instants per second and that the signal is thus completely defined by such sample values.

A formal proof of the theorem follows by considering the nature of the signal after sampling and by showing that the original waveform can be recovered.

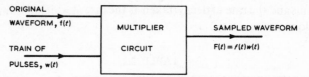

Fig. 2.12. Sampling process.

The sampling process can be carried out by multiplying the original waveform, $f(t)$ by a waveform $w(t)$ consisting of a series of short pulses at the sampling rate, f_s to produce a sampled waveform, $F(t)$ where

$$F(t) = f(t) \, w(t) \tag{2.15}$$

This process is indicated in Fig. 2.12.

The waveforms $f(t)$, $g(t)$ and $F(t)$ are shown in Fig. 2.13. The shape of the sampling pulse does not influence the final result.

We now consider the frequency spectra of the waveforms $f(t)$, $w(t)$ and $F(t)$, say $g(\omega)$, $v(\omega)$ and $G(\omega)$ respectively. Since $f(t)$ contains frequencies up to W Hz only, $g(\omega)$ will be zero if $|\omega|$ exceeds $2\pi W$. The waveform $w(t)$ is periodic and is represented by a Fourier series, the spectrum thus being of the discrete variety as shown earlier in Fig. 2.8. Such a discrete spectrum may be represented mathematically as an infinite sum of impulse functions (see section A.3.4);

$$v(\omega) = \sum_{n=-\infty}^{\infty} v_n \delta(\omega - 2\pi n f_s) \qquad (2.16)$$

The values of the constants v_n depend on the shape of the pulses used in the pulse train.

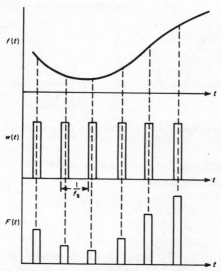

Fig. 2.13. Sampling process waveforms.

Since $F(t)$ is given by equation (2.15), its spectrum can be found by using the convolution theorem discussed in Section A.3.5, whence:

$$G(\omega) = \frac{1}{2\pi} \int_{-\infty}^{\infty} g(\omega - x) v(x) dx \qquad (2.17)$$

(see equation A.35; the integration variable has been changed from w in equation A.35 to x above).

When the expression for $v(\omega)$ is substituted in this integral, each term in the series can be integrated.

$$\int_{-\infty}^{\infty} g(\omega - x) \delta(x - 2\pi n f_s) dx = g(\omega - 2\pi n f_s)$$

and so

$$G(\omega) = \frac{1}{2\pi} \sum_{n=-\infty}^{\infty} v_n g(\omega - 2\pi n f_s) \qquad (2.18)$$

The spectrum of the sampled waveform therefore consists of a succession of repetitions of the spectrum of the original waveform

displaced along the frequency axis as shown in Fig. 2.14. This figure has
been drawn for the case in which the upper frequency limit, W, of the
waveform, $f(t)$, is less than $f_s/2$ and shows that no overlap occurs between
any of the repetitions which form $G(\omega)$. Finally, it follows from Fig. 2.14
that $g(\omega)$ and hence $f(t)$ is recovered if the waveform $F(t)$ is applied to a
filter which passes all frequencies up to W and rejects all frequencies
above $f_s - W$.

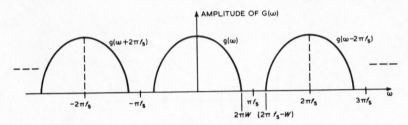

Fig. 2.14. Spectrum of sampled waveform: $g(\omega)$, the original waveform, is restricted
to the frequency range, 0 to W Hz.

It is therefore theoretically possible to recover the original waveform
from the sampled waveform provided

$$f_s - W \geqq W$$

i.e.   (2.19)

$$f_s \geqq 2W$$

The limiting value for f_s, equal to $2W$ is usually called the Nyquist
sampling rate, since the proof of this result was first provided by
H. Nyquist in 1924. In practice, operation at the Nyquist rate is not
possible since a filter with a sharp cut-off at W is required. Filters can
however be constructed with a sufficiently rapid transition from pass
band to stop band to allow satisfactory operation at sampling rates of
about 20 per cent above the Nyquist rate.

The action of the filter in reconstituting, $f(t)$, from the sampled waveform $F(t)$ will be examined a little further, for the case in which each
pulse is very short and may be approximated by an impulse. $F(t)$ can
thus be expressed as

$$F(t) = \Sigma f(n\tau)\delta(t - n\tau)$$ (2.20)

i.e. as a train of impulses, the strength of the impulse at the sampling
instant $n\tau$ being $f(n\tau)$.

The reconstituting filter is assumed to be ideal, *i.e.* it has a uniform
response up to the Nyquist frequency, $2/\tau$ and zero response for all
higher frequencies. The output from such a filter corresponding to a
unit impulse input is $\sin(\pi t/\tau)/(\pi t/\tau)$ and so the reconstituted waveform
is:

$$f_1(t) = \Sigma f(n\tau)\frac{\sin\left[\pi(t-n\tau)/\tau\right]}{\pi(t-n\tau)/\tau} \tag{2.21}$$

The (sin x)/x function has unit value when $x = 0$ and vanishes when x equals any multiple of π. It is now obvious that $f_1(t)$ equals $f(t)$ at each of the sampling instants, while the summation on the right of equation (2.21) is a standard mathematical form for interpolation from a series of equally spaced samples. Fig. 2.15 illustrates the reconstruction process.

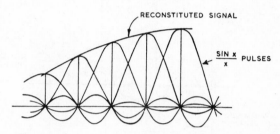

Fig. 2.15. Illustration of reconstruction of sampled signal.

The sampling theorem provides the possibility of replacing a continuous function by a series of values at discrete times. We shall find several examples of its importance in communication applications.

2.6. The Permissible Distortion in Received Signals

It is part of the communication engineer's task to ensure that the signal to be transmitted is delivered to the receiving end with the minimum of distortion. However, this usually becomes a problem of economics, because the cost may increase out of proportion as the distortion is reduced to small values.

In the case of digital signals the problem is often amenable to calculation. Consider, for example, the transmitted binary double current signal in Fig. 2.16(a). (The actual signal would be pre-filtered before transmission to prevent interference with other signals. See Chapter 4.) At the receiving end, after the influence of limited bandwidth and noise interference the signal might appear as in Fig. 2.16(b). Correct regeneration of the signal can be carried out so long as the envelope does not pass to the wrong side of the point where the selection instant crosses the zero axis. If the signal followed the dotted line in Fig. 2.16(b) an error would occur.

It is assumed that means always exist for deriving the best selection instants as described in the section on synchronizing. If noise is the main

cause of distortion of the envelope it is possible to calculate the probability of pulses being in error as will be shown in Chapter 6. The retimed
signal is shown in Fig. 2.16(*c*). Owing to the finite time for signal build-up
in the channel, the retimed signal is delayed according to the selection
instants.

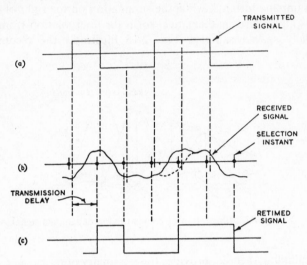

Fig. 2.16. Permissible distortion in binary signals.

For multi-level digital signals the problem is somewhat more involved
but calculation of error rates from signal to noise ratio is still feasible.
When selection or sampling instants are subject to some uncertainty,
as in the start-stop system, the problem is more complicated although a
first-order estimate can often be made.

With analogue signals, of the telemetering and telecontrol type, the
permissible distortion is usually an error in the value communicated at
any instant. This error is usually expressed as a percentage of the full
scale range. An error of 5 or 10 per cent would be a low accuracy system,
an error of 1 per cent would be medium accuracy and 0·1 per cent would
be high accuracy.

With communication systems whose output is perceived by human
senses such as telephony and television, the permissible distortion is
often a matter of opinion. This topic will be discussed further at the end
of Chapter 4.

Chapter 3

Amplitude and Angle Modulation

3.1. Introduction

The need for modulation in a communication system arises because the range of frequencies contained in the baseband signal to be transmitted is not in general the same as the range of frequencies which can be transmitted by the communication channel. This is most obvious for radio systems which must operate with frequencies of say 30 kHz and upwards, whereas the baseband signal will usually contain frequencies in the audio frequency range and upwards. This suggests that the modulation process is basically one of frequency translation. At the same time, however, there are often advantages to be gained by using a more complicated modulation process, in which changes other than a simple frequency translation are incorporated. These changes are designed to overcome the limitations on any system imposed by noise and distortion and the reasons for their use will be discussed later. In this chapter a number of the most useful types of modulation will be discussed and examples given of their applications.

A simple introduction to the possibilities of different forms of modulation may be given by considering the ways in which an r.f. signal may be used to transmit information. An r.f. oscillator delivers an output which can be expressed as

$$f_c(t) = A_c \cos (\omega_c t + \phi_c) \tag{3.1}$$

where the suffix c is attached to all the quantities to indicate that we are dealing with a *carrier wave*, this being the term normally applied to the basic unmodulated r.f. wave. The other symbols are defined as follows:

A_c is the carrier amplitude.

ω_c is the carrier angular frequency (this is expressed in terms of radians per second and is equal to 2π times the frequency in Hz).

ϕ_c is the phase angle of the carrier wave at $t = 0$.

If the three quantities A_c, ω_c, ϕ_c, which specify the carrier wave, remain unaltered, then no information is communicated. The communication of information requires that one or more of these quantities must vary with time in some way related to the base-band signal to be communicated. There are three obvious possibilities:

(a) The carrier amplitude is made to vary in sympathy with the baseband signal,

(b) The carrier frequency is made to vary in sympathy with the baseband signal,

(c) The carrier phase is made to vary in sympathy with the baseband signal.

These three possibilities are referred to as *amplitude modulation* (A.M.), *frequency modulation* (F.M.), and *phase modulation* (P.M.) respectively. The last two have many similarities. Both may be regarded as changing the angle $(\omega_c t + \phi_c)$; they are often taken together and called *angle modulation*.

Other types of modulation can be developed by starting with a different form of carrier signal in place of the simple continuous wave. One such carrier, which is often used, consists of a train of rectangular pulses of equal amplitudes and lengths, and occurring at a constant repetition frequency. Modulation can be applied in a number of ways. For example, by altering the amplitude of the pulse (pulse amplitude modulation, P.A.M.), altering the length or duration of the pulse (pulse duration modulation, P.D.M.), altering the repetition frequency (pulse frequency modulation, P.F.M.), or altering the time of occurrence of each pulse relative to a mean position (pulse position modulation P.P.M.). The combination of the train of pulses with a coding technique leads to *pulse code modulation* (P.C.M.), to be discussed in Chapter 7.

3.2. Amplitude Modulation

Suppose that the baseband signal which is to be used to modulate the carrier is given by the time function $f_m(t)$. It will be convenient to refer to this as the modulating signal and this is the reason for the choice of the suffix m. The simplest form of amplitude modulated wave should therefore be

$$f(t) = Kf_m(t) \cos (\omega_c t + \phi_c) \qquad (3.2)$$

A little reflection shows that this may lead to difficulties if the modulating signal can take both positive and negative values. The carrier amplitude in equation (3.1) is essentially a positive quantity and a negative value of $f_m(t)$ in the above equation would imply that the phase of the carrier wave is shifted by π radians. To avoid this difficulty we therefore add to equation (3.2) a carrier wave of constant amplitude and choose this amplitude to be sufficiently large so that it will always exceed the largest negative value of $Kf_m(t)$. The constant K is simply an amplitude factor to cover any effective amplification or attenuation in the modulating process. At the same time we can eliminate the phase ϕ_c by a slight change in the time origin. The change required will only be a fraction of the r.f. period and since the highest frequency in the modulating signal will in general be much less than the r.f. frequency, this change will have no appreciable effect on the modulating signal. We are therefore led to consider as our basic amplitude modulated wave the expression:

$$f(t) = [A_c + Kf_m(t)] \cos(\omega_c t) \tag{3.3}$$

in which

$$A_c \geqslant Kf_m(t) \quad \text{for all values of } t \tag{3.4}$$

It was shown in in the last chapter that any signal can be regarded as being made up from an assembly of continuous waves of different frequencies and this idea will be applied to the amplitude modulated carrier. The modulating signal can be expressed in terms of its frequency spectrum, $g_m(\omega)$ as:

$$f_m(t) = \frac{1}{2\pi} \int_{-\omega_1}^{\omega_1} g_m(\omega) \exp(j\omega t) \, d\omega \tag{3.5}$$

where the integration range is restricted to frequencies below ω_1, the largest significant frequency component in $f_m(t)$. A typical curve showing the form of the magnitude of $g_m(\omega)$ is shown in Fig. 3.1(a). If $\cos(\omega_c t)$ is expressed in terms of exponential functions, equation (3.3) can be written as:

$$f(t) = \tfrac{1}{2} \int_{-\omega_1}^{\omega_1} [2\pi A_c \delta(\omega) + K g_m(\omega)] \exp(\overline{j\omega + \omega_c}t) \, d\omega/2\pi$$

$$+ \tfrac{1}{2} \int_{-\omega_1}^{\omega_1} [2\pi A_c \delta(\omega) + K g_m(\omega)] \exp(\overline{j\omega - \omega_c}t) \, d\omega/2\pi \tag{3.6}$$

where the terms corresponding to the carrier A_c have been included under the integration by use of the Dirac delta function introduced in Appendix A.3. This equation can be transformed by slight changes in the integration variables to a form which shows the frequency spectrum of the modulated carrier. This form is:

$$f(t) = \tfrac{1}{2} \int_{\omega_c - \omega_1}^{\omega_c + \omega_1} [2\pi A_c \delta(\omega - \omega_c) + K g_m(\omega - \omega_c)] \exp(j\omega t) \, d\omega/2\pi$$

$$+ \tfrac{1}{2} \int_{-\omega_c - \omega_1}^{-\omega_c + \omega_1} [2\pi A_c \delta(\omega + \omega_c) + K g_m(\omega + \omega_c)] \exp(j\omega t) \, d\omega/2\pi \tag{3.7}$$

If the carrier frequency ω_c exceeds ω_1, then the first integral contains only positive frequency components and the second integral contains only negative frequency components. Further, the form of the spectrum

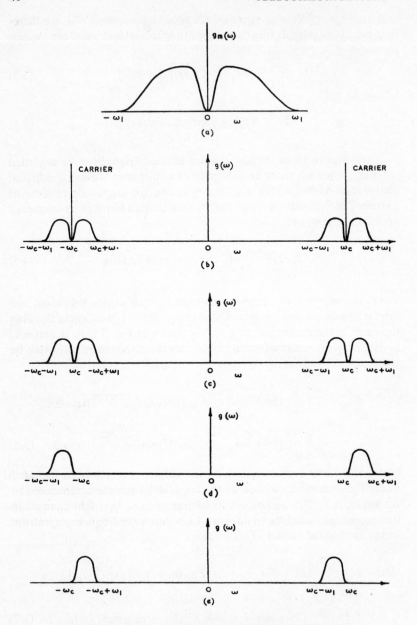

Fig. 3.1.(a) Frequency spectrum of modulation signal
(b) Frequency spectrum of amplitude modulated carrier (D.S.B.)
(c) Frequency spectrum of suppressed carrier modulation
(d) Frequency spectrum of single side band modulation (upper sideband)
(e) Frequency spectrum of single side band modulation (lower sideband)

is given by the integrands and can be used to give the curve in Fig. 3.1(b). This shows the frequency translation property of amplitude modulation: the spectrum of the modulated carrier is simply the spectrum of the modulating waveform displaced bodily by an amount ω_c both upwards and downwards. It may also be seen that the presence of positive and negative frequency components in the spectrum of the modulating signal leads to components both above and below the carrier frequency in the spectrum of the modulated carrier. This last result can be proved without the introduction of the idea of positive and negative frequencies by using the properties of the trigonometrical functions sine and cosine, but the derivation given here illustrates the Fourier Transform. The trigonometric approach will be given presently.

The spectrum of the modulated carrier has three components: a continuous wave at the carrier frequency, an *upper side band* with frequency components between ω_c and $\omega_c + \omega_1$, and a *lower sideband* with frequency components between $\omega_c - \omega_1$ and ω_c. The spectral distribution in the lower sideband is a mirror image with respect to the carrier frequency of the spectral distribution of the upper sideband. It is therefore clear that either sideband alone contains a replica of the modulating signal and this suggests that it is unnecessary to transmit more than one sideband. This is in fact true but practical difficulties arise in both generation and reception if only a single sideband is used. However, these difficulties can be overcome by suitable techniques which will be described in the next sections. There are thus a number of possible variations in the use of amplitude modulation. The case which we have so far examined, that illustrated in Fig. 3.1(b) is called *double sideband modulation* (DSB). Since the carrier is not essential in the transmission process, it can be eliminated, giving the spectrum shown in Fig. 3.1(c): this is referred to as suppressed carrier modulation. If only one sideband is transmitted, as in Figs. 3.1(d) and 3.1(e), the system is said to operate with *single sideband modulation* (SSB). The prefix, upper or lower, is used to indicate which sideband is being transmitted. An obvious advantage of single sideband operation is that the bandwidth required in the transmission medium is only one half of that required by a double sideband system.

The double sideband system has advantages in that it requires the minimum of modulation and detecting equipment and so we will consider its properties in some further detail. It will suffice to examine the behaviour when the modulating signal is a single frequency component so that:

$$f_m(t) = A_m \cos(\omega_m t) \qquad (3.8)$$

where A_m is the amplitude of the modulation and ω_m is its frequency.

For a complete discussion we should add an arbitrary phase angle in the expression for $f_m(t)$ but its presence does not alter any of our conclusions and so it will be omitted. The spectrum can be found either by direct substitution in the general result given by equation (3.7) or by working from the expression in equation (3.3). We begin by considering the latter. The modulated carrier can be written as:

$$f(t) = A_c[1 + m \cos(\omega_m t)] \cos(\omega_c t) \qquad (3.9)$$

in which KA_m has been replaced by mA_c.

The product of the two cosines can be expressed as a sum of two continuous waves by using the trigonometric relation:

$$\cos \alpha \times \cos \beta = \tfrac{1}{2} \cos(\alpha + \beta) + \tfrac{1}{2} \cos(\alpha - \beta) \qquad (3.10)$$

so that:

$$
\begin{aligned}
f(t) = A_c \cos(\omega_c t) + \tfrac{1}{2}A_c m \cos\overline{(\omega_c + \omega_m t)} \\
+ \tfrac{1}{2}A_c m \cos(\omega_c - \omega_m t)
\end{aligned}
\qquad (3.11)
$$

This shows a carrier of amplitude A_c, an upper sideband of frequency $(\omega_c + \omega_m)$ and amplitude $\tfrac{1}{2}mA_c$ and a lower sideband of frequency $(\omega_c - \omega_m)$ and amplitude $\tfrac{1}{2}mA_c$. The same result can be obtained from equation (3.7). The frequency spectrum corresponding to $f_m(t)$ is:

$$g_m(\omega) = \pi A_m \delta(\omega - \omega_m) + \pi A_m \delta(\omega + \omega_m) \qquad (3.12)$$

When this is substituted in equation (3.7) and the integrals evaluated we obtain:

$$
\begin{aligned}
f(t) = {} & \tfrac{1}{2}A_c[\exp(j\omega_c t) + \exp(-j\omega_c t)] \\
& + \tfrac{1}{4}KA_m[\exp(j\overline{\omega_c + \omega_m}\,t) + \exp(j\overline{\omega_c - \omega_m}\,t) \\
& + \exp(-j\overline{\omega_c + \omega_m}\,t) + \exp(-j\overline{\omega_c - \omega_m}\,t)]
\end{aligned}
\qquad (3.13)
$$

If the exponentials are combined to give trigonometrical functions and mA_c is substituted for KA_m, it is easily seen that the above expression is identical to that in equation (3.11).

The time function $f(t)$ will usually be a voltage or a current and in either case the power delivered to a resistive load by each single frequency component will be proportional to the square of its amplitude. The power involved in the amplitude modulated wave is thus made up of carrier power of relative amplitude A_c^2 and two sidebands, each of relative power $\tfrac{1}{4}m^2 A_c^2$. The ratio of the power contained in the sidebands to the total power is therefore $\tfrac{1}{2}m^2/(1 + \tfrac{1}{2}m^2)$ and depends only on the quantity m. This quantity m is called the modulation depth: it is usually expressed numerically as a percentage, and is then referred to as percentage modulation. We may note that the amplitude of either sideband relative to the carrier is $\tfrac{1}{2}m$ for the case of a single frequency modulating signal. Further, if we apply the restriction, expressed by equation (3.4) to this case we have:

$$A_c \geqslant KA_m = mA_c$$

i.e.
$$m \leqslant 1 \tag{3.14}$$

This means that if we are to avoid the complication of a 'negative amplitude' the modulation depth must be kept below unity, *i.e.* the percentage modulation must be below 100. If 100 per cent modulation is used, the total power in the two sidebands is half that of the carrier and therefore one-third of the total power. This points to a weakness of this form of modulation, namely that only a fraction of the total transmitted power is affected by the modulating signal.

The performance of systems using amplitude modulation can often be analysed very simply by an extension of the phasor method used for a.c. circuits. The carrier may be represented by a phasor of length A_c which rotates at the uniform angular velocity ω_c. The projection of this phasor upon a fixed line coincident with its position at $t = 0$ is the instantaneous value $A_c \cos(\omega_c t)$. In applying phasor methods, we normally assume that the phasor diagram is viewed by an observer also rotating at the angular frequency ω_c so that the phasor representing the carrier appears fixed in position. Such an observer will see the phasors corresponding to the sidebands rotating at the difference angular velocities. If the usual convention of positive angular velocity in the counter clockwise direction is retained, the upper sideband will appear to rotate in the counter clockwise direction with angular velocity ω_m and the lower sideband in the clockwise direction with angular velocity ω_m. This is illustrated in Fig. 3.2. The relative positions of the phasors are found by noting from equation (3.11) that all the components have positive maxima at time zero.

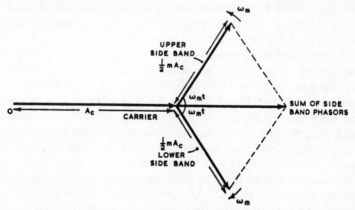

Fig. 3.2. Phasor diagram for an amplitude modulated carrier

The diagram in Fig. 3.2 is drawn for an arbitrary time instant and it is seen that the resultant of the two sideband phasors has an amplitude $2 \times (\frac{1}{2} mA_c \cos \omega_m t)$ and has the same direction as the carrier phasor. The instantaneous amplitude of the modulated carrier is therefore $A_c[1 + m \cos (\omega_m t)]$ as would be expected. The phasor diagram may be used to find the effects of changes in the phases of either the carrier or the modulating signal. In general an algebraic approach provides more effective solutions, but phasor diagrams are often helpful in interpreting such solutions.

3.3. The Generation and Detection of Amplitude Modulated Waves

The advantages and limitations of the various types of amplitude modulation can only be appreciated if the techniques available for modulation and detection are considered. We will therefore examine the principles of these techniques on a block diagram basis. We begin by considering the problem of the detection of the double sideband amplitude modulated wave defined by equation (3.3). The relation of this wave to the modulating signal $f_m(t)$ is shown in Fig. 3.3. Two situations are illustrated: in the first, Fig. 3.3(b), the condition $A_c \geqslant Kf_m(t)$ always holds while in the second, Fig. 3.3(c), this condition is violated for part of the time interval. In the first case the envelope of the amplitude modulated wave is a faithful reproduction of the modulating signal and this suggests that the modulating signal can be recovered by using a detector whose output at any time is directly proportional to the instantaneous amplitude of the modulated carrier. Such a detector can be made very easily by using a diode rectifier in conjunction with a resistor and a capacitor. For obvious reasons this combination is referred to as an envelope detector. The envelope of the waveform in Fig. 3.3(c) is shown in Fig. 3.3(d) and differs appreciably from the original modulating signal. If an envelope detector is used to extract information from the amplitude modulated signal in Fig. 3.3(c) it is clear that considerable distortion will occur. One possible method of overcoming this is to use a narrow band filter to extract the carrier signal, then to amplify this carrier to a level such that when it is recombined with the sidebands, the waveform will revert to the form shown in Fig. 3.3(b). Such a procedure requires a sharply tuned filter centred on the carrier frequency, an extra amplifier and suitable recombining circuits and so leads to a considerable increase in the complexity of the receiver. A similar procedure can in principle be applied to the detection of a suppressed carrier double sideband system. In this case the carrier must be generated in the receiver and then be combined with the sidebands. The frequency of this locally generated carrier must be identical with that used at the transmitter and its phase must be such as to preserve the correct relationship with the frequency

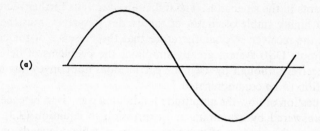

(a)

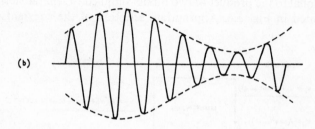

(b)

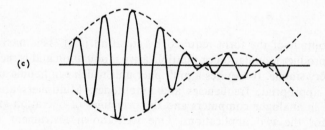

(c)

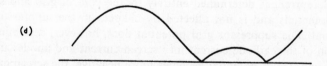

(d)

Fig. 3.3. Waveform for amplitude modulated waves.
(a) Modulating waveform
(b) Modulated carrier when *m* is less than unity
(c) Modulated carrier when *m* is greater than unity
(d) Envelope of (c)

components in the sidebands. The difficulties are thus further increased in that a highly stable oscillator at the carrier frequency must be provided in the receiver. We see therefore that the presence of the carrier in the simple DSB system greatly simplifies the problems of detection at the receiver although by itself the carrier does not convey any of the information to be communicated.

The question of how the amplitude modulated wave is to be generated can be answered by examining the expression in equation (3.3). This expression is the product of two factors, one of which depends only on the modulating signal, while the other is the unmodulated carrier. One possible modulator is therefore a device which produces an output proportional to the product of two inputs supplied to separate terminals as indicated in Fig. 3.4. Appropriate selection of the constants leads

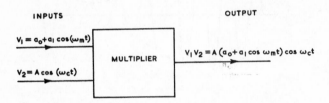

Fig. 3.4. Principle of operation of product modulator.

to an output of the form required by equation (3.3). This method is called product modulation or multiplicative modulation and is theoretically very simple. It requires a multiplier unit which can handle signals of the appropriate frequencies and amplitudes. Multipliers are also needed in analogue computers and the principles of operation are the same for the two applications. One well-known instrument which operates as a multiplier is the dynamometer wattmeter but the frequency range is much too restricted for application to communications. A more promising alternative is the pentode valve. In an ideal pentode, the cathode current is determined entirely by the control grid and screen grid potentials and is not affected by changes in the suppressor grid potential. The suppressor grid potential does, however, determine the division of the cathode current into screen current and anode current. As the suppressor potential becomes more negative, the screen current rises and the anode current falls, until eventually all the cathode current flows via the screen grid. The characteristics of such an ideal valve are shown in Fig. 3.5: the dependence of the cathode current (i_c) on the control grid voltage (v_g) is shown in Fig. 3.5(a) on the assumption that the screen potential is constant. The effect of the suppressor potential

(v_s) in splitting the cathode current into screen current (i_s) and anode current (i_a) is illustrated by Fig. 3.5(b), it being assumed that the anode potential is constant. If in the absence of input signals, the control grid and the suppressor grid are biassed to the mid-points of their control

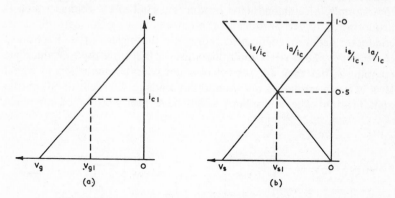

Fig. 3.5. Characteristics of ideal pentode (anode and screen voltages constant) V_g : control grid voltage; V_s : suppressor grid voltage; i_c : cathode current; i_a : anode current; i_s : screen current.

ranges, $i.e.$ to v_{g1} and v_{s1} respectively, then the cathode current will be i_{c1} and the anode current will be $\frac{1}{2}i_{c1}$. Suppose now that a carrier voltage $E_c \cos(\omega_c t)$ is also applied to the control grid. The cathode current will now be:

$$i_c = i_{c1} + g_m E_c \cos(\omega_c t) \tag{3.15}$$

where g_m is the mutual conductance defined by the slope of the cathode current-grid voltage characteristics. If, further, the modulating signal $f_m(t)$ is applied to the suppressor grid, the anode current will be given by the expression:

$$i_a = [\tfrac{1}{2} + kf_m(t)]i_c \tag{3.16}$$

where k is a constant defined by the slope of the anode current-suppressor voltage characteristic. Combining equations (3.15) and (3.16) gives:

$$i_a = [\tfrac{1}{2} + kf_m(t)][i_{c1} + g_m E_c \cos(\omega_c t)] \tag{3.17}$$

The part of the right-hand side of this equation which involves i_{c1} contains only the frequency components in the modulating signal. If, therefore, the current is filtered by a network which passes only frequencies in the vicinity of the carrier frequency, we are left with

$$i_a = [\tfrac{1}{2} + kf_m(t)]g_m E_c \cos(\omega_c t) \tag{3.18}$$

which is identical in form to the basic amplitude modulated wave defined by equation (3.3). An ideal pentode can thus act as a product modulator. In practice pentodes depart quite considerably from the ideal characteristics postulated in Fig. 3.5, the principal difference being that the anode current-suppressor grid potential characteristic is not sufficiently linear. There are other devices which can act as multipliers, for example, a semiconductor produces a Hall effect voltage which is proportional to the product of the current flowing through the semiconductor and a magnetic field acting at right angles to the direction of current flow. Hall effect multipliers are in fact now used in analogue computers, but they are subject to a considerable reduction in signal level between input and output and this leads to the need for additional amplification. The multipliers which have been mentioned are ideal

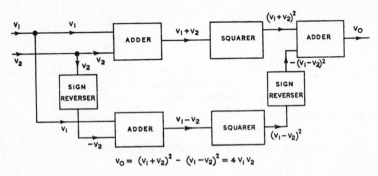

$$v_O = (v_1+v_2)^2 - (v_1-v_2)^2 = 4\,v_1 v_2$$

Fig. 3.6. Multiplier circuit using non-linear devices as squaring elements

devices in that they do not rely on any non-linear behaviour. Until the advent of the Hall multiplier, analogue computers relied on the use of non-linear elements to give a product indirectly by the arrangement sketched in Fig. 3.6. The key components in this circuit are the squaring elements, which give an output which is proportional to the square of the input. A similar technique can be used for modulation, but now that the need for a non-linear device is seen an obvious question is whether the complexity of the arrangement in Fig. 3.6 is necessary. Is there a simpler possibility, using only one non-linear element? Valves and transistors are inherently non-linear in their behaviour, a typical anode current-grid voltage characteristic for a valve being as shown in Fig. 3.7. This characteristic may be expressed algebraically by writing the anode current in the form of a series of powers of the grid voltage:

$$i_a = i_{a0} + b_1 v_g + b_2 v_g{}^2 + b_3 v_g{}^3 \qquad (3.19)$$

where i_{a0} is the quiescent anode current at the working point selected

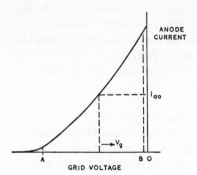

Fig. 3.7. Anode current-grid voltage characteristic of a valve (anode voltage constant).

(see Fig. 3.7) and v_g is the displacement of the grid voltage from the bias voltage defining the working point. The number of terms used in the series determines the accuracy with which the expression in equation (3.19) satisfies the actual characteristic. Over a limited range of grid voltages such as the range A to B in Fig. 3.7, the first few terms suffice. In the calculations which follow we will retain terms up to the cubic. Suppose now that the grid voltage v_g is the sum of a carrier voltage $E_c \cos(\omega_c t)$ and a modulating waveform $f_m(t)$, i.e.:

$$v_g = E_c \cos(\omega_c t) + f_m(t) \tag{3.20}$$

Then:

$$
\begin{aligned}
i_a = \; & i_{a0} + b_1[E_c \cos \omega_c t + f_m(t)] + b_2[E_c \cos \omega_c t + f_m(t)]^2 \\
& + b_3[E_c \cos \omega_c t + f_m(t)]^3 \\
= \; & i_{a0} + \underline{b_1 E_c \cos \omega_c t} + b_1 f_m(t) + b_2 E_c^2 \cos^2 \omega_c t + \underline{2b_2 f_m(t) E_c \cos \omega_c t} \\
& + b_2 f_m^2(t) + \underline{b_3 E_c^3 \cos^3 \omega_c t} + 3b_3 E_c^2 f_m(t) \cos^2 \omega_c t \\
& + \underline{3b_3 E_c f_m^2(t) \cos \omega_c t} + b_2 f_m^3(t) \tag{3.21}
\end{aligned}
$$

The part of this expression which is of interest is the one with frequency components in the vicinity of the carrier frequency. This can be extracted physically by passing the current through a filter with a response centred on the carrier frequency and a bandwidth great enough to accommodate the sideband frequencies. The terms involved can be selected from equation (3.21) by remembering that $f_m(t)$ is a slowly varying function of time and by replacing powers of $\cos(\omega_c t)$ by their expansions in terms of $\cos(\omega_c t)$, $\cos(2\omega_c t)$, $\cos(3\omega_c t)$. The only terms which will contribute to the filtered output are underlined and when the trigonometrical powers are replaced, we obtain:

$$i_a = [b_1 E_c + \tfrac{3}{4} b_3 E_c^3 + 2b_2 E_c f_m(t) + 3b_3 E_c f_m^2(t)] \cos(\omega_c t) \tag{3.22}$$

The envelope is therefore $[b_1 + \tfrac{3}{4} b_3 E_c^2 + 2b_2 f_m(t) + 3b_3 f_m^2(t)] E_c$ and is a distorted replica of the modulating signal, the distortion arising from the presence of the term in $f_m^2(t)$. The output from an envelope detector will be $2b_2 f_m(t) + 3b_3 f_m^2(t)$. The distortion can be kept to an acceptable value by making the coefficient b_3 much less than b_2.

Modulation by the additive process described above is simple and requires little equipment. It is used in practice for low-level modulation. High-level modulation is generally effected by impressing the modulation signal as a voltage on the anode of the carrier frequency amplifier valve. If the anode voltage versus carrier amplitude is linear then true AM is produced.

3.3.1. SINGLE SIDEBAND GENERATION AND DETECTION

It has already been seen that a single sideband transmission requires only half the bandwidth of a double sideband transmission. Further, all the transmitted power is used in the sideband and so there is no unnecessary wastage in a carrier. These advantages make it clear that the single sideband system is potentially much more economical both in power and bandwidth than double sideband transmission. We will therefore examine the possible methods of generation and detection with a view to seeing what difficulties arise. The starting-point for generation is the product type of modulator discussed in the previous section. This can easily produce a suppressed-carrier double-sideband transmission and in principle this can be converted to a single sideband transmission simply by filtering out one sideband. The difficulty in this method is simply one of producing a filter with the necessary characteristics. The changeover from a passband to a stopband must occur sufficiently rapidly for the lowest frequencies in the transmitted sideband to be passed with little attenuation while the lowest frequencies in the rejected sideband must be attenuated strongly. This is illustrated by Fig. 3.8. We see that the changeover from pass to stop conditions should occur in a frequency interval of the order of 50 Hz for sound transmission and this means that elaborate filters are required.

The usual form of modulator to produce the suppressed-carrier

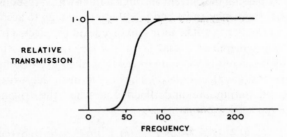

Fig. 3.8. Filter characteristic to produce D.S.B.
to S.S.B. conversion.

double-sideband signal is the product type based on non-linear elements as shown in Fig. 3.6. A typical arrangement is shown in Fig. 3.9 and is often referred to as a balanced modulator. The first non-linear element,

A, operates on the sum of the carrier $E_c \cos (\omega_c t)$ and the modulating signal $f_m(t)$, and the output after filtering in the vicinity of the carrier frequency is given by equation (3.22) as:

$$i_A = [b_1 + \tfrac{3}{4}b_3 E_c^2 + 2b_2 f_m(t) + 3b_3 f_m^2(t)]E_c \cos (\omega_c t) \quad (3.23)$$

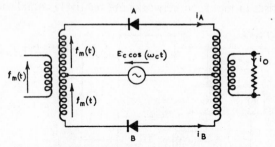

Fig. 3.9. Balanced modulator designed to generate suppressed carrier D.S.B. modulation.
A and B are non-linear elements (diodes).

The second non-linear element, B, operates on the difference between the carrier and the modulating signal. Its output is therefore identical to that of A except that the sign of $f_m(t)$ is reversed, *i.e.*

$$i_B = [b_1 + \tfrac{3}{4}b_3 E_c^2 - 2b_2 f_m(t) + 3b_3 f_m^2(t)] \, E_c \cos (\omega_c t) \quad (3.24)$$

The net output from the circuit is the difference of i_A and i_B and so is:

$$i_0 = 4b_2 f_m(t) \, E_c \cos (\omega_c t) \quad (3.25)$$

This has the required form of a suppressed carrier double sideband signal and it may also be noted that the distortion term in $f_m^2(t)$ is eliminated. A more complete treatment would show that distortion would arise from the even terms in the power series (equation (3.19)) for the non-linear element. These higher-order distortion terms are usually negligible. A single sideband output can be obtained by connecting a filter to the output of the balanced modulator.

An alternative approach, designed to eliminate the need for the filter to separate the sidebands, is illustrated by Fig. 3.10. The principle of this circuit is that two double sideband outputs are produced, such that when they are combined one sideband is cancelled and the other is reinforced. The operation is most easily seen for a sinusoidal modulating signal, $\sin (\omega_m t)$. If the balanced modulators are assumed to be perfect, the output from I is $E_c \cos (\omega_c t) . \sin (\omega_m t)$ and that from II is $E_c \sin (\omega_c t) . \cos (\omega_m t)$. Adding these two outputs together, we obtain:

$$I_0 = E_c \cos (\omega_c t) . \sin (\omega_m t) + E_c \sin (\omega_c t) . \cos (\omega_m t)$$

$$= E_c \sin (\overline{\omega_c + \omega_m t}) \quad (3.26)$$

This output is the appropriate term for the upper sideband only. Since any modulating signal can be resolved into a spectrum of sinusoidal waves, the same result will hold in general, provided that the phase shifting network N_1 produces a phase shift of exactly 90 degrees for any frequency contained in the modulating signal. Once again a practical difficulty arises in that such networks are relatively complicated.

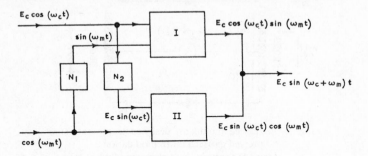

Fig. 3.10. Single side band modulator.
I and II are balanced modulators as shown in Fig. 3.9. N_1 shifts the phase of each component of the modulating signal by 90 degrees. N_2 shifts the phase of the carrier by 90 degrees.

The detection of a single sideband can be regarded as a modulation process in reverse. The single sideband modulator simply shifts the frequencies of the components of the modulating signal upwards by an amount ω_c. The converse process of shifting the frequencies downwards by an amount ω_c will convert an upper single side band back into the modulating signal. This frequency shift can be effected by using circuits similar to those described for modulation. There must of course be available at the receiver a c.w. signal of the correct carrier frequency and this can be provided either by using a highly stable oscillator or by transmitting a 'pilot carrier' in addition to the sideband. For most transmissions the phase of the local carrier is unimportant.

One important advantage of a single sideband system is that it is less subject to one type of fading than is a double sideband system. When radio communication is used over long distances, the transmitted signals are reflected by the ionosphere. The reflecting properties of the ionosphere are frequency dependent and this often leads to unwanted differential phase shifts for the frequency components in the signal. If a single sideband system is used, these phase shifts cause corresponding phase shifts in the detected wave form but do not cause any changes in the relative amplitudes of the components. If the system is being used for speech communication, these phase shifts are unimportant since the ear does not detect them. Satisfactory single sideband communication can thus be made, even if this type of distortion is present. Such phase shifts do, however, lead to changes in the amplitudes of the frequency components in a double sideband system, as will be seen for a simple example considered below. The amplitude changes take the form of a reduction in level of certain of the frequency components and lead to what is called 'selective fading'. This can be sufficiently serious to render a double sideband speech communication unintelligible. As an example of this effect, we will consider a single frequency modulating signal so that the transmitted double sideband signal can be written as:

$$V = E_c \cos(\omega_c t) + \tfrac{1}{2} m E_c \cos\overline{(\omega_c + \omega_m t)} + \tfrac{1}{2} m E_c \cos\overline{(\omega_c - \omega_m t)}$$

$$(3.27)$$

Suppose that each of the three frequency components in this signal undergoes a phase shift during transmission and that because of the nature of ionospheric reflection, these phase shifts are unrelated. The received signal will therefore be:

$$V_r = E_c \cos(\omega_c t + \phi_c) + \tfrac{1}{2} m E_c \cos\overline{(\omega_c + \omega_m t + \phi_u)}$$

$$+ \tfrac{1}{2} m E_c \cos(\omega_c - \omega_m t + \phi_l) \quad (3.28)$$

and this expression can be manipulated to show the form of the envelope:

$$
\begin{aligned}
V_r &= E_c \cos(\omega_c t + \phi_c) + \tfrac{1}{2} m E_c \cos(\omega_c t + \phi_c)\cos(\omega_m t + \phi_u - \phi_c) \\
&\quad - \tfrac{1}{2} m E_c \sin(\omega_c t + \phi_c)\sin(\omega_m t + \phi_u - \phi_c) \\
&\quad + \tfrac{1}{2} m E_c \cos(\omega_c t + \phi_c)\cos(\omega_m t - \phi_l + \phi_c) \\
&\quad + \tfrac{1}{2} m E_c \sin(\omega_c t + \phi_c)\sin(\omega_m t - \phi_l + \phi_c) \\
&= [1 + \tfrac{1}{2} m \cos(\omega_m t + \phi_u - \phi_c) \\
&\qquad\qquad + \tfrac{1}{2} m \cos(\omega_m t - \phi_l + \phi_c)] E_c \cos(\omega_c t + \phi_c) \\
&\quad + [\tfrac{1}{2} m \sin(\omega_m t - \phi_l + \phi_c) \\
&\qquad\qquad - \tfrac{1}{2} m \sin(\omega_m t + \phi_u - \phi_c)] E_c \sin(\omega_c t + \phi_c) \\
&= V(t) . E_c \cos(\omega_c t + \phi_c + \psi) \qquad\qquad\qquad (3.29)
\end{aligned}
$$

where

$$
V^2(t) = \{[1 + \tfrac{1}{2} m \cos(\omega_m t + \phi_u - \phi_c) + \tfrac{1}{2} m \cos(\omega_m t - \phi_l + \phi_c)]^2
$$
$$
+ \frac{m^2}{4}[\sin(\omega_m t - \phi_l + \phi_c) - \sin(\omega_m t + \phi_u - \phi_c)]^2\}^{\tfrac{1}{2}}
$$

$$(3.30)$$

Expansion of the squares and simplification of the resulting expression gives:

$$
V^2(t) = 1 + 2m \cos\left(\omega_m t + \frac{\phi_u - \phi_l}{2}\right)\cos\left(\phi_c - \frac{\phi_u + \phi_l}{2}\right)
$$
$$
+ m^2 \cos^2\left(\omega_m t + \frac{\phi_u - \phi_l}{2}\right)
$$

$$(3.31)$$

If the receiver has an envelope detector, the demodulated output is $V(t)$. Examination of eqn. (3.31) shows that $V(t)$ reduces to $1 + m \cos [\omega_m t + (\phi_u - \phi_l)/2]$ if $2\phi_c = \phi_u + \phi_l$ and in this case the modulation is recovered with its original amplitude and without distortion. If however the phases ϕ_c, ϕ_u, ϕ_l do not satisfy the above condition (or any equivalent one with addition of multiples of 2π), the demodulated output contains harmonics of the modulating frequency and the amplitude of the term of frequency ω_m is reduced, becoming zero if $\phi_c - \frac{1}{2}(\phi_u + \phi_l)$ is an odd multiple of $\pi/2$.

In practice the phase shifts may vary rapidly with frequency, and so a speech modulation will have some frequency components completely eliminated and others severely attenuated. The particular frequencies which are attenuated change with time as the reflecting properties of the ionosphere change and this gives a fading characteristic to the detected output from the receiver. The possibility of eliminating this selective fading by using a single sideband system has led to a significant improvement in long distance radio communication circuits.

3.3.2. VESTIGIAL SIDEBAND SYSTEMS

The advantages of single sideband transmission are only obtained at the expense of using more complicated transmitting and receiving equipment than is required for double sideband transmission. A compromise can be effected by using a vestigial sideband system in which one sideband is transmitted completely and the other is partially filtered. This means that the bandwidth required for the transmission is intermediate between that required for single and double sideband systems and the saving can be appreciable if modulating signals with large bandwidths are being handled. This arises in television and the type of filter characteristic used is shown in Fig. 3.11. Such a

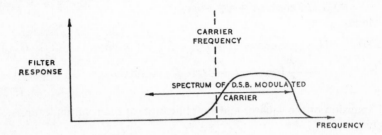

Fig. 3.11. Vestigial side band filter.

filter is much less elaborate than that required if one sideband is to be completely suppressed. A simple envelope detector can still be used in the receiver although some distortion may arise. A full analysis is complicated by the fact that the presence of the low-frequency components in both sidebands leads to a reduction in the high-frequency distortion caused by suppressing one sideband. We will consider only the worst case (which is never likely to arise in practice), that of a single frequency modulating signal, the frequency lying in the range for which one sideband is appreciably attenuated. The transmitted signal is therefore:

$$V(t) = E_c \cos(\omega_c t) + \tfrac{1}{2}mE_c \cos(\overline{\omega_c + \omega_m t})$$
$$+ \tfrac{1}{2}maE_c \cos(\overline{\omega_c - \omega_m t}) \quad (3.32)$$

where a is a constant, less than unity, and represents the attenuation of the lower sideband. The envelope of the waveform given by equation (3.32) is:

$$V(t) = E_c[\{1 + \tfrac{1}{2}m(1 + a)\cos(\omega_m t)\}^2 + \tfrac{1}{4}m^2(1 - a)^2 \sin^2(\omega_m t)]^{\frac{1}{2}}$$
$$(3.33)$$

The form of this expression suggests that it may be approximated by a Fourier series involving terms up to the second harmonic, *i.e.* by:

$$V(t) = p + q\cos(\omega_m t) + r\cos(2\omega_m t) \quad (3.34)$$

Approximations to p, q, r may be obtained by equating the two expressions for three values of t, a convenient choice being $\omega_m t$ equal to 0, $\pi/2$ and π. This gives:

$$p + q + r = E_c\{1 + \tfrac{1}{2}m(1 + a)\} \quad (3.35)$$

$$p \quad - r = E_c[1 + \tfrac{1}{4}m^2(1 - a)^2]^{\frac{1}{2}} \quad (3.36)$$

$$p - q + r = E_c\{1 - \tfrac{1}{2}m(1 + a)\} \quad (3.37)$$

from which we find:

$$q = \tfrac{1}{2}m(1 + a)E_c \quad (3.38)$$

$$r = -\tfrac{1}{2}[\{1 + \tfrac{1}{4}m^2(1 - a)^2\}^{\frac{1}{2}} - 1]E_c \quad (3.39)$$

The value of r relative to q is a measure of the distortion produced by the detector. If a equals unity, the situation is the same as in a normal double sideband system: r is then zero so that there is no distortion. As a is decreased from unity towards zero, r steadily increases and q decreases, so that the worst distortion occurs, as would expected, when one sideband is completely attenuated. Further, the magnitude of this distortion increases as m increases. The worst possible case is therefore given by letting a equal zero and m equal unity in equations (3.38) and (3.39). The values are then $q = \tfrac{1}{2}E_c$ and $r = -0.06\,E_c$, so that the second harmonic output is approximately 12 per cent of the fundamental.

As has already been mentioned, this case will never arise in practice because practical signals will always contain frequency components over the whole frequency range of the sidebands; this results in a receiver output which is a more faithful replica of the original modulating signal.

The vestigial sideband system is a useful compromise in that it needs only the detecting equipment of a double sideband system but it does lead to a reduction in the bandwidth required.

3.4. Angle Modulation

The basic modulated carrier wave, defined by equation (3.1), has an instantaneous value determined by the amplitude A_c and the angle $(\omega_c t + \phi_c)$. In the amplitude modulation systems which have been discussed, A_c is made to vary in sympathy with the modulating signal while the angle is kept the same as in the unmodulated carrier. In angle modulation, the converse situation holds, A_c being unaltered and the angle being varied according to the modulating signal. We will define the instantaneous value of this angle as $\theta(t)$, so that for an unmodulated carrier

$$\theta(t) = \omega_c t + \phi_c \tag{3.40}$$

There are an infinite number of ways in which $\theta(t)$ may be made to vary in sympathy with the modulating signal but only two of them need be considered in detail. In the first of these, ϕ_c is made instantaneously proportional to the modulating signal $f_m(t)$ so that we have:

$$\theta(t) = \omega_c t + K_p \cdot f_m(t) \tag{3.41}$$

where K_p is a constant. In this system, the phase angle is proportional to the modulating signal and the name phase modulation is applied. The second possibility appears a little more complicated, but it results in a simplification of the techniques required for modulation and detection and is in fact more widely used. In this system, the instantaneous frequency of the modulated carrier is made proportional to the modulating signal. The term frequency is defined with reference to a sinusoidal wave and is a constant. We must therefore extend our definition to cover the situation in which the frequency can change with time. This is most easily done by representing the wave by a rotating phasor. The sinusoidal wave has an instantaneous value which is equal to the projection of a phasor of length A_c rotating at constant angular velocity ω_c. The angular frequency of the wave is equal to the angular velocity of the phasor and this suggests that if the angular velocity is varying with time we should define our instantaneous angular frequency as being equal to the instantaneous angular velocity. The angular velocity is the rate of change of the angle $\theta(t)$, so that we are led to the definition of instantaneous angular frequency as $d\theta(t)/dt$. This gives the constant value ω_c for the angular frequency of the unmodulated carrier and so is consistent with the definition for a sinusoidal wave. We now introduce a modulation system, frequency modulation, in which the instantaneous frequency of the modulated carrier has a component proportional to the modulating signal, *i.e.* the instantaneous angular frequency of the modulated carrier is:

$$\omega(t) = \frac{d\theta(t)}{dt} = \omega_c + K_f f_m(t) \tag{3.42}$$

where K_f is a constant. The form of this expression is such that the instantaneous frequency has the constant value ω_c in the absence of a modulating signal. Equation (3.42) may be integrated with respect to time to give:

$$\theta(t) = \omega_c t + K_f \int_0^t f_m(t)\,dt + \phi_0 \tag{3.43}$$

in which ϕ_0 is an integration constant, equal to the phase of the modulated carrier at $t = 0$. The value of this constant is unimportant and it can be taken as zero without affecting any of our conclusions. A comparison of equations (3.41) and (3.43) shows that they differ only in the form of the added time dependent function. Equation (3.43) may be regarded as a phase modulated wave in which the modulating signal is $\int_0^t f_m(t)\,dt$ in place of $f_m(t)$. This means that a frequency modulated wave can be produced by first integrating the modulating signal and then using it as the input to a phase modulator, as indicated in Fig. 3.12. Conversely,

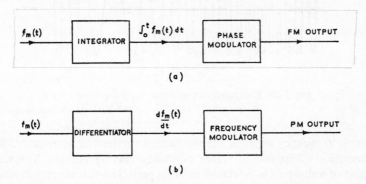

Fig. 3.12. Relations between frequency and phase modulation.

a phase modulated wave can be obtained by first differentiating the modulating signal and then applying it to a frequency modulator. This argument shows that we may deduce all the properties of phase modulated waves from those of frequency modulated waves and conversely. The other forms of angle modulated waves referred to above can similarly be regarded as systems in which some other operation is carried out on the modulating signal before it is applied to either a phase modulator or a frequency modulator. They need not therefore be considered further.

3.4.1. FREQUENCY MODULATION

We will now concentrate on the frequency modulated wave and examine its properties in detail. The expression for such a wave is:

$$f(t) = A_c \cos \left[\omega_c t + K_f \int_0^t f_m(t) \, dt \right] \tag{3.44}$$

and it has the characteristics illustrated by Fig. 3.13. If the modulating signal is changing with time at a rate which is slow compared with the

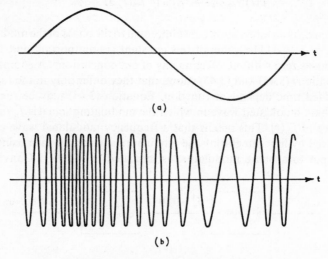

(a)

(b)

Fig. 3.13. Characteristics of a frequency modulated wave.
(a) Waveform of modulating signal;
(b) Waveform of modulated carrier.

carrier frequency ω_c, then the modulated carrier can be regarded as a succession of sinusoidal waves of slowly varying periods. The r.m.s. value of a sinusoid is independent of its period so that we may conclude that the r.m.s. value of the modulated carrier is the same as that of the unmodulated carrier. This means that the power transmitted by a frequency modulated oscillator is not altered by the modulation. It will be recalled that the transmitted power increases if the oscillator is amplitude modulated.

In order to determine the bandwidth required by a frequency modulated wave, we must resolve the modulated carrier into its sidebands. This is a much more difficult piece of analysis than for an amplitude modulated carrier and it is virtually impossible if the modulating wave is not restricted in some way. We will therefore take the simplest possibility, that of a sinusoidal modulating signal of angular frequency ω_m. The modulated carrier is then:

$$f(t) = A_c \cos\left[\omega_c t + K_f \int_0^t \cos(\omega_m t)\, \mathrm{d}t\right]$$

$$= A_c \cos\left[\omega_c t + (K_f/\omega_m)\sin(\omega_m t)\right] \qquad (3.45)$$

in which the amplitude of the modulating signal is absorbed into the constant K_f. By differentiating within the brackets, the instantaneous frequency of this wave is $(\omega_c + K_f \cos \omega_m t)/2\pi$ Hz and oscillates between $(\omega_c - K_f)/2\pi$ and $(\omega_c + K_f)/2\pi$ Hz. The peak value of the difference between the instantaneous frequency and the carrier frequency is defined as the *frequency deviation* and in the present example equals $K_f/2\pi$ Hz. Since K_f is proportional to the amplitude of the modulating signal, the frequency deviation is also proportional to this amplitude. The maximum frequency deviation is the value corresponding to the largest amplitude of the modulating signal to be handled by the particular system considered. The ratio of the actual frequency deviation to the maximum frequency deviation is called the *deviation ratio* (usually expressed as a percentage). The deviation ratio is the parameter of a frequency modulated signal which corresponds to the modulation depth of an amplitude modulated wave. If f_d is the maximum frequency deviation and f_a is the actual frequency deviation, then the deviation ratio, m, is given by:

$$m = f_a/f_d \qquad (3.46)$$

The deviation ratio for the modulated carrier given by equation (3.45) is therefore $K_f/2\pi f_d$, since $f_a = K_f/2\pi$.

Examination of equation (3.45) shows that the constant K_f only appears in the ratio (K_f/ω_m) and it is the value of this ratio which determines the sideband structure. Replacing K_f by $2\pi f_a$ and ω_m by $2\pi f_m$, where f_m is the frequency of the modulating signal, we have:

$$K_f/\omega_m = f_a/f_m = mf_d/f_m \qquad (3.47)$$

This ratio, *i.e.* the ratio of the actual frequency deviation to the modulating frequency, is defined as the *modulation index*. Care should be taken that this is not regarded as the equivalent of the modulation depth of an amplitude modulated signal: the true equivalent of modulation depth is the deviation ratio. If the modulation index is denoted by m_p, equation (3.45) can be written as:

$$f(t) = A_c \cos \left[\omega_c t + m_p \sin (\omega_m t) \right] \tag{3.48}$$

and can be expanded into a series of sidebands by the method explained in Appendix B. The result of this expansion is

$$f(t) = A_c \sum_{n=-\infty}^{\infty} J_n(m_p) \cos (\omega_c + n\omega_m)t \tag{3.49}$$

where the functions $J_n(m_p)$ are Bessel functions, whose properties are discussed in Appendix B.

The most striking feature of the expression in equation (3.49) is that

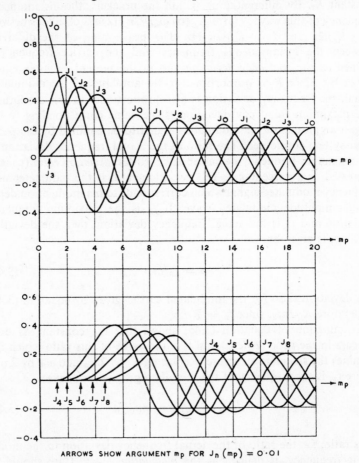

ARROWS SHOW ARGUMENT m_p FOR $J_n(m_p) = 0 \cdot 01$

Fig. 3.14. Bessel functions of order 1 to 8 for arguments up to 20.

there is an infinite number of sidebands, spaced at intervals equal to the modulation frequency. In principle therefore the modulated carrier has a frequency spectrum extending over all possible frequencies. This is in marked contrast to the amplitude modulated carrier which has only a single pair of sidebands. In practice, only a finite number of these sidebands has significant amplitudes and we will show that the frequency modulated carrier can be transmitted with little distortion through a finite bandwidth. This result also follows from the properties of the Bessel functions, some of which are derived in Appendix B. The form of the first few of these functions is shown in Fig. 3.14: they have an oscillatory character with decreasing values at successive maxima. If m_p is zero, all the functions are zero except $J_0(0)$ which equals unity: this means that the expression in equation (3.49) reduces to the unmodulated carrier if m_p is zero. We are most interested in the number of terms required in the series for a specified value of m_p. If m_p is kept constant at a value greater than unity and n is varied, it is found that the Bessel function with the largest amplitude has a value for n just below m_p and that if n increases above this value the succeeding terms decrease rapidly in amplitude. Such terms do not contribute greatly to the waveform but may be considered significant if they have amplitudes greater than some selected value. A convenient choice for this value is 0·01 which means that we will ignore all sidebands whose amplitudes are less than one per cent of the amplitude of the unmodulated carrier. It should be stressed that the rapid decrease of the amplitudes of the Bessel functions for values of n above that which gives the greatest amplitude, enables us to ignore the possibility that there may be a significant contribution from a large number of very small amplitude sidebands. The Bessel function $J_{-n}(m_p)$ equals $(-1)^n J_n(m_p)$ so that the amplitudes of corresponding upper and lower sidebands are equal. Table 3.1 shows the number of significant sidebands, calculated on the one per cent basis explained above. The numbers given in the table include both the upper and lower sidebands. The bandwidth required for the modulated carrier is given by the product of the number of the sidebands and the modulation frequency. An equivalent expression can be found in terms of the frequency deviation by using the definition of the modulation index. The values for the bandwidth, expressed in each of these two ways, are also given in Table 3.1. An interesting and most important property emerges from this table: for small values of m_p, i.e. values less than 0·3, the bandwidth is twice the modulating frequency, whereas for large values of m_p, i.e. values greater than about 5, the bandwidth is rather more than twice the deviation frequency. These two cases are sometimes referred to as narrow band frequency modulation (m_p less than 0·3) and wideband frequency modulation, this being taken to cover all the values of m_p for which the bandwidth is appreciably larger than twice the modulation frequency. We

Carlson's rule

will be much more concerned with the wideband case. If the frequency deviation is kept constant and the modulating frequency is increased, the modulation index m_p will decrease and we see from the final column of Table 3.1 that the bandwidth required will also increase. This enables

TABLE 3.1

Bandwidth Required for a Frequency Modulated Signal

Modulation Index m_p	No. of significant sidebands (N)	Bandwidth*	
		(a) As multiple of modulation frequency	(b) As multiple of frequency deviation
0·1	2	2	20
0·3	4	4	14
0·5	4	4	8
1·0	6	6	6
2·0	8	8	4
5·0	16	16	3·2
10·0	28	28	2·8
20·0	50	50	2·5
30·0	70	70	2·3

* The bandwidth equals $Nf_m = Nf_a/m_p$
where f_a = frequency deviation; f_m = modulating frequency

us to predict the bandwidth required for a frequency modulated transmitter, for we see that the maximum bandwidth is associated with the highest modulating frequency and with the largest frequency deviation being used. As an example we may consider the B.B.C. V.H.F. service which uses frequency modulation with a maximum frequency deviation of 75 kHz. The highest modulating frequency transmitted is 15 kHz and the corresponding modulation index at maximum frequency deviation is 5. The bandwidth required is therefore 3·2 times the frequency deviation, *i.e.* 240 kHz. There are eight pairs of significant sidebands under these conditions and it is this feature which leads to the advantage of using frequency modulation. If a noise signal is superimposed on the modulated carrier it is improbable that the various noise sidebands will have the precise relationship required for them to act in the same way as a frequency modulated signal. This means that if a detector responds only to frequency modulation, it is not likely to give an output from the superimposed noise. This topic will be explored in more detail in Chapter 6. The advantage of comparative freedom from noise interference is only gained if a wideband frequency modulated system is used.

The frequency spectrum of a frequency modulated wave becomes very complicated if the modulating signal is not sinusoidal. It cannot be obtained by a linear combination of the spectra for the sinusoidal components of the modulating signal. An example is given in the Appendix for a modulating signal containing two sinusoidal terms of frequencies ω_1 and ω_2, and it is shown that the spectrum contains all frequencies $n\omega_1 + m\omega_2$, where n,m are any integers. An important feature of the spectrum of a frequency modulated wave is that there is in general no simple relation between the upper and lower sidebands. Single sideband operation therefore leads to considerable complications when frequency modulation is being used, and will not be discussed.

3.4.2. METHODS OF GENERATING AND DETECTING FREQUENCY MODULATION

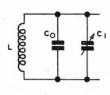

Fig. 3.15. Simple frequency modulator

The most direct method of generating a frequency modulated wave is to vary the resonant frequency of an oscillator in sympathy with the modulating signal. Suppose, for example, that the resonant frequency of the oscillator is controlled by the parallel resonant circuit, shown in Fig. 3.15. The capacitance of this circuit is shown in two parts, C_0 being fixed and C_1 being capable of varying proportionally to the modulating signal $f_m(t)$, i.e.

$$C_1 = kf_m(t) \tag{3.50}$$

where k is a proportionality constant. The resonant frequency of the tuned circuit is:

$$\omega_r^2 = 1/[L\{C_0 + kf_m(t)\}] \tag{3.51}$$

and if the magnitude of $kf_m(t)$ is always much less than C_0, we have approximately:

$$\omega_r = \omega_c - \frac{k\omega_c}{2C_0} f_m(t) \tag{3.52}$$

where

$$\omega_c = 1/(LC_0)^{\frac{1}{2}} \tag{3.53}$$

The output from the oscillator is a frequency modulated wave whose instantaneous frequency is given by equation (3.52) and comparison with equation (3.42) shows that:

$$K_f = -k\omega_c/2C_0 \tag{3.54}$$

The instantaneous frequency deviation is $K_f f_m(t)/2\pi$ Hz and therefore equals $-kf_c f_m(t)/2C_0$ Hz. Since the magnitude of $kf_m(t)$ must be much less than C_0, we see that the approximation used in deriving equation (3.52) implies that the frequency deviation must always be much less than the carrier frequency. In practice, frequency modulation is only used in the V.H.F. and higher frequency bands and this restriction is always satisfied. It is often convenient to modulate an oscillator operating at a much lower frequency and then to multiply up to the final frequency. If this is done the multiplier increases the instantaneous frequency by a constant ratio and so the frequency deviation is multiplied by the same factor as is the carrier frequency.

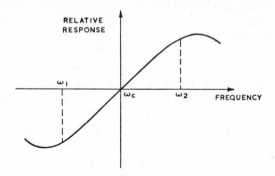

Fig. 3.16. Characteristic of frequency discriminator.
The useful operating range is between frequencies ω_1
and ω_2

A frequency-modulated signal is demodulated by using a frequency discriminator with a characteristic as shown in Fig. 3.16. The discriminator output is proportional to the instantaneous frequency deviation and so has an amplitude modulation which is a replica of the frequency modulation. The amplitude modulation is then detected by a diode.

3.4.3. PHASE MODULATION

It has already been pointed out that all the properties of phase modulation can be deduced from those of frequency modulation by making an appropriate change in the form of the modulating signal. We will use this argument to show that phase modulation uses a specified bandwidth less effectively than does frequency modulation. It is seen from Table 1 that the bandwidth required for a frequency modulated carrier is approximately a constant multiple of the frequency deviation, independently of the value of the modulating frequency. This statement, which is strictly true for large values of the modulation index, means that the bandwidth required for a sinusoidal modulating frequency of maximum amplitude does not depend on the modulating frequency. The available band is thus used equally effectively, irrespective of the modulating frequency. A different situation arises for a phase modulated carrier. Consider a sinusoidal modulating signal of constant amplitude, corresponding to the largest value to be handled, and of frequency f_m. The phase modulated carrier can be produced by using the derivative of the modulating signal as the input to a frequency modulator. The amplitude of this derivative is proportional to the modulating frequency f_m and so the frequency deviation, corresponding to a constant value of the amplitude of the phase modulation, will increase proportionally to f_m. This means that the bandwidth required by the phase modulated carrier also increases as the modulating frequency increases and so the available bandwidth for the transmission cannot be used economically. It is for this reason that frequency modulation is much more widely. used than any other form of angle modulation.

The bandwidth of the frequency modulated carrier is only strictly proportional to the frequency deviation if the modulation index is large. For example, we see from Table 1 that the bandwidth tends to twice the frequency deviation if m_p is large but is $3 \cdot 2$ times the frequency deviation if m_p equals 5. This leads to an increase in the bandwidth required for a modulating signal with a high modulation frequency. It can be corrected by passing the modulating signal through a filter which progressively reduces amplitude as the frequency rises before applying it to the modulator. This process is known as de-emphasis and can be compensated by a complementary filter in the receiver. In practice this process is not used because it is most improbable that a single high-frequency component will ever occur in a broadcast transmission. In broadcast transmitters, the high-frequency components are in fact accentuated by a pre-emphasis filter since this leads to an improvement in signal-noise performance. This will be discussed in Chapter 6.

3.4.4. NARROW BAND FREQUENCY MODULATION

Although narrow band frequency modulation is not often used for communication, it does arise in certain measuring techniques. The properties can be very easily discussed with the aid of a phasor diagram similar to that used for examining amplitude modulation and such a discussion brings out some important differences between frequency and amplitude modulation. If the modulation index is small, say less than $0 \cdot 3$, $J_0(m_p)$ is not appreciably different from unity and only the first two sidebands are of significant amplitude (see Table 1). Equation (3.49) thus reduces to:

$$f(t) = A_c[J_0(m_p)\cos(\omega_c t) + J_1(m_p)\cos(\omega_c + \omega_m)t$$
$$+ J_{-1}(m_p)\cos(\omega_c - \omega_m)t] \quad (3.55)$$

We now use the approximations:

$$J_0(m_p) = 1 : J_1(m_p) = -J_{-1}(m_p) = \tfrac{1}{2}m_p$$

which are valid for small m_p and equation (3.55) becomes:

$$f(t) = A_c[\cos(\omega_c t) + \tfrac{1}{2}m_p \cos(\omega_c + \omega_m)t - \tfrac{1}{2}m_p \cos(\omega_c - \omega_m)t]$$
$$(3.56)$$

This expression is similar to the corresponding one for amplitude modulation (equation (3.10)) but has the important difference that the sign of the lower sideband is reversed. The phasor diagram for the frequency modulated carrier is shown in Fig. 3.17 from which it can be seen that the resultant of the two sideband phasors is always at right angles to the carrier phasor. This result should be contrasted with the diagram for amplitude modulation (Fig. 3.2). It is readily verified from Fig. 3.17 that the instantaneous value of the angle ϕ is $m_p \sin(\omega_m t)$, provided that m_p is small. To the accuracy being used, the phasors in Fig. 3.17 therefore combine to give the resultant $A_c \cos(\omega_c t + m_p \sin \omega_m t)$ as required.

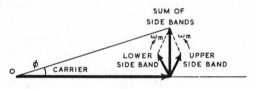

Fig. 3.17. Phasor diagram for narrow band F.M.

Since amplitude and narrow-band frequency modulation differ only in the phases of the sidebands, it is possible to convert one to the other by a network with a suitable phase-frequency response. An examination of the sideband structure of a frequency modulated carrier with sinusoidal modulation of large modulation index shows that all the odd order sidebands contribute a component in quadrature with the carrier while the even order sidebands contribute a component in phase with the carrier. The time variation of the phase angle is thus due to the odd order sidebands.

3.5. Other Modulation Systems

The other modulation systems which were mentioned in the introduction to this chapter all stem from the use of a train of radio frequency pulses as the carrier wave. The modulating signal is used to control one of the characteristics of the pulses. This process may be analysed by supposing that the carrier is a continuous wave and that the modulating signal is first coded into a series of pulses before being applied to the modulator. This approach, which is used throughout this volume, makes a distinction between coding and modulation and means that all the modulation theory required is contained in the preceding sections. The alternative is to analyse the properties of modulated trains of pulses but this does not lead to any significant new results.

Chapter 4

The Properties of Communication Channels

4.1. Introduction

The function of a communication channel is to convey a signal of a specified type from some transmitting point to a distant receiving point. We will be concerned in this chapter with an examination of the properties of a channel which determine its suitability or otherwise for use in a specified application. Ideally, the output from the channel should be an exact replica of the input but in practice this ideal can never be realized and the output will always be modified or distorted to some extent. The problem of assessing the suitability of a channel for a specified application thus reduces to first specifying the amount of distortion which can be tolerated, and then determining from the physical characteristics of the channel the amount of distortion which will arise. The specification of the permissible amount of distortion is essentially a matter of experience and in many cases, *e.g.* international telephony circuits, agreed specifications have been drawn up by the appropriate international bodies. These bodies, such as C.C.I.R. (Comité Consultatif International de Radio) meet at regular intervals to revise their specifications in the light of new requirements and of improvements in the performance of available equipment.

A large number of parameters is necessary in order to define fully the performance of any particular channel. These parameters may not all be equally important in any specified application and a logical approach to the classification of channels is first to specify the type of signal which is to be handled. The major distinction is between continuous (or analogue) signals and discrete signals. Most of our consideration will be given to channels designed for continuous signals and these may be further classified into linear and non-linear channels. Most practical channels are linear for a certain range of input signal level and a detailed study of linear continuous channels will provide much of the information we require. This is complemented by an examination of the effects of non-linearity in so far as they modify the performance of practical channels.

The classification outlined above is summarised in Table 4.1, and the principal characteristics required to give a knowledge of the behaviour of each kind of channel are listed. By approaching the problem in this way, we do not need to make any specific distinctions between baseband and carrier channels or between radio and line channels, other than to note that some of the characteristics may only apply to particular kinds of channel, *e.g.* fading is only encountered in radio channels. Similarly, for the present purpose we may consider active and passive networks together: once again, the only difference relevant to the performance as a telecommunications channel appears in the characteristics listed.

This chapter contains a general discussion of the characteristics of channels and the methods by which they may be specified and will be followed in the next chapter by a more detailed account of the mathematical methods of analysis.

TABLE 4.1

Classification and Main Characteristics of Communication Channels

Classification of channels			Main characteristics
Continuous (analogue) Baseband or Carrier Channels	Linear	Active or Passive	Amplitude-frequency characteristic Phase-frequency characteristic Delay characteristic Multipath effects (stable) Non-reciprocal effects
	Non-linear	Active or Passive	Input-Output non-linearity Overload characteristic (Dynamic range) Multipath effects (non-stable) Fading
Discrete (digital)	2-level		Telegraph distortion Error rate
	Multi-level		

4.2. Linear Channels

For the purposes of our discussion, any channel can be regarded as a four-terminal electrical network. The input is applied to one pair of terminals and the output is obtained at a second pair of terminals. An obvious example is the uniform transmission line as used in telephony. A radio link may equally well be looked on as a four-terminal network, the terminals of the transmitting aerial being the input pair and the terminals of the receiving aerial being the output pair. The network may also include amplifiers or repeaters, modulators and demodulators. In every case, we are concerned primarily with the relations between the output and input voltages and currents and the bulk of our study can be carried through without any need to know the details of the components contained within the network. It is not even necessary to restrict the discussion to purely electrical components. Transducers may be allowed for by an appropriate change in the physical quantities considered. In every case, a pair of quantities, *e.g.* pressure and velocity, can be found which will play the same essential role as voltage and current. Alternatively, we may say that any transducer can be represented by an equivalent electrical network.

A number of definitions, taken from electrical network theory, are used to describe the essential features of any channel. The most important definition is that of linearity. A channel is *linear* if the output voltage and current are directly proportional to the input voltage and current. Linear networks are very much simpler to analyse than non-linear ones, largely because the Superposition Theorem can be used. In the present context, this theorem states that if two (or more) input signals are simultaneously applied, the output signal is the sum (or superposition) of the two (or more) outputs obtained if the inputs are applied individually. At a given frequency the input and output impedances of linear networks are constants, independently of the magnitudes of the signals which are being transmitted. Any circuit comprising only resistance, inductance and capacitance is a linear network. A network containing valves or transistors can often be regarded as linear, provided the signal magnitudes are not excessive, *i.e.* provided that the 'small signal equivalent circuits' may be used.

A *passive* network is one in which there are no internal sources of power (except noise). If there are internal sources of power, *e.g.* in amplifiers, the network is said to be *active*.

There is further distinction between *reciprocal* and *non-reciprocal* networks. Reciprocal networks obey the reciprocity theorem which states that if an e.m.f. E acting in one branch of a network causes a current I to flow in a second branch, then the same e.m.f. E acting in the second branch will cause a current I in the first branch. Most passive networks are reciprocal but there are a few important exceptions, which do occur in telecommunications work.

Some examples of networks classified according to the above definitions are given below.

1. Linear, passive, reciprocal networks
(*a*) networks containing only resistance, inductance, capacitance
(*b*) transmission lines and waveguides
(*c*) networks defined between the input to a transmitting aerial and the output from a receiving aerial, provided the transmission path does not pass through the ionosphere
2. Linear, passive, non-reciprocal elements:
(*a*) Networks as in 1(*c*) with transmission paths through the ionosphere may be non-reciprocal
(*b*) Waveguides and co-axial lines containing magnetised ferrite elements ('gyrators' and 'isolators')

3. Linear, active, non-reciprocal networks:
    Valve and transistor amplifiers working under 'small signal' conditions
4. Linear, active, reciprocal networks:
    Symmetrical amplifiers specially designed for two-way operation
5. Non-linear, active, non-reciprocal networks:
    Networks containing valves and transistors working under 'large signal' conditions

Most communication links are designed to operate in a linear manner and the next section, which deals with linear four-terminal networks, is applicable. The inevitable departures from linearity which arise in valves and transistors can be regarded as a source of distortion and can be examined by the methods discussed in a later section.

4.3. Characteristics of Four-Terminal Networks

Many of the essential characteristics of a four-terminal network are illustrated by the simple resistance attenuator shown in Fig. 4.1(a). It is clear that any signal voltage applied to the input terminals AB will cause a reduced voltage to appear at the output terminals CD, *i.e.* the signal is attenuated in passing through the network. If the resistors are pure, so that there is no inductance or capacitance in the network, the attenuation will be the same for all frequencies. Any signal $f(t)$ applied to the input will appear at the output with the same time waveform but to a reduced scale. The output will be $a \cdot f(t)$ where a is a scale factor, less than unity, determined by the attenuation introduced by the network. The value of this attenuation depends not only on the network between AB and CD but also on the input device or transducer which supplies the voltage to AB and on the output device or transducer to which the output voltage is applied.

Thevenin's Theorem[1] enables us to represent any input device in terms of a zero impedance voltage generator in series with a source impedance. Similarly, any load device can be represented in terms of the load impedance it presents to the output terminals CD. This division of any electrical system into source voltage, source impedance, four-terminal network, and load impedance can always be effected and is convenient for the analysis of the performance. In the example of the resistance attenuator, both the source impedance and the load impedance must be purely resistive if a true scale relation between output and input is required.

The simple network of Fig. 4.1(a) will pass all frequencies from zero to infinity so that it clearly cannot represent a practical communication channel. In any practical system, there are always shunt capacitances and series inductances which cause the four-terminal network to be less efficient in the transmission of high frequencies. Thus Fig. 4.1(b) is a better representation of a practical channel. As the frequency of an applied sinusoidal signal is increased, the series impedance will increase and the shunt impedance will fall, both effects combining to reduce the signal. The network shown is, in fact, a form of filter which will pass all frequencies up to a certain value, and will cause attenuation for higher frequencies. It is often called a low-pass filter.

The network of Fig. 4.1(a) is capable of transmitting a direct current,

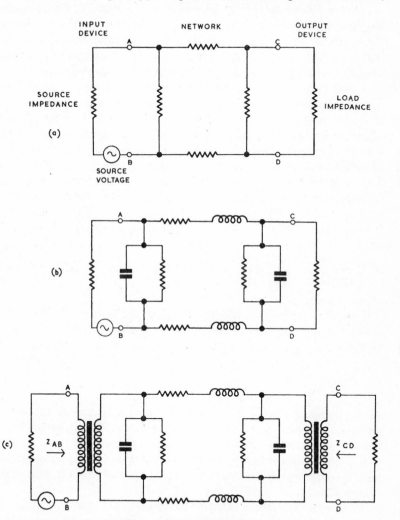

Fig. 4.1. Simple examples of communication channels.
(a) Resistance attenuator (transmits all frequencies).
(b) Channel (a) modified to include reactances operative at low and high frequencies.
(c) Channel which does not transmit d.c.

but many communication channels are incapable of passing d.c. If transformers are added in tandem, as in Fig. 4.1(c), we have a network which will not pass d.c. or very low frequencies and which will cut off signals with frequencies above a certain value. Such a network has similar properties to a large class of communication channels particularly those intended for speech or audio signals.

A complete analysis of the behaviour of the circuit in Fig. 4.1(c) requires a knowledge of the impedances Z_{AB} and Z_{CD}, which would be measured if the network were cut at sections AB or CD and measurements made in the directions of the arrows. It should be noticed that the input impedance Z_{AB} depends on the load impedance. Communication networks are usually designed so that impedances such as Z_{AB} and Z_{CD} are mainly resistive within the frequency bands to be transmitted and have equal values. Further, the source and load impedances are chosen to be equal to Z_{AB} and to Z_{CD}. The behaviour of the network can then be described sufficiently by considering the ratio of the output voltage developed across CD to the input voltage applied to AB. Suppose that a sinusoidal voltage of frequency f and r.m.s. output value V_{AB} is applied to AB and that the r.m.s. output voltage is V_{CD}. The voltage amplification factor F is now defined by the ratio:

$$F_{AB-CD} = \frac{\text{r.m.s. output voltage}}{\text{r.m.s. input voltage}} = \frac{V_{CD}}{V_{AB}} \qquad (4.1)$$

For passive networks, such as those in Fig. 4.1, this ratio will be less than unity, but we may use the same definition for active networks containing amplifiers. The network may then have gain and the ratio will be greater than unity.

If two four-terminal networks are connected in cascade, the output of the first network is the input of the second network and the overall amplification factor is given by the product of the separate factors. Thus, if F_1 and F_2 are the first and second factors respectively, then the overall factor is

$$F = F_1 \times F_2 \qquad (4.2)$$

In this formula, the factors F_1 and F_2 must be defined for the appropriate impedance terminations. The impedance Z_{AB} of network 2 is the load impedance of network 1 and the impedance Z_{CD} of network 1 is the source impedance of network 2. There is clearly a great simplification in operating with a common value for the source and load impedances Z_{AB} and Z_{CD}.

For several networks in tandem the separate factors are all multiplied together to give the overall factor representing the overall gain or loss. In practice it is usual to take the logarithm of each factor so as to convert the product into an addition. This leads to the conventional way of defining the gain or loss of a four-terminal system. To avoid difficulties

associated with networks having different impedance values this definition is based on the ratio of the power flowing past CD into the load
to the power entering the network at AB. Thus

$$\text{Transmission gain or loss} = \log_{10} \frac{P_{CD}}{P_{AB}} \text{(Bels)} \qquad (4.3)$$

where P_{AB} and P_{CD} are the powers entering and leaving the network
respectively.

For a simple network in which the impedance Z_{AB} equals the load
impedance, we have:

$$\text{Transmission gain or loss} = \log_{10} \left(\frac{V_{CD}}{V_{AB}}\right)^2 \text{(Bels)} \qquad (4.4)$$

The unit, called the Bel after the inventor of the telephone, is rather
large for most problems and a unit one-tenth the size of the Bel is
normally used. This is known as the decibel (abbreviation—dB). Thus
we have finally:

$$\text{Transmission gain or loss} = 10 \log \left(\frac{P_{CD}}{P_{AB}}\right) \quad \text{decibels}$$

$$= 20 \log \left(\frac{V_{CD}}{V_{AB}}\right) \quad \text{decibels} \qquad (4.5)$$

This definition is of great importance and it is desirable to know the
approximate logarithms of certain integers for everyday calculations.
A summary of the most useful power and voltage ratios and the corresponding number of decibels is given in Table 4.2 (see next page).

In the above definitions the output power or voltage appears in the
numerator so that if the network has gain the power or voltage ratio is
greater than unity and the logarithm is positive. If the ratio is less than
unity corresponding to a loss, the logarithm is negative. This convention
should be preserved so that, when carrying out calculations involving
gains and losses in tandem, it is only necessary to carry out simple addition and subtraction.

An example of the type of calculations which arise is given below.

A transmitter delivers 1 watt to an aerial. The transmission loss
between the transmitting aerial and a distant receiving aerial is 56 dB.
The receiving aerial is connected to a receiver with 33 dB gain. What is
the output power from the receiver? Calculate the additional gain
required to increase the output power to 30 mW.

The total 'gain' of the system is the sum of -56 dB for the transmission
path and 33 dB for the receiver, *i.e.* -23 dB.

From Table 4.2, we find that 20 dB$+3$ dB is equivalent to a power
ratio of 100×2.

Since the gain in dB is negative, the input power is divided by the
ratio to give the output power as $1/200$ W, *i.e.* 5 mW.

To increase the output power to 30 mW, a power ratio of 6, *i.e.* 2×3, is needed. From Table 4.2 we find that this corresponds to $3+4\cdot8$, *i.e.* $7\cdot8$ dB.

TABLE 4.2

Power ratio	Voltage ratio	Decibels (dB)
1	1	0
1·25	1·12	1
2	1·41	3
3	1·73	4·8
4	2	6
5	2·24	7
8	4	9
10	3·16	10
100	10	20
1000	31·6	30
10^4	100	40
10^5	316	50
10^6	1000	60

A further justification for taking the logarithm of power or intensity ratios as a measure of gains and losses in communication problems is the fact that the human senses respond according to a logarithmic scale of values. This is an example of an old-established law in psychology which states that the sensation is proportional to the logarithm of the stimulus. This is known as the Weber-Fechner Law[2]. In the case of sound intensity, tests on many observers show that the increase of intensity δI that can just be discerned is roughly proportional to the intensity I. Thus

$$\frac{\delta I}{I} = \text{constant } k$$

If we consider a series of intensities $I_1, I_2, I_3, \ldots I_n, I_{n+1}$ each just discernible from the previous value, then

$$\frac{I_2}{I_1} = \frac{I_3}{I_2} = \ldots \frac{I_{n+1}}{I_n} = (k+1)$$

Then

$$\frac{I_{n+1}}{I_1} = (k+1)^n$$

$$\log_{10} \frac{I_{n+1}}{I_1} = n \log_{10} (k+1)$$

Although the value of k varies between observers, and also varies with frequency, a rough order of magnitude is $0\cdot1$. If we then convert to decibels the $(n + 1)$th intensity can be expressed as a number of decibels above the first intensity, thus:

$$20 \log_{10} \frac{I_{n+1}}{I_1} = n \cdot 20 \cdot \log_{10} (1.1)$$

$$\doteqdot n \, \text{dB}$$

In fact 1 dB is about a 12% increase in intensity or voltage but the agreement between the unit and the least discernible step is good enough for most purposes. That is, the unit of power or intensity difference in communications, the decibel, is roughly the least discernible step to the ear.

If V_{AB} is taken to be 1 volt, then the voltage gain or loss factor is simply the received voltage V_{CD}. If the input signal of 1 volt is varied in frequency over a range which embraces the frequency band a particular channel has to transmit we can plot V_{CD} against frequency as in Fig. 4.2(a). If the channel passes d.c. the factor starts from a certain value at zero frequency and as the frequency is increased the factor may vary somewhat, but eventually it must cut off in a manner similar to a low pass filter as shown. If the channel has tuned or bandpass filter components in the circuit, the curve, which is called the amplitude-frequency characteristic, will appear as shown to the right of Fig. 4.2(a).

The amplitude-frequency characteristic alone is not sufficient to define all the external properties of a four-terminal network. It is also necessary to consider the phase shift between the input and output voltages, and this may be taken into account by the network or transfer function. If V_{CD} and V_{AB} are regarded as vectors or complex numbers, containing both amplitude and phase information the ratio

$$\frac{V_{CD}}{V_{AB}} = A(\omega) \exp\left[-j\phi(\omega)\right] \tag{4.6}$$

is defined to be the transfer function. A and ϕ are functions of the angular frequency $\omega = 2\pi f$ to indicate that the amplitude and phase vary with frequency. A is the amplitude used for the amplitude characteristic in Fig. 4.2, ϕ is the phase shift between input and output voltage and it is shown on the same figure by the dotted curves. If we take natural logarithms, the transfer function is expressed in real and imaginary parts as:

$$\text{Log}_e \frac{V_{CD}}{V_{AB}} = \log_e A(\omega) - j\phi(\omega) \tag{4.7}$$

The real part giving the attenuation in nepers and the imaginary part the phase shift in radians. The attenuation in decibels is given by $20 \log_{10} A(\omega)$ and it is readily shown from the relation between natural logarithms and logarithms to the base 10, that 1 neper is equivalent to 8·68 dB. The phase may of course be measured in radians or degrees.

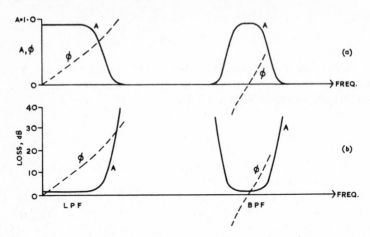

Fig. 4.2. Typical channel transfer functions. Those on the left apply to a low pass filter (LPF) and those on the right to a band pass filter (BPF). The solid curves (A) show the dependence of the voltage loss factor on frequency and the dotted curves show the dependence of phase on frequency. In (b) the loss factor is expressed in decibels.

The amplitude characteristic may be plotted logarithmically as shown in Fig. 4.2(b). For filters it is usual to plot the graph with attenuation or loss increasing in the positive direction as shown.

The overall transfer function for a number of networks or channels connected in tandem is given by multiplying together the individual transfer functions, provided that the correct impedance values are maintained at each junction. We therefore have for the combination

$$\frac{V_{out}}{V_{in}} = A(\omega) \exp [-j\phi(\omega)]$$

$$= A_1(\omega) \exp [-j\phi_1(\omega)] \times A_2(\omega) \exp [-j\phi_2(\omega)]$$
$$\times A_3(\omega) \exp [-j\phi_3(\omega)] \times \ldots \tag{4.8}$$

where

$$A_i(\omega) \exp [-j\phi_i(\omega)]$$

is the transfer function of the ith network and

$$A(\omega) \exp [-j\phi(\omega)]$$

is the transfer function of the combination.

Equation (4.8) is equivalent to the two separate equations:

$$A(\omega) = A_1(\omega) A_2(\omega) A_3(\omega) \ldots \tag{4.9}$$

and

$$\phi(\omega) = \phi_1(\omega) + \phi_2(\omega) + \phi_3(\omega) + \ldots \tag{4.10}$$

i.e. the amplification factors are multiplied together, as has already been seen, while the phases are added together. If logarithms are taken, we find from equation (4.8) that:

$$20 \log_{10} A(\omega) = 20 \log_{10} A_1(\omega) + 20 \log_{10} A_2(\omega) + \ldots \tag{4.11}$$

showing that the overall gain in decibels is the sum of the decibel gains of the individual networks.

The transfer function has been introduced above on the assumption that the signal has not undergone any change, such as modulation or coding during its passage through the channels considered. The concept of a transfer function can often be used even if modulation and de-modulation occurs within the channel. The ratio of the output voltage to the input voltage is again evaluated for a sinusoidal signal of angular frequency ω. It is clear that this approach is only permissible if there is no change in the sinusoidal waveform, i.e. if the output is a pure sine wave of the same frequency as the input sine wave. This requirement is equivalent to saying that the modulation and demodulation processes are linear. Any non-linearity will result in the appearance of the har-monics of an input sine wave, and an exact analysis then becomes extremely difficult. If the non-linearity is not excessive, it is, however, usually possible to apply the transfer function method as a first approx-imation and then to estimate the effects of the non-linearity at a later stage.

The behaviour of a linear channel passing a signal which is not sinus-oidal can be predicted by combining the ideas of frequency analysis, introduced in Chapter 2, with the transfer function discussed above. This will be done in the next section.

4.4. Signal Distortion in Linear Channels

If the input to a linear channel is a sinusoidal wave of angular frequency ω, the output will be a sinusoidal wave of the same frequency and its amplitude and phase can be determined with the aid of the transfer function. We have shown in Chapter 2 that any time waveform can be analysed into component sinusoidal waves, the number of components possible being infinitely large. The output from a linear channel can thus be calculated by carrying through the following three steps:

(a) Analysing the input waveform into its component frequencies.

(b) Changing the amplitude and phase of each component according to the value of the transfer function appropriate to its frequency, thus obtaining the corresponding output components.

(c) Summing the output component sinusoidal waves.

The simplest possibility arises when the transfer function is a real number which does not change with frequency. In this case, the amplitude of each input component is multiplied by a constant and its phase is unchanged. The summation in step (c) then reproduces the input waveform, apart from a change in amplitude. One circuit which has this type of transfer function is the ideal resistance attenuator shown in Fig. 4.1(a). The output from such a network is a faithful replica of the input.

In carrying out step (b) above, if the amplitude of the transfer function varies with frequency, or if the phase of the transfer function is not proportional to frequency, the output waveform is not identical in shape to the input waveform. The waveform is then distorted in passing through the channel. It is usual to distinguish between distortion caused by a frequency dependent amplitude characteristic and that caused by a frequency dependent phase characteristic.

4.4.1. AMPLITUDE DISTORTION

To illustrate the effect of amplitude distortion, we consider a square wave of fundamental frequency f, as shown in Fig. 4.3(a). This wave can be analysed by Fourier's Theorem into an infinite number of frequency components, given by the series:

$$E_1 \sin(\omega t) + \tfrac{1}{3}E_1 \sin(3\omega t) + \tfrac{1}{5}E_1 \sin(5\omega t) + \dots \qquad (4.12)$$

Suppose now that this square wave is transmitted through a channel which has an amplitude characteristic of the type shown in Fig. 4.4 by the curve AD. All components with frequencies of $5f$ and greater are attenuated to such an extent that their amplitudes may be taken as zero. The components of frequencies f and $3f$ are subject to equal attenuation, and so the channel output is simply proportional to the sum of the first two terms of equation (4.12). This output waveform is shown in Fig. 4.3(b) and is a distorted version of the square wave input. This kind of distortion is sometimes called 'frequency distortion', although 'amplitude distortion' is to be preferred. Further examples of such distortion are shown by the waveforms in Figs. 4.3(c) and 4.3(d). The one in Fig. 4.3(c) results if the waveform in Fig. 4.3(a) is transmitted through a channel with a characteristic similar to the curve AB in Fig. 4.4. The channel loss is greater for the third harmonic than for the fundamental. Similarly, if the channel characteristic is like the curve AC in Fig. 4.4, the third

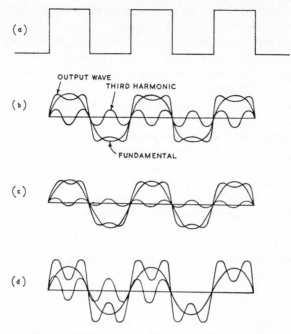

Fig. 4.3. Waveforms to show the effects of amplitude or frequency distortion: (a) Square wave, (b) Waveform obtained by suppressing all the harmonics in (a) except the first and the third, (c), (d) Waveform obtained by passing the square wave through a channel with characteristics similar to AB, AC in Fig. 4.4. respectively

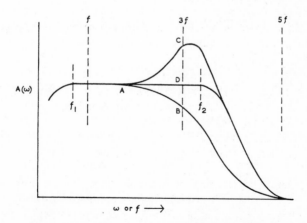

Fig. 4.4. Channel amplitude characteristics used to illustrate amplitude or frequency distortion.

harmonic is increased relative to the fundamental and the output wave-form has the shape shown in Fig. 4.3(*d*).

If the signal in Fig. 4.3(*b*) is to be received with no distortion of this kind, the amplitude relation of the third harmonic to the fundamental must be the same at the receiving end as at the transmitting end, and it is clear that the transmission loss (or gain) at *f* and 3*f* must be the same. In other words, the amplitude characteristic between *f* and 3*f* should be flat with frequency as shown in AD.

If a channel has to carry a signal having frequency components within the range f_1 to f_2, then the condition for no amplitude distortion of the signal is that the amplitude characteristic should be flat from f_1 to f_2. Generally speaking, it is possible to correct frequency distortion by connecting a suitable 4-terminal network of resistors, inductors and capacitors, known as an equalizer, in tandem with the channel. The equalizer has an amplitude characteristic which is complementary to that of the channel over the band f_1 to f_2 and the two added together give a flat characteristic. Sometimes this equalizer may be a simple capacitor or inductor connected in series or shunt at a suitable point in the channel.

4.4.2. PHASE AND GROUP DELAY AND PHASE DISTORTION

Before we can examine the effect of a frequency-dependent phase characteristic it is necessary to consider the relation between the phase characteristic and the time of transmission of a signal through the channel. Suppose a steady sinusoidal signal of angular frequency ω radians per second is being carried by a channel which has a total phase shift of β radians at that frequency. This phase shift may or may not be greater than 2π, but it is important to note that β is the total phase shift. If the input voltage and the received voltage are represented by two vectors, the received vector will lag the input vector by β radians. The time for the received voltage vector to sweep out this angle is simply:

$$t = \beta/\omega \qquad\qquad (4.13)$$

This time is known as the phase delay of the channel. Where it is associated with a distance of transmission, as, for example, in a line or a radio path, it can be used to calculate the phase velocity. However, when there are circuits in the sending and receiving equipment with filter character-istics, which cause delay, distance in metres or miles has no specific meaning.

It is essential to realize that the phase delay is not necessarily the signal delay. A steady sinusoidal signal cannot carry any intelligence, so that it is incorrect to deduce from the above reasoning that the phase delay is the signal delay. A signal can only be transmitted by making some change in the sinusoidal wave. If a very low frequency modulation is applied to the sine wave, <u>an observation of the delay of the modulation envelope provides the true signal delay</u>. If this envelope is observed at the input and output, the relationship between the envelope and the sine wave may in general differ at the two points. We wish to determine the delay of the envelope, rather than that of the enclosed sine wave. A signal with a superimposed low frequency modulation contains a narrow group of frequencies, and the delay of the envelope is therefore called the group or envelope delay.

Suppose that the signal consists of two infinitely long duration sine waves of equal amplitudes, one having an angular frequency of ω, the other having a slightly larger angular frequency ($\omega + \delta\omega$). These two sine waves, if plotted and added, are found to produce beats as shown in Fig. 4.5. The envelope has an angular frequency of $\delta\omega/2$, as shown by the following argument.

$$\text{Input signal} = \cos\omega t + \cos(\omega + \delta\omega)t$$
$$= 2\cos\left(\frac{\delta\omega}{2}t\right)\left[\cos\left(\omega + \frac{\delta\omega}{2}\right)t\right] \qquad (4.14)$$

The term in square brackets is the mean frequency of the two sine waves

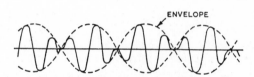

Fig. 4.5. Modulated waveform obtained by the superposition of two sinusoidal waves.

and is the frequency of the full line curve in Fig. 4.5. The term $2\cos(\delta\omega/2\ t)$ shows how the amplitude of this curve varies, *i.e.* it gives the envelope, shown by the dotted curve. Its angular frequency is ($\delta\omega/2$).

We now suppose that this signal passes through a channel which has the phase characteristic of Fig. 4.6. If β is the channel phase shift at angular frequency ω, and ($\beta + \delta\beta$) the phase shift at ($\omega + \delta\omega$) we have

$$\text{Output signal} = \cos(\omega t - \beta) + \cos\{(\omega + \delta\omega)t - (\beta + \delta\beta)\}$$
$$= 2\cos\left(\frac{\delta\omega}{2}t - \frac{\delta\beta}{2}\right)\left[\cos\left\{\left(\omega + \frac{\delta\omega}{2}\right)t - \left(\beta + \frac{\delta\beta}{2}\right)\right\}\right]$$
$$(4.15)$$

We are only concerned with the change produced in the envelope and this can be obtained from the first term, $2\cos(\frac{1}{2}t\delta\omega - \frac{1}{2}\delta\beta)$. The

envelope of the output therefore lags the input envelope by the phase
difference $\delta\beta/2$ and since the envelope frequency is $\delta\omega/2$ we find

$$\text{Envelope or group delay} = \frac{\frac{1}{2}\delta\beta}{\frac{1}{2}\delta\omega} = \frac{\delta\beta}{\delta\omega} \qquad (4.16)$$

In this argument, we have assumed that $\delta\omega$ is a very small frequency
difference and we now allow this difference to tend to zero so that the
group delay becomes $d\beta/d\omega$, the derivative of the channel phase shift
with respect to angular frequency. The group delay can therefore be
interpreted geometrically as the slope of the phase characteristic.

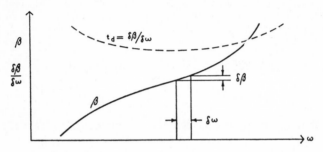

Fig. 4.6. Phase characteristic and group delay time of a channel.
The solid curve shows the variation of the phase shift, with frequency, and the dotted curve shows the corresponding group
delay time.

The group delay corresponding to the phase characteristic in Fig. 4.6
is shown by the dotted curve and, in the case illustrated, is seen to vary
with frequency. If a signal containing components throughout the frequency band shown is divided into a set of narrow band signals, each
of these will be subject to a different time delay in passing through the
channel. The output signal must therefore have a different wave form
from the input and distortion again appears. This type of distortion is
called phase or delay distortion. The difference between the maximum
and minimum values of the group delay over a prescribed operating
frequency band is called the differential delay and provides a convenient
numerical measure of the delay distortion.

A necessary condition for absence of delay distortion is that the group
delay time should be constant. This requires that

$$\frac{d\beta}{d\omega} = C \qquad (4.17)$$

where C is a constant.

Integration gives

$$\beta = C\omega + D \qquad (4.18)$$

where D is an integration constant.

To avoid delay distortion, we must therefore have a linear phase-frequency characteristic. This condition is not, however, sufficient, for the integration constant D may result in distortion. This may be seen by considering an input signal consisting of two sinusoids of different frequencies, say:

$$V_{in} = a_1 \sin(\omega_1 t) + a_2 \sin(\omega_2 t) \qquad (4.19)$$

If the network phase characteristic is defined by equation (4.18), the output waveform is:

$$V_{out} = a_1 \sin(\omega_1 t - \omega_1 C + D) + a_2 \sin(\omega_2 t - \omega_2 C + D)$$

and this may be expressed as:

$$V_{out} = a_1 \sin(\omega_1 t' + D) + a_2 \sin(\omega_2 t' + D) \qquad (4.20)$$

if t' is defined as $t - C$.

The dependence of V_{out} on t' is identical to the dependence of V_{in} on t, if the constant D is zero and clearly there is then no distortion. Further, if D is a multiple of π, V_{out} as a function of t' is the same as V_{in} as a function of t, apart from a possible sign change and again there is no distortion. The effect of any other value of D will be to displace the second sinusoid along the time axis relative to the first sinusoid and the output waveform is no longer a replica of the input waveform. This type of distortion is known as intercept distortion, and it arises if D has any value other than $n\pi$. D is the phase shift through the network for a sinusoid of frequency arbitrarily close to zero, *i.e.* it is the intercept of the phase-frequency characteristic on the phase axis.

4.5. Transient Response of Linear Channels

The properties of linear channels have so far been considered in relation to the steady-state behaviour specified by sinusoidal inputs. An alternative approach is to examine the behaviour of the channel when it is subjected to an input of a transient character. The transient selected depends on the type of network being investigated. For low frequency networks, either a unit step or a unit impulse is taken. For band pass networks, the transient is a carrier wave of mid-band frequency, modulated by a unit step or a unit impulse. Equivalent results must of course be obtained by the steady state and transient methods and this will be demonstrated in the next chapter.

If the unit step shown in Fig. 4.7(a) is applied to the input of a channel of the type shown in Fig. 4.1(c), the output signal will have the form of the curve ABC in Fig. 4.7(*b*). The time of build-up of the network is closely related to the upper cut-off frequency, f_2, of the frequency response. The decay time, t_1, is correspondingly related to the lower cut-off frequency, f_1. A channel which transmits d.c. will have an output similar to ABD, there now being no decay. In more complicated networks, the output waveform may include ripples, related as in all cases to the nature of the amplitude and phase characteristics.

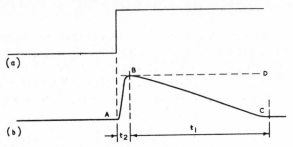

Fig. 4.7. (a) Unit step of voltage as used in testing low frequency networks. (b) Typical output waveform for unit step input

A unit step is a mathematical idealization and cannot be realized exactly in practice. Any practical test signal must have a finite build-up time, during which the signal rises from zero to unity. For satisfactory results, the build-up time of the test signal should be not greater than one-tenth of the build-up time of the output from the network under test. Present-day techniques are such that step waveforms with build-up times of a few nanoseconds (1 nanosecond $= 10^{-9}$ sec) can be achieved and these are adequate for testing networks with upper cut-off frequencies of up to about 100 MHz.

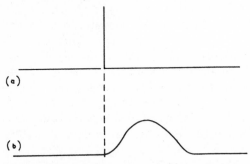

Fig. 4.8. (a) Idealized unit impulse of voltage.
(b) Typical ouput waveform for unit impulse input.

An alternative form of transient is the unit impulse, shown in its ideal form by the sharp spike in Fig. 4.8(*a*). A typical output waveform is shown in Fig. 4.8(*b*). The unit impulse is also a mathematical idealization and practical signals consist of pulses of finite duration. The input pulse length must be an order of magnitude less than the expected output pulse length for satisfactory results. The spectrum of a short impulse is flat over a wide range of frequencies and if the pulse is applied to a low pass filter whose cut-off frequency is within this range, the spectrum of the output pulse is equal to the frequency response of the filter (see Appendix A). This is an important result which will be used in a later chapter.

Band-pass channels can be tested by similar methods, using a carrier wave of mid-band frequency and modulating it either by a unit step (Fig. 4.9(*a*)) or by a short pulse (Fig. 4.9(*b*)). The envelopes of the output waveforms contain the required information regarding the channel properties. The considerations relating to the choice of pulse length in the

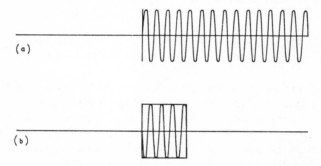

Fig. 4.9. (a) Carrier wave modulated by unit step, (b) Carrier wave modulated by short pulse.

second case are similar to those discussed above, and lead to the conclusion that the pulse length must be appreciably less than the reciprocal of the bandwidth of the channel under test. The phase of the carrier at switching in either of the cases illustrated in Fig. 4.9 has a slight effect on the response and in some networks, for example, those used in radar it is preferable to make the switching instant coincide with a zero of the carrier waveform. The pulsed-carrier signal of Fig. 4.9(*b*) is particularly well suited for investigating a type of distortion arising from multi-path effects and this is examined in the next section.

4.6. Channels with Multipath Effects

A further type of distortion which can occur in linear channels is one which is caused by multipath effects. Within the channel the signal can reach the receiving terminals by two or more paths of different delays, these different paths being associated with reflections from discontinuities. For example, if a cable system has two impedance irregularities as in Fig. 4.10(*a*) the signal can reach the receiving end via the two paths shown, one being direct, the other having an additional delay corresponding to twice the distance between the irregularities. In a radio system using very high frequencies, it is possible to receive signals by direct transmission between two aerials and sometimes by reflection from other objects such as hills, buildings, etc., as shown in Fig. 4.10(*b*). This causes the signal to consist of two components, one the direct component, the other a slightly delayed echo. In a long-distance radio system using high frequencies which are reflected from the ionosphere the same effect can take place owing to the existence of one hop and two hop paths as shown in Fig. 4.10(*c*). In some cases there may be

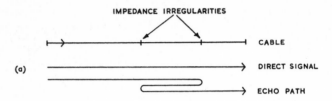

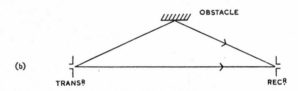

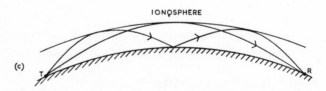

Fig. 4.10. Effect of discontinuities in producing multipath effects. (a) Cable system with impedance irregularities. (b) Reflection of radio waves from an obstacle. (c) Multiple paths involving reflections from the ionosphere.

more than two paths or echoes present, and so the general term multi-path effect is used.

When there are two paths, as in the above examples, the channel performance can be represented as in Fig. 4.11 by two paths in parallel. Up to the point B the delay t of the two paths is the same, so that at this point the two signals are still in phase and the envelopes are in step. Between B and C there is an additional delay Δt in the second path. It is assumed that each path is dispersionless, *i.e.* each has a linear phase characteristic. If the channel is carrying a steady-state sine wave of frequency f_1 the number of periods n_1 in the excess delay is given by

$$n_1 = \Delta t \cdot f_1 \qquad (4.21)$$

If, now, the frequency is varied there will be a simple case of interference between two sine waves at the receiving end. If n_1 is a whole number the two waves will add. When there are $n_1 + \frac{1}{2}$ periods in Δt they will subtract. The frequency f_2 at which the two waves will again add is given by

$$\Delta t \cdot f_2 = n_1 + 1 = \Delta t \cdot f_1 + 1$$

or

$$f_2 - f_1 = \frac{1}{\Delta t} \qquad (4.22)$$

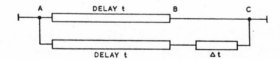

Fig. 4.11. Representation of the multipath effect by parallel paths.

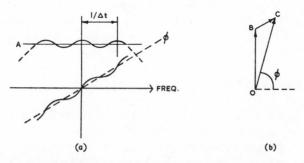

Fig. 4.12. (a) Amplitude and phase characteristics for a channel showing multipath effect. (b) Phasor diagram for multipath effect.

If the signal in the second path is rather weaker than the direct signal, then the amplitude-frequency curve will show ripples as in Fig. 4.12(a), the frequency difference between consecutive ripples being $1/\Delta t$. The phase characteristic will also have ripples as can be seen from the phasor diagram in Fig. 4.12(b). As the frequency is varied the small interfering signal phasor BC rotates on the end of the phasor OB representing the direct signal. The resultant vector OC will vary in phase about the mean phase of the direct signal. In one interference cycle the phase-frequency curve will show one cycle of ripple as in Fig. 4.12(a).

This is an example of an important theorem which controls the relationship between the amplitude and phase characteristics in physical networks. It is important to realize that the amplitude and phase characteristics cannot be chosen entirely independently, a point which is considered in more detail in Chapter 5.

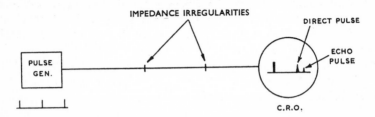

Fig. 4.13. Experimental arrangement to display multipath effect.

If the two-path line in Fig. 4.10(a) has an input signal consisting of a very narrow pulse, or at least a pulse narrower than Δt the received signal will consist of the direct pulse, followed by a weaker echo pulse. (In the radio case a pulse of carrier would be used.) If the input pulse is repetitive, the received signal can be displayed on a cathode ray oscillograph having a synchronized time base, as shown diagrammatically in Fig. 4.13. This is an example of the close relationship between steady state and transient testing, and between amplitude/phase characteristics and impulse response.

4.7. Non-linear Distortion

It has so far been assumed that the channel under consideration is linear, *i.e.* that the output voltage or current is directly proportional to the impressed input voltage. A passive network assembled from resistors, inductors and capacitors, each obeying Ohm's Law, has this linear behaviour, but any active network must inevitably show non-linear behaviour if the output is increased beyond a limit determined by the power which the network is capable of supplying. In such a situation, the network is said to be overloaded and an increase in the input voltage will not cause a corresponding increase in the output voltage. The transition from linear to overloaded operation usually occurs gradually and a typical output-input voltage characteristic is shown in Fig. 4.14. Any active circuit, *e.g.* any circuit using valves or transistors, has such a characteristic and certain passive circuits, *e.g.* those in which the coils are wound on ferro-magnetic materials, may behave similarly. A communication channel in general has a number of active devices connected in cascade and each of these devices will have its own overload characteristic. We will be concerned in this section with the overall behaviour of the channel obtained by plotting the characteristic for the channel output against the channel input. The presence of a non-linear characteristic leads to signal distortion, known as non-linear distortion. A general analysis of non-linear behaviour is extremely complicated but, fortunately, in most communication problems, we are interested primarily in those parts of the characteristic which are approximately linear. The departures from linearity can then be assumed small and approximate methods of calculation can be used. We are interested in finding out what effects the non-linearity produces on the signal and in obtaining numerical measures by which the non-linear behaviour can be assessed.

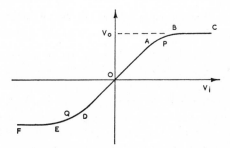

Fig. 4.14. Typical non-linear input-output characteristic.

The characteristic shown in Fig. 4.14 shows only the effect on the amplitude of the signal. In practice, we will be concerned with baseband channels for which the phase characteristic is virtually linear throughout the frequency range being used and we may therefore assume that the channel introduces no phase shift. It is then sufficient to concentrate our attention on the amplitude characteristic. The curve in Fig. 4.14 has a reasonably linear portion between D and A, two 'knees' between A and B and D and E, and two regions BC and EF in which the output is unaffected by the input. In these latter two regions, the network is fully overloaded or saturated. The useful range of operation will be defined by points P,Q lying somewhere between A and B and between D and E, the precise position of P and Q being determined by the degree of non-linear distortion which can be tolerated. P,Q can be regarded as overload points, but it is clear that they are not well defined and one of our objectives is to obtain a criterion for defining them.

One possible method of assessing the channel behaviour is to use the graphical method illustrated in Fig. 4.15. The input signal waveform is drawn on the input axis, and the output waveform obtained by point-to-point projection for selected equal time intervals. In practice, the degree of distortion with which we are concerned is so small that it is barely perceptible on a graph of this kind. The effect is deliberately exaggerated in Fig. 4.15 by taking the input amplitude sufficiently large to involve operation in the saturated regions. Recourse is therefore made to an analytical method and the characteristic is expressed mathematically by writing the output voltage v_0 as a power series of the input voltage v_i:

$$v_o = av_i + bv_i^2 + cv_i^3 \ldots \qquad (4.23)$$

The first term represents the linear response of the channel, the constant a representing the gain or loss. The second term represents a lack of symmetry between the positive and negative portions of the response curve and the magnitude of the asymmetry is determined by the constant b. (This constant would be zero for the curve shown in Fig. 4.15.) The third term gives a first approximation to the flattening of the curve

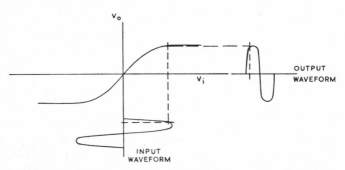

Fig. 4.15. Graphical method of assessing non-linear distortion.

caused by overloading and produces a symmetrical effect. In a typical case, such as that in Fig. 4.15, the corresponding constant c is negative and its magnitude is a measure of the extent to which the curve is flattened. Terms involving higher powers of v_i in equation (4.23) may usually be neglected when attention is concentrated on the region of the curve which approximates to linear operation.

The analysis of the effect of the non-linear terms is in principle similar to that used in Chapter 3 to discuss the operation of detectors and modulators. The emphasis is, however, changed in that non-linear operation is the essential feature of a detector or modulator, whereas here we are concerned with small departures from linearity. We begin by considering the input signal to be a sinusoidal wave, *i.e.*

$$v_i = E \sin(\omega t) \qquad (4.24)$$

and substitution in equation (4.23) gives:

$$v_o = aE \sin(\omega t) + bE^2 \sin^2(\omega t) + cE^3 \sin^3(\omega t) \ldots$$
$$= aE \sin(\omega t) + \tfrac{1}{2}bE^2(1 - \cos 2\omega t) + \tfrac{1}{4}cE^3(3 \sin \omega t - \sin 3\omega t) + \ldots \qquad (4.25)$$

The second term causes the appearance of both d.c. and second harmonic terms, while the third introduces a change in the fundamental amplitude and a third harmonic term. Since we are concerned primarily in distortion, *i.e.* changes in the shape of the waveform, we may ignore the d.c. contribution. Further, since we have postulated that the departures from linearity are small, we can neglect the contribution of the third term to the fundamental. The output waveform then contains the following components, with amplitudes as shown:

Fundamental : aE
Second harmonic: $\tfrac{1}{2}bE^2$
Third harmonic : $\tfrac{1}{4}cE^3$

One effect of non-linearity is thus to cause the appearance of harmonics in the output if the input is a pure sinusoid. This is referred to as harmonic distortion. The amplitudes of the harmonics are given by the expressions above and it should be noted that an increase in the input signal E causes relatively greater increases in the harmonic amplitudes. For example, if the amplitude of the input signal is doubled, the second and third harmonics are increased by factors of 4 and 8 respectively. A 6 db increase in the input thus causes a 12 db increase in the second harmonic and an 18 db increase in the third harmonic. If the maximum tolerable values of the harmonic amplitudes in the output relative to the fundamental in the output are specified, the above results can be used to determine the largest input signal which can be used.

A second effect caused by non-linearity can be illustrated by supposing that the input voltage contains two sine waves, *i.e.*

$$v_i = E_1 \sin(\omega_1 t) + E_2 \sin(\omega_2 t) \qquad (4.26)$$

The output voltage is calculated by substituting this expression in equation (4.23) and to simplify the working, we will examine the individual terms. We have:

$$bv_i{}^2 = b(E_1 \sin \omega_1 t + E_2 \sin \omega_2 t)^2$$
$$= bE_1{}^2 \sin^2 \omega_1 t + 2bE_1 E_2 \sin \omega_1 t \cdot \sin \omega_2 t + bE_2{}^2 \sin^2 \omega_2 t$$
$$= \tfrac{1}{2}(bE_1{}^2 + bE_2{}^2) - \tfrac{1}{2}bE_1{}^2 \cos(2\omega_1 t) - \tfrac{1}{2}bE_2{}^2 \cos(2\omega_2 t)$$
$$+ bE_1 E_2 \cos(\omega_1 - \omega_2 t) - bE_1 E_2 \cos(\omega_1 + \omega_2 t) \quad (4.27)$$

In addition to the d.c. and second harmonic terms corresponding to the single frequency input, we have terms at the sum and difference frequencies. Such terms are referred to as intermodulation products. The term $cv_i{}^3$ gives a similar result, in that, in addition to the expected fundamental and third harmonic terms, there will be further intermodulation products. Concentrating our attention on these, we have:

$$cv_i{}^3 = c(E_1 \sin \omega_1 t + E_2 \sin \omega_2 t)^3$$
$$= c(E_1{}^3 \sin^3\omega_1 t + E_2{}^3 \sin^3 \omega_2 t) + 3cE_1 E_2(\sin \omega_1 t \times \sin \omega_2 t)$$
$$\times (E_1 \sin \omega_1 t + E_2 \sin \omega_2 t)$$
$$= \text{terms in } \omega_1, 3\omega_1, \omega_2, 3\omega_2,$$
$$+ \frac{3cE_1 E_2}{2}[\cos(\omega_1 - \omega_2)t - \cos(\omega_1 + \omega_2)t]$$
$$\times [E_1 \sin \omega_1 t + E_2 \sin \omega_2 t]$$
$$= \text{terms in } \omega_1, 3\omega_1, \omega_2, 3\omega_2,$$
$$+ \frac{3cE_1{}^2 E_2}{4}(\sin \overline{2\omega_1 - \omega_2}t + \sin \omega_2 t)$$
$$+ \frac{3cE_1 E_2{}^2}{4}(\sin \omega_1 t + \sin \overline{2\omega_2 - \omega_1}t)$$
$$- \frac{3cE_1{}^2 E_2}{4}(\sin \overline{2\omega_1 + \omega_2}t - \sin \omega_2 t)$$
$$- \frac{3cE_1 E_2{}^2}{4}(\sin \overline{2\omega_2 + \omega_1}t - \sin \omega_1 t) \quad (4.28)$$

Equation (4.28) shows that the output contains intermodulation products of frequencies $2\omega_1 \pm \omega_2$ and $2\omega_2 \pm \omega_1$. The products arising from the term in $v_i{}^2$ are referred to as second-order products and those from the term in $v_i{}^3$ as third-order products. The intermodulation products can thus be summarized as follows:

	Frequency	Amplitude
Second order	$\omega_1 + \omega_2$	$bE_1 E_2$
	$\omega_1 - \omega_2$	$bE_1 E_2$
Third order	$2\omega_1 + \omega_2$	$\tfrac{3}{4}cE_1^2 E_2$
	$2\omega_1 - \omega_2$	$\tfrac{3}{4}cE_1^2 E_2$
	$2\omega_2 + \omega_1$	$\tfrac{3}{4}cE_1 E_2^2$
	$2\omega_2 - \omega_1$	$\tfrac{3}{4}cE_1 E_2^2$

The above table shows the way in which the amplitude of each component is influenced by changes in the amplitudes of either constituent of the input waveform.

The amplitudes of the harmonics and the intermodulation products can be measured experimentally by using a harmonic analyser. This consists of a very narrow band tunable bandpass filter which can be connected to the output of the channel. The amplitudes of each frequency component in the output can thus be measured by a suitable valve voltmeter and the relations between the output and input components can be verified. Such experiments confirm that the three-term approximation to the output-input characteristic is adequate to describe the non-linear behaviour of most practical communication channels.

If the input signal contains more than two sinusoidal components, a more complicated situation results and intermodulation and harmonic distortion can be particularly objectionable in multichannel telephony, where independent signals are conveyed by adjacent frequency bands. Harmonics produced by components in one frequency band can have frequencies which lie in an adjacent band and can thus cause interference between the two independent signals. Similarly, intermodulation products caused by components in two different bands can have frequencies lying in a third band. To avoid such complication, the equipment used in multichannel systems must have a high degree of linearity.

Owing to the gradual onset of overload it is difficult to specify precisely the overload level, and an arbitrary definition must be employed. This is sometimes defined as the input level of a sine wave which causes the output sine wave to be accompanied by a certain amount of harmonic (generally a total harmonic r.m.s. voltage which is 5 per cent of the fundamental output voltage). Sometimes it is defined as the input level of a sine wave which produces a fundamental frequency output which is 1 db below that which it should be if linear. This is sometimes referred to as 1 db compression. The ratio of this overload level to the r.m.s. noise level (Chapter 6) is called the dynamic range of the channel.

If a four-terminal non-linear network which has the non-linear characteristic of Fig. 4.16(a) is connected in tandem with a network having the reciprocal characteristic of Fig. 4.16(b) it follows that the overall characteristic produces no distortion. This can be seen by carrying out the graphical construction of Fig. 4.15 and then carrying out a similar construction, applying the distorted wave to the reciprocal characteristic. In theory at least non-linear distortion can be eliminated or reduced, provided a network with a sufficiently accurate reciprocal characteristic can be constructed. Of course the reciprocal characteristic can only apply up to a certain input level, beyond which it must exhibit overload phenomena as shown by the dotted portion in Fig. 4.16(b). This level can be kept above the range of interest. In practice, any precise cancellation of non-linear distortion is, with present knowledge, out of the question, but these principles are often applied in telephony to match a signal widely varying in amplitude range between overload and noise. Thus a device with a characteristic such as Fig. 4.16(a)

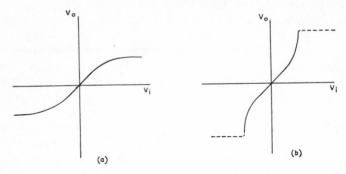

Fig. 4.16. Examples of reciprocal characteristics used to eliminate non-linear effects.

included in the sending equipment may be used to compress the amplitude range of a signal, and by working sufficiently close to the channel overload, a better margin over noise is achieved. The reciprocal device or expander in the receiving equipment has the characteristic of Fig. 4.16(b) and restores the original range of the signal. Such a device might be called an instantaneous compressor, and such principles are applied in automatic modulation control and companders (compressor-expanders).

4.8. Miscellaneous Channel Characteristics

The characteristics considered so far apply to some degree to any form of practical channel. There are several other properties which may be encountered in particular types of channel and this section contains some comments on the more important of these.

It has been assumed that the channel characteristics remain constant with time but in practice changes may occur. The extent to which these changes affect the transmission of a signal is known as the stability characteristic of the channel. Changes may arise in the equipment used, for example, if the power supply is subject to fluctuations. It is, however, possible by good design to ensure that equipment changes are kept to a sufficiently small level to avoid any significant change in the signal. The stability of equipment can be assumed adequate for all practical purposes.

In certain radio channels, however, changes in channel performance may arise because of changes in propagation conditions. For example, long-distance radio channels use transmission paths in which the radio waves are reflected from the ionosphere, and the reflecting properties of the ionosphere do change for a number of reasons. Changes in the amplitude of a signal arising from such changes in propagation conditions are referred to as fading. The ratio of the maximum signal to the minimum signal at the receiver input is defined as the fading ratio and may be expressed in decibels. In many practical cases, the statistics of the received signal, when a steady carrier signal is being transmitted, are similar to those of a noise signal at the output of a filter of very narrow bandwidth. The effect of fading may thus be likened to the introduction of noise sidebands at frequencies in the immediate vicinity of the carrier. One method of reducing fading is to apply automatic gain control (AGC) to the receiving amplifier. This is a form of negative feedback in which the signal level at the output of the amplifier is used to control the amplifier gain in such a way that the output level remains constant despite changes of the input level. The operation of the AGC removes the noise sidebands near the carrier and this plays the part of a narrow band rejection filter. The bandwidth of the filter is determined by the time constant of the negative feedback loop. Any signal components lying within the bandwidth of the equivalent rejection filter will also be removed and so in such systems very low modulation frequencies are not used.

Fading may sometimes be strongly frequency dependent and may affect the different frequency components of a modulated waveform to different degrees. The term selective fading is then applied. One advantage of single sideband operation is that it is less affected by selective fading than is a double sideband system.

In radio channels and in some other channels, the signals are subjected to frequency changes in modulators and demodulators. If the local oscillators used at the modulator and the demodulator are not in exact synchronism, the frequencies in the output signal from the channel will be displaced from those of the input. Such a frequency shift is equivalent to a waveform distortion and must be kept within tolerable limits. It is particularly prone to occur in single sideband systems and a high degree of stability in the local oscillators, used in such systems, is required to avoid excessive distortion. A frequency shift can also arise if a radio wave is reflected from a moving object because of the Doppler effect.

4.9. Multi-channelling

It is frequently necessary to obtain more than one channel from a communication facility such as a cable or radio link. Two methods are available, corresponding to the time and frequency forms of a message. The frequency division method is the more generally known because it

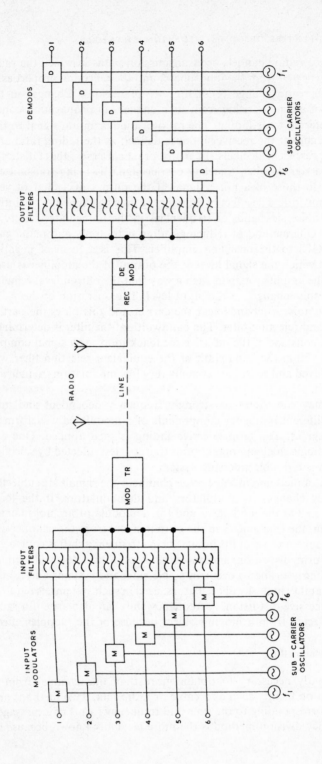

Fig. 4.17. Essentials of a Frequency Division Multiplex (FDM) system.

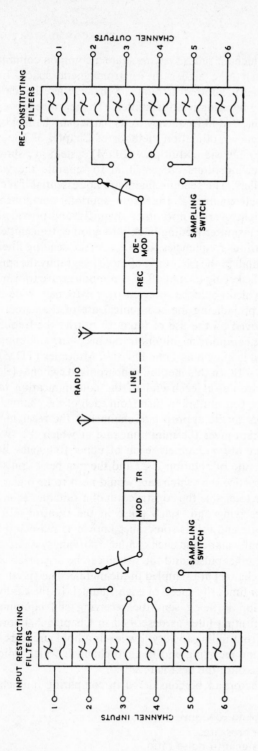

Fig. 4.18. Essentials of a Time Division Multiplex (TDM) system.

is the method by which all radio systems share a common communication medium, namely, the aether or electromagnetic space. In this method each channel is allotted a frequency band, these bands being stacked in frequency by using bandpass filters. The message for a particular channel is translated in frequency to its particular channel by one of the carrier frequency modulation methods of Chapter 3. The essentials of a Frequency Division Multiplex (FDM) system are shown in Fig. 4.17. The receiving filters are essential to separate the wanted channel from all others. The sending filters are not essential if it can be guaranteed that each channel is free from spurious components of frequency which might overlap into other channels, and provided that the power of one transmitter feeding back into another transmitter will not cause other spurious frequencies. In practice, the sending filters are almost always essential. In the early days of radio telegraphy the spurious frequencies due to the keying on and off of continuous wave transmitters caused key clicks which could be received over extremely wide bands of frequency, thus prejudicing the economical use of the aether. This situation was improved by the use of filters to restrict the frequencies radiated by radio telegraph transmitters to the necessary minimum.

The other method is known as Time Division Multiplex (TDM), and is illustrated in Fig. 4.18. In this method synchronized switches (electromechanical or electronic) at each end of the communication facility enable samples to be transmitted in turn from each of the channels and delivered as samples to the appropriate channel at the receiving end. The sampling theorem gives the minimum rate at which the switches must rotate. If there are n channels, each of upper frequency limit f, then the minimum rate of rotation is $2f$ and the number of pulses per second is $2nf$. In practice the switch rate would need to be higher than the minimum figure to enable the construction of a suitable reconstituting filter at the receiving end. Each pulse in the transmission path conveys an amplitude and this is sometimes known as pulse amplitude modulation. If the information in each channel is of binary digital form, the timing of the channel pulses and the switch can be organized so that the 0's and 1's in a channel are sampled in the middle. The speed on the line is then simply n times the speed of each channel. In the case where analogue information is being sent the receiving end equipment is basically a re-constituting filter as described in Chapter 2. In order to prevent the signal from falling to very low levels a hold circuit is sometimes provided at the receiving end in each channel, which holds the voltage of one pulse until the next is received.

The important factors to be considered in comparing multichannel systems are:

1. Frequency band economy.
2. Crosstalk or crossfire.
3. Power and signal to noise ratio.

If ideal filters and techniques are assumed, both FDM and TDM systems require the same total bandwidth. Thus, with FDM assuming ideal frequency translation (requiring the use of single sideband modulation), and ideal rectangular filter frequency characteristics, n channels, each of bandwidth 0 to f would require a total band of nf. The TDM system would have to transmit $2nf$ pulses per second. If these pulses were of the $(\sin x)/x$ shape, the peak amplitude representing the sample, then an ideal filter of frequency cut-off equal to nf would pass the pulses. This follows from the unique feature of this shape of pulse. In Fig. 4.19

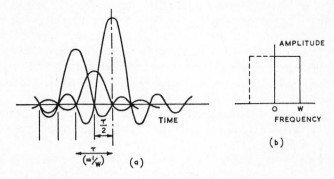

Fig. 4.19. Sequence of pulses in an ideal TDM system. Number of pulses per second $= 2/\tau = 2W = 2nf$ where $n =$ no. of channels and $f =$ cut-off frequency of each channel.

a number of successive pulses are shown. If the maximum amplitude of a particular pulse represents the quantity to be conveyed, then all other pulses are zero at that instant. The frequency transform of such a pulse waveform is rectangular. If the time to the first zero of a pulse is $1/2nf$ seconds, the cut-off of the spectrum is nf cycles per second and the pulse will pass through an ideal filter of the same cut-off without change. This ideal argument is not practicable because filters of infinitely sharp cut-off and linear phase characteristic have an infinite group delay (infinite lag filters).

In the FDM system the channel filters require a frequency band between channels where the filters can develop their attenuation, sometimes called the cross-over band or guard band. The width of this band may be reduced by using more elaborate and expensive filters but it can rarely be less than about 20 per cent of a channel bandwidth. In the TDM system the main channel pulses must be of finite duration and the main channel frequency cut-off must extend well beyond the ideal limit given above. The actual amounts by which the bandwidths of FDM and TDM systems exceed the above ideal limits will depend upon several factors, the most important of which is perhaps crosstalk between channels.

Crosstalk in FDM may be investigated by considering adjacent filters of two different types, as illustrated in Fig. 4.20. The first pair in Fig. 4.20(*a*) are smooth cut-off filters such as might be produced by simple resonant or coupled circuits. The second pair shown in Fig. 4.20(*b*) have

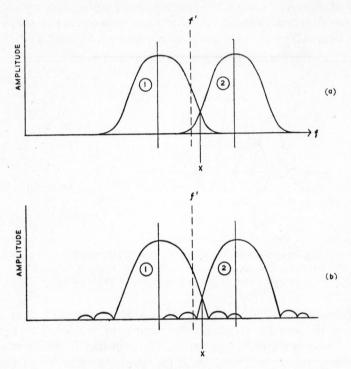

Fig. 4.20. Adjacent filters in an FDM system.

frequency characteristics in which there are zeros of amplitude response or 'attenuation peaks'. If now we consider a frequency f' which falls within channel 1, it will cause leakage into channel 2 with both types of filter. The amount of leakage depends entirely upon the filter design. If the system has similar sending filters, a cross attenuation measurement from the input of channel 1 to the output of channel 2 over the whole band of the two filters should indicate sufficient loss to keep crosstalk down to the desired value. This generally means that the filter attenuation in decibels at the middle of the guard band (point X) should be at least half the desired crosstalk attenuation.

In the TDM case, the main channel frequency characteristic will cause each channel pulse to have a rise time and a decay time. If the decay time of one pulse extends into the sampling time of the succeeding pulse it is clear that crosstalk will be produced. This situation can be dealt with if the system is handling a binary digital signal, but if an analogue signal is being used, difficulties can arise. It is generally necessary to shape the frequency characteristic of the main channel so as to provide pulses which are reasonably symmetrical and monotonic, *i.e.* free from ringing. This requires a filter of gradual cut-off, generally something approaching the Gaussian filter. It is difficult to generalize, but it is usually more difficult to provide high crosstalk attenuations with the TDM system, than with the FDM system.

The last important factor to consider in comparing FDM and TDM methods of multi-channelling is that of power and signal to noise ratio. If we return to the two systems with hypothetical ideal filters, the noise power from the main channel will be the same in both cases and the noise power per channel will also be the same. If the transmitter has a common amplifier, for the same received signal to noise ratio, this transmitter will require to work at the same mean power level for TDM and FDM systems. In the TDM case the transmitter peak power is the same as the channel peak power. In the FDM case the transmitter must be capable of handling n^2 times the channel peak power because the peak voltages in each channel may all add in phase, although the probability of this is low. This consideration tends to favour TDM when the number of channels is large, but it must be stressed that every case must be considered on its merits.

Intermediate or terminal amplifiers in an FDM multi-channel system must be capable of handling the sum of the instantaneous voltages in each channel. Consider a system of n channels, having a signal of a certain amplitude in each channel. It is assumed that when these signals are translated into line frequencies they will result in a signal on the line consisting of n uniformly spaced frequencies. Such a state of affairs is represented by a voice frequency telegraph system or a single sideband carrier telephone system carrying steady signals in each channel. It may may be represented graphically by a frequency group as in Fig. 4.21.

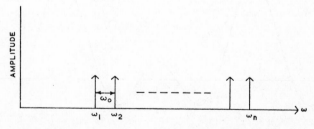

Fig. 4.21. Frequency group in an FDM telegraph system.

If we make a further assumption that at time $t = 0$ all frequencies have zero instantaneous amplitude and are all increasing positive, then the total signal on the line is given by:

$$e = \sum_{k=1}^{n} \sin(\omega_k t) \tag{4.29}$$

Suppose that the frequencies are equally spaced at frequency interval ω_0 so that $\omega_k = \omega_1 + k\omega_0$. Then:

$$e = \text{Im.} \sum_{k=1}^{n} \exp j(\omega_1 + k\omega_0)t$$

$$= \text{Im.} \exp(j\omega_1 t) \frac{[1 - \exp(jn\omega_0 t)]}{1 - \exp(j\omega_0 t)}$$

$$= \text{Im.} \exp \left(j\omega_1 t + j\frac{n}{2}\omega_0 t - j\frac{\omega_0}{2}t \right) \cdot \frac{\exp(\frac{1}{2}jn\omega_0 t) - \exp(-\frac{1}{2}jn\omega_0 t)}{\exp(\frac{1}{2}j\omega_0 t) - \exp(-\frac{1}{2}j\omega_0 t)}$$

$$= \text{Im.} \overline{\exp(j\omega_1 + \frac{1}{2}n\omega_0 - \frac{1}{2}\omega_0 t)} \frac{\sin(\frac{1}{2}n\omega_0 t)}{\sin\frac{1}{2}\omega_0 t}$$

$$= \sin(\omega_m t) \cdot \frac{\sin(\frac{1}{2}n\omega_0 t)}{\sin\frac{1}{2}\omega_0 t} \tag{4.30}$$

where

$$\omega_m = \omega_1 + \frac{1}{2}(n - 1)\omega_0 \tag{4.31}$$

The frequency ω_m is the mean of the frequencies of the group and equation (4.30) can be interpreted as implying that the resultant wave is a sinusoid of frequency ω_m with an envelope given by

$$A(t) = \frac{\sin(\frac{1}{2}n\omega_0 t)}{\sin\frac{1}{2}\omega_0 t} \tag{4.32}$$

Examination of the properties of $A(t)$ shows that its largest magnitude is n and that this occurs whenever $\omega_0 t$ is a multiple of 2π. A typical curve is shown in Fig. 4.22 for $n = 5$.

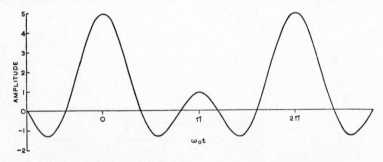

Fig. 4.22. Envelope of five frequency group in FDM system.

Any line amplifier, either sending, intermediate or terminal, must be capable of handling this signal. It should not overload with a signal amplitude of n times the amplitude of the signal per channel. This is an important feature and must be considered in the choice of transistors, the design of transformers, the rating of resistors and other components. However, in practical systems, relaxations are usually possible because peak amplitudes are unlikely to occur simultaneously in all channels.

Another important requirement of the line amplifier is that for signals smaller than the designed maximum, the production of harmonics and combination tones must be as small as possible. If a signal in a lower frequency channel, in passing through an amplifier, produces harmonics which fall in an upper frequency channel then interference of an unintelligible crosstalk nature will occur in the latter channel. Likewise, if frequencies in·two or more channels are mixed in passing through a line amplifier so as to produce combination frequencies within the band of the carrier system, then crosstalk between channels will occur. These phenomena are the result of non-linearity in the system below the overload level.

4.10. Objective and Subjective Testing of Communication Channels

Channels, including their transducers, for personal communications (*e.g.* telephony, broadcast radio, television), may be tested by two methods, objective and subjective. By objective testing we mean the use of instruments to measure specific properties of the channel such as the amplitude-frequency characteristic, or harmonic distortion. Several different instruments may be necessary to determine all the relevant properties. The comparison of channels when only quantities obtained by such measurements are available may present some difficulty. Nevertheless, such objective measurements are an essential requirement for the development and close control of the performance of channels.

For the ultimate comparison of channels there is a need for an overall type of test which includes the human beings at the ends of the channel, a test which proves the overall ability of the channel to provide the kind of communication concerned. This is known as subjective testing and in the case of telephony it has played an important part in improving the quality of telephone connections. One method of carrying out such tests is to set up a test telephone circuit with acoustically treated rooms at each end in order to be independent of such conditions, and to pass large numbers of meaningless words. The percentage of words correctly recorded by the listener is a measure of the quality of the channel with its transducers. This is known as articulation testing and it provides a severe criterion of performance. A slightly less severe method is to employ meaningful short sentences. In order to obtain repeatable results it is necessary to control very closely the test procedures employed. The test card employed for television system testing is a form of subjective test. The use of alternate black and white lines carefully spaced provides a test of the high-frequency response of the whole channel including the receiver.

For telegraph systems, test messages and error counts provide a form of subjective test which augments objective tests such as telegraphic distortion measurements.

Chapter 5

The Response of Linear Channels

5.1. Introduction

The specification of linear channels by the steady-state input method and by the transient method was dealt with in general terms in the last chapter. The spectra of certain signals were deduced in Chapter 2 and methods of modulating signals or translating them to higher frequencies were discussed in Chapter 3. This chapter to some extent correlates these topics by considering the response of a linear channel to an arbitrary applied signal. Just as there are two ways of specifying the properties of a channel, so, also, are there two ways of calculating the general response. These will be discussed. Certain assumptions are made regarding the characteristics of channels in order to simplify the mathematics involved. These assumptions mean that such channels are not obtainable in practice and they are referred to as idealised channels. The limitations on the characteristics of practical channels are discussed and it is shown that a consideration of idealised channels is useful in giving an insight into the behaviour of the practical channels.

5.2. Determination of the Response

5.2.1. THE FOURIER INTEGRAL METHOD

This method stems directly from the steady-state method of specifying the behaviour of a network or channel. Consider the network in Fig. 5.1,

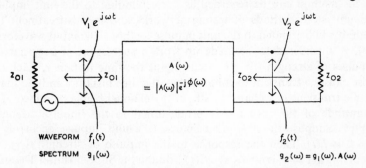

Fig. 5.1. Network under steady state test; the source and load are matched to be input and output of the network respectively.

which is correctly matched at both terminations, and suppose it is under steady-state test at the angular frequency ω. If the input voltage is V_1

$\exp(j\omega t)$ and the output voltage is $V_2 \exp(j\omega t)$, then, considering V_1 and V_2 as phasors, we have:

$$V_2/V_1 = A(\omega) \tag{5.1}$$

where $A(\omega)$ is the complex transfer function of the channel. The properties of the transfer function are discussed in Section 4.3.*

If the input signal has the general time waveform $f_1(t)$, we may obtain the input spectrum by calculating the Fourier transform $g_1(\omega)$. As we have seen in Chapter 2, $g_1(\omega)$ will in general be a complex function of frequency. Since the channel is a linear one, the output spectrum is obtained at every frequency by multiplying the input spectrum by the transfer function and so we have:

$$g_2(\omega) = A(\omega) \cdot g_1(\omega) \tag{5.2}$$

where $g_2(\omega)$ is the output spectrum. The output waveform, $f_2(t)$, obtained by using the inverse transform, is therefore:

$$f_2(t) = \frac{1}{2\pi} \int_{-\infty}^{\infty} A(\omega) \cdot g_1(\omega) \cdot \exp(j\omega t) \cdot d\omega \tag{5.3}$$

In many practical problems, it is impossible to evaluate this integral explicitly, but nevertheless the general principle of this method is of fundamental importance. If numerical information is required for a particular case, the integral can always be evaluated by using a digital or an analogue computer. A suitable form of analogue computer is, of course, a network which has the same transfer function as the channel being considered.

5.2.2. THE DUHAMEL INTEGRAL METHOD

This method can be regarded as corresponding to the unit impulse and unit step methods of testing networks and in this treatment, the method will be related to the unit impulse method. The input waveform, $f_1(t)$, is considered to be made up from a succession of very narrow impulses of strength $f_1(\tau)\,d\tau$, occurring at the time instant τ. Each impulse causes a transient to commence at the time instant τ and the shape of this transient is a characteristic of the channel being examined. The magnitude of the response is proportional to the impulse strength, $f_1(\tau)\,d\tau$. Suppose the network response to a unit impulse occurring at $t = 0$ is $h(t)$. Then the response to the impulse of strength $f_1(\tau)\cdot d\tau$ occurring at $t = \tau$, is $h(t - \tau) \cdot f_1(\tau) \cdot d\tau$ and the Superposition Theorem shows that the output waveform is the sum of all these contributions.

* $A(\omega)$ is now regarded as a complex number giving information on both the amplitude and phase response of the network. In Section 4.3, $A(\omega)$ was used to indicate amplitude response only and corresponds to $|A(\omega)|$ in the notation of the present Chapter.

The sum becomes an integral if the element $d\tau$ is infinitesimally small. This principle is illustrated in Fig. 5.2. If the input waveform is zero up to the time zero, then the integration will commence at τ equal to zero. Clearly, a physical network cannot give an output in advance of the input signal and this implies that the impulse response, $h(t)$, is zero for all negative values of t. Similarly, $h(t - \tau)$ is zero if τ exceeds t and so,

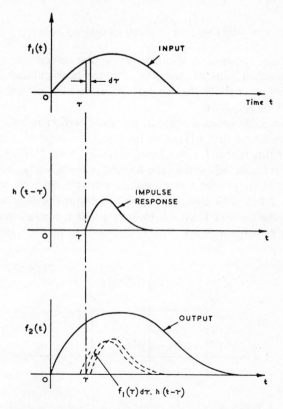

Fig. 5.2. Illustration of the DuHamel integral method: (a) Input waveform resolved into a succession of unit impulses. (b) Unit impulse response. (c) Output waveform obtained by summation of impulse responses.

in summing the contributions to the output at time t, we need only include values of τ less than or equal to t. The upper limit of the integration with respect to the variable τ is therefore t and so, under the conditions postulated, we obtain for the output waveform:

$$f_2(t) = \int_0^t f_1(\tau) \cdot h(t - \tau) \cdot d\tau \qquad (5.4)$$

A more mathematical derivation of this result is given in Section A.5, where it is also proved that the expressions for the output waveforms in equations (5.3) and (5.4) give identical results.

The relation between the Fourier and Duhamel methods stems from the result (proved as example 3 in Section A.4.1) that the impulse response $h(t)$ is the Fourier transform of the network transfer function, $A(\omega)$, i.e.

$$h(t) = \frac{1}{2\pi} \int_{-\infty}^{\infty} A(\omega) \cdot \exp(j\omega t) \cdot d\omega \qquad (5.5)$$

This equation shows that, if either the impulse response or the channel transfer function is known, then the other can be calculated. This is why it is possible to specify the network characteristic in terms of one or other of these functions.

Equation (5.5) raises a most important question in relation to the assumption above that $h(t)$ is zero for negative values of t: the question is whether this is true for any transfer function $A(\omega)$? We will see from a simple example below that the answer is no. Clearly, however, the answer must be yes for any physically possible network and we are, therefore, faced with determining the consequent restrictions on the nature of the transfer function, to ensure that it may correspond to a physically possible network. This is discussed in the next section.

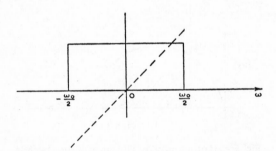

Fig. 5.3. Ideal rectangular channel:
————— amplitude characteristic: --------- phase
characteristic.

As an example of the need to restrict the nature of the transfer function, we consider the idealized low pass filter with constant amplitude and linear phase shift, as shown in Fig. 5.3, i.e., we assume:

$$\left. \begin{array}{ll} A(\omega) = \exp(-jk\omega): & \text{for} \quad -\tfrac{1}{2}\omega_0 < \omega < \tfrac{1}{2}\omega_0 \\[2mm] \quad\quad = 0 & : \text{for} \quad \omega > \tfrac{1}{2}\omega_0 \text{ and } \omega < -\tfrac{1}{2}\omega_0 \end{array} \right\} \qquad (5.6)$$

The response has been defined for positive and negative frequencies as required in using Fourier transforms, and satisfies the condition derived in Section A.4.1, that

$$A(-\omega) = A^*(\omega) \tag{5.7}$$

which ensures that a real input waveform produces a real output waveform. The bandwidth of the filter is $\frac{1}{2}B$ Hz, where B is equal to $\omega_0/2\pi$, and the phase shift changes linearly at a rate determined by the constant, k. The amplitude of $A(\omega)$ has the constant value unity throughout the passband.

The impulse response is given by equation (5.5) as:

$$h(t) = \frac{1}{2\pi} \int_{-\frac{1}{2}\omega_0}^{\frac{1}{2}\omega_0} \exp(-jk\omega + j\omega t) . \, d\omega \tag{5.8}$$

in which the integration limits have been reduced to the frequency range within which $A(\omega)$ does not vanish. Equation (5.8) is readily integrated to give the result:

$$h(t) = \frac{\sin[\frac{1}{2}\omega_0(t-k)]}{\pi(t-k)} = \frac{B \cdot \sin x}{x} \quad \text{if} \quad x = \pi B(t-k) \tag{5.9}$$

The impulse response is a $(\sin x)/x$ pulse of peak amplitude B centred on time k as shown in Fig. 5.4. The pulse length between the first zeros

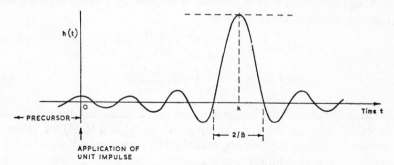

Fig. 5.4. The impulse response for the rectangular channel.

is $2/B$ and the build-up time of the pulse from the zero at the beginning of the main lobe to the peak value is $1/B$. It is obvious from the figure that there is some output from the network before the time $t = 0$ at which the input impulse is applied. This shows that an apparently simple transfer function leads to a result which cannot occur in practice and we can only conclude that no physical network can have a transfer function of this type. An exception to this statement arises if k becomes infinitely large, for then the output is delayed by an infinite time and the amplitude of $h(t)$ is vanishingly small for t less than zero.

5.3. Physically realisable transfer Functions

The condition that the impulse function must be zero for all negative values of t can be used in conjunction with equation (5.5) to determine the condition that the transfer function may correspond to a physically possible network. Such a transfer function is said to be physically realisable, *i.e.*, it can be realised in practice by a suitably chosen network. We begin by taking the inverse transform of equation (5.5), giving:

$$A(\omega) = \int_0^\infty h(t) \cdot \exp(-j\omega t) \cdot dt \qquad (5.10)$$

the lower limit being taken as zero, since $h(t)$ vanishes for negative t. The impulse function $h(t)$ must be a real function of t and so, by taking real and imaginary parts of equation (5.10), we obtain:

$$\text{Re.}[A(\omega)] = \int_0^\infty h(t) \cdot \cos(\omega t) \cdot dt \qquad (5.11)$$

$$\text{Im.}[A(\omega)] = -\int_0^\infty h(t) \cdot \sin(\omega t) \cdot dt \qquad (5.12)$$

The impulse function can be eliminated from these equations to yield a relation between the real and imaginary parts of $A(\omega)$.

A similar problem is examined in Section A.8 and following the method described there, we obtain:

$$\text{Re.}[A(\omega)] = -\frac{2}{\pi} \int_0^\infty \frac{\zeta \cdot \text{Im}[A(\zeta)]}{\zeta^2 - \omega^2} \cdot d\zeta \qquad (5.13)$$

$$\text{Im.}[A(\omega)] = \frac{2}{\pi} \int_0^\infty \frac{\omega \cdot \text{Re}[A(\zeta)]}{\zeta^2 - \omega^2} \cdot d\zeta \qquad (5.14)$$

If either the real or the imaginary part is specified, then the other can be calculated. The integrals in equations (5.13) and (5.14) are Hilbert transforms and are evaluated by excluding the region $\omega - \delta < \zeta < \omega + \delta$, where δ is a vanishingly small number. This removes the difficulty which arises as a result of the integrand becoming infinite when $\zeta = \omega$.

Although the above equations provide the required restriction on the nature of the transfer function, they are not convenient to use in practice, since we are more interested in the amplitude and phase characteristics of the channel. A pair of equations, similar in form to those above, is satisfied by $\log_e |A(\omega)|$ and the phase, $\phi(\omega)$, provided that a restriction is placed on the phase characteristic. If we denote $-\log_e |A(\omega)|$, *i.e.*, the attenuation of the channel in nepers, by $N(\omega)$, the equations are:

$$N(\omega) = -\frac{2}{\pi} \int_0^\infty \frac{\zeta \cdot \phi(\zeta)}{\zeta^2 - \omega^2} \cdot d\zeta \qquad (5.15)$$

$$\phi(\omega) = \frac{2}{\pi} \int_0^\infty \frac{\omega \cdot N(\zeta)}{\zeta^2 - \omega^2} \cdot d\zeta \qquad (5.16)$$

The derivation of these equations (see Section A.8) shows that they are only valid if the network has the least possible change in phase with frequency for the given amplitude·characteristic. Networks with this property are known as minimum phase networks and the majority of networks used in communications are of this type. A non-minimum phase network can always be regarded as equivalent to a minimum phase network in tandem with an all-pass network, *i.e.*, a network with a constant attenuation but a frequency dependent phase characteristic. This aspect of network theory is too elaborate for discussion here but comprehensive treatments are available in the literature.

From equations (5.15) and (5.16), we may deduce the phase character-istic associated with a given amplitude characteristic, or conversely. One example of the use of these equations is given in Section A.8.2, where it is shown that the ideal rectangular low pass filter, with no attenuation in the pass band and infinite attenuation in the stop band (Fig. 5.3(a)), has infinite group delay.

Another example occurs in Section 5.5.3.

5.4 Application to Bandpass Channels

5.4.1. EQUIVALENCE TO BASEBAND CHANNELS

The results which have been given in this chapter have applied to baseband signals and we now consider how the theory can be applied to modulated carriers.

The proof of this result is quite straightforward but is complicated by the restrictions placed on the possible forms of network transfer functions by the need for them to be physically realisable. We recall (Section 5.2.2.) that a complex transfer function $A(\omega)$ corresponds to a physically realisable network only if

$$A(-\omega) = A^*(\omega) \qquad (5.17)$$

There must always be a response in the negative frequency region. Sup-pose now that we have a bandpass circuit whose response is restricted to the frequency interval $\omega_c - \frac{1}{2}\omega_1$ to $\omega_c + \frac{1}{2}\omega_1$. The mid-band frequency has been taken as ω_c, the frequency of the input carrier. In view of equation (5.17), the complex transfer function of such a network must have the form:

$$\begin{aligned} A_1(\omega) &= A(\omega) \quad \text{if} \quad \omega_c - \tfrac{1}{2}\omega_1 < \omega < \omega_c + \tfrac{1}{2}\omega_1 \\ &= A^*(-\omega) \quad \text{if} \quad -\omega_c - \tfrac{1}{2}\omega_1 < \omega < -\omega_c + \tfrac{1}{2}\omega_1 \\ &= 0 \quad \text{for all other values of } \omega. \end{aligned} \qquad (5.18)$$

This transfer function is shown in Fig. 5.5, an asymetrical pass band being chosen to emphasise the relation between the positive and negative frequency contributions.

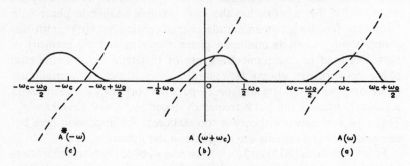

Fig. 5.5. Transfer characteristics used in analysing bandpass channels
———————— amplitude ------------ phase
(a) positive frequency transfer function of bandpass channel. (b) equivalent lowpass channel. (c) related negative frequency transfer function for the bandpass channel. The amplitude characteristics are shown as asymmetric curves to emphasise the relations between the three sets.

Suppose that a modulated carrier is applied to the input of this network. The only signals which can occur in practice are real functions of time and the most general form of modulated carrier can be written:

$$v_i(t) = f(t) \cdot \cos[\omega_c t + \phi(t)] \qquad (5.19)$$

where the time dependence of $f(t)$ corresponds to amplitude modulation and the time dependence of $\phi(t)$ corresponds to angle modulation. As in steady-state problems, it is more convenient to work with complex functions and we therefore consider the complex input,

$$v_i'(t) = f(t) \cdot \exp[j\{\omega_c t + \phi(t)\}] \qquad (5.20)$$

The actual input $v_i(t)$ is the real part of $v_i'(t)$.

We will calculate $v_o'(t)$, the output corresponding to $v_i'(t)$ and then, exactly as in steady-state analysis, the output corresponding to $v_i(t)$ will be Re $\cdot v_o''(t)$. (It should be noted that this statement is only true if the network is physically realisable and this is the reason for stressing the presence of the negative frequency contribution in the transfer function.)

The analysis now proceeds by the normal Fourier method. Let

$$f(t) \exp[j\phi(t)] = \frac{1}{2\pi} \int_{-\infty}^{\infty} g(\omega) \exp(j\omega t) \cdot d\omega \qquad (5.21)$$

This complex time waveform gives complete information about the modulation of the input wave and will be referred to as the input modulation function. We now express $v_i'(t)$ as a Fourier spectrum:

$$v_i'(t) = \frac{1}{2\pi} \int_{-\infty}^{\infty} g(\omega) \exp[j(\omega + \omega_c)t]d\omega$$

$$= \frac{1}{2\pi} \int_{-\infty}^{\infty} g(\omega - \omega_c) \exp(j\omega t)d\omega \qquad (5.22)$$

so that the spectrum of $v_i'(t)$ is simply the spectrum of the input modulation displaced upwards in frequency by ω_c. The output corresponding to $v_i'(t)$ follows by including the transfer function $A_1(\omega)$ under the integral, *i.e.*,

$$v_o'(t) = \frac{1}{2\pi} \int_{-\infty}^{\infty} A_1(\omega) \cdot g(\omega - \omega_c) \exp(j\omega t)d\omega \qquad (5.23)$$

The transfer function $A_1(\omega)$ is given by equation (5.18) and so,

$$v_o'(t) = \frac{1}{2\pi} \int_{\omega_c - \frac{1}{2}\omega_1}^{\omega_c + \frac{1}{2}\omega_1} A(\omega) \cdot g(\omega - \omega_c) \exp(j\omega t)d\omega$$

$$+ \frac{1}{2\pi} \int_{-\omega_c - \frac{1}{2}\omega_1}^{-\omega_c + \frac{1}{2}\omega_1} A^*(-\omega) \cdot g(\omega - \omega_c) \exp(j\omega t)d\omega \quad (5.24)$$

In most practical problems, the spread of the spectrum $g(\omega - \omega_c)$ will be such that there is no contribution from the second integral. For example, suppose $g(\omega)$ is vanishingly small if ω is less than $-\frac{1}{2}\omega_2$. Then $g(\omega - \omega_c)$ is vanishingly small it ω is less than $-\frac{1}{2}\omega_2 + \omega_c$ and the integrand in the second integral is negligible provided:

$$-\omega_c + \frac{1}{2}\omega_1 < -\frac{1}{2}\omega_2 + \omega_c$$

i. e. if
$$\frac{1}{2}(\omega_1 + \omega_2) < 2\omega_c$$

Since the carrier frequency ω_c is usually much higher than either ω_1 the channel bandwidth, or ω_2, the modulation bandwidth, this condition is generally satisfied.

We then have:

$$v_o'(t) = \frac{1}{2\pi} \int_{\omega_c - \frac{1}{2}\omega_1}^{\omega_c + \frac{1}{2}\omega_1} A(\omega)g(\omega - \omega_c) \exp(j\omega t)d\omega$$

$$= \frac{1}{2\pi} \int_{-\frac{1}{2}\omega_1}^{\frac{1}{2}\omega_1} A(\omega + \omega_c)g(\omega) \exp[(j\omega + \omega_c)t]d\omega$$

$$= f_0(t) \exp[j\phi_0(t) + j\omega_c t] \qquad (5.25)$$

The output waveform is thus expressed in the form of a complex modulation function and a carrier wave and taking the real part gives an expression for the output in the same form as the input (equation (5.19)). The output modulation function is

$$f_0(t)\exp[j\phi_0(t)] = \frac{1}{2\pi}\int_{-\frac{1}{2}\omega_1}^{\frac{1}{2}\omega_1} A(\omega + \omega_c)g(\omega)\exp(j\omega t)\,d\omega \quad (5.26)$$

and, therefore, equals the output which would be obtained if the input modulation function (of spectrum $g(\omega)$) is passed through a baseband channel with a passband in the frequency range $-\frac{1}{2}\omega_1$ to $+\frac{1}{2}\omega_1$. We have thus proved the desired result.

The transfer function of this equivalent baseband channel is $A(\omega+\omega_c)$ i.e., the positive frequency bandpass transfer function displaced downwards by ω_c. This baseband may not be physically realizable but this does not affect the validity of the above approach. Since, however, all our previous results are obtained for physically realizable baseband networks, the usefulness of this approach is greatest when the baseband channel can be realized.

The condition for this is,

$$A(-\omega + \omega_c) = A^*(\omega + \omega_c)$$

and this implies that the amplitude characteristic of the bandpass circuit is symmetrical with respect to the carrier frequency ω_c. The phase characteristic must be zero at the carrier frequency and is an odd function of the frequency departure from the carrier frequency.

The results of the above argument may be summarised as follows:

(a) Any bandpass function must have a transfer function with responses at both positive and negative frequencies, related as shown in Fig. 5.5.

(b) If a modulated carrier is passed through such a channel, the output can be deduced by considering the behaviour of the modulation, expressed as a complex function to preserve amplitude and phase information, in passing through the related baseband channel.

It follows that there is no need to give bandpass channels any particular treatment and, further, that the above procedure emphasizes the fundamental nature of the baseband concept.

5.4.2. DESIGN OF BANDPASS FILTERS FROM BASEBAND FILTERS

The previous section shows that the behaviour of signals in bandpass channels can be deduced from the corresponding baseband situation. Suppose that this leads to a choice of a suitable baseband filter in some particular situation. We are then faced with providing a bandpass filter with a transfer function related to the baseband transfer function, in the way shown in Fig. 5.5. To facilitate the discussion, we denote the base-band transfer function by $A_b(\omega)$ and assume, as before, that the bandwidth is $\frac{1}{2}B$ Hz. The transfer function for the corresponding bandpass channel can thus be expressed as:

$$
\begin{aligned}
A(\omega) &= A_b(\omega - \omega_c) \quad \text{if} \quad \omega_c - \tfrac{1}{2}\omega_0 < \omega < \omega_c + \tfrac{1}{2}\omega_0 \\
&= A_b(\omega + \omega_c) \quad \text{if} \quad -\omega_c - \tfrac{1}{2}\omega_0 < \omega < \omega_c + \tfrac{1}{2}\omega_0 \qquad (5.27) \\
&= 0 \quad \text{for all other values of } \omega
\end{aligned}
$$

We will consider the situation in which the baseband filter is assembled from passive elements, *i.e.*, resistors, inductors and capacitors. The frequency dependent characteristic arises only from the variation of reactance with frequency, each inductor having a reactance ωL and each capacitor having a reactance $-1/\omega C$. We wish to modify the circuit so that the baseband behaviour near $\omega = 0$ is reproduced by the passband circuit near the frequencies ω_c and $-\omega_c$. At $\omega = 0$, the reactances of the inductors and capacitors are zero and infinity respectively: this suggests that we replace each inductor by a series resonant circuit giving zero reactance at ω_c and $-\omega_c$, and each capacitor by a parallel resonant circuit giving infinite reactance at ω_c and $-\omega_c$. Each L is replaced by the series combination L_1, C_1 (Fig. 5.6) giving the reactance $(\omega L_1 - 1/\omega C_1)$, which can be written as $k(\omega - \omega_c^2/\omega)L$ if $L_1 = kL$ and $L_1 C_1 = 1/\omega_c^2$. Similarly, each capacitor C is replaced by a parallel combination L_2, C_2 with reactance $-\omega L_2/(\omega^2 L_2 C_2 - 1)$ and this equals

$$-1/[k(\omega - \omega_c^2/\omega)C] \quad \text{if } C_2 \text{ is taken as } kC, \text{ and } L_2 C_2 = 1/\omega_c^2$$

The constant k has been introduced to allow for the possibility of a change in bandwidth. The effect of each of these changes is equivalent

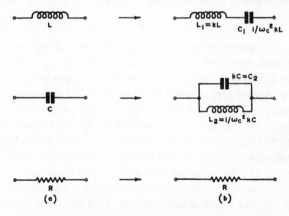

Fig. 5.6. The design of a bandpass filter from the equivalent lowpass filter. The low pass elements in (a) are replaced by the combinations in (b) to give the bandpass filter.

to replacing ω in the baseband case by the combination $k(\omega - \omega_c{}^2/\omega)$ in the passband case. Suppose that each resistor in the baseband circuit is repeated in the new circuit. Then, the behaviour of the new circuit is given simply by replacing ω in the baseband circuit by $k(\omega - \omega_c{}^2/\omega)$ and its transfer function is therefore:

$$A(\omega) = A_b\left[k\left(\omega - \frac{\omega_c{}^2}{\omega}\right)\right] \tag{5.28}$$

The function $A_b(x)$ is zero if $|x|$ exceeds $\tfrac{1}{2}\omega_0$ and so $A(\omega)$ will be zero if $|k(\omega - \omega_c{}^2/\omega)|$ exceeds $\tfrac{1}{2}\omega_0$. In most practical cases, the carrier frequency ω_c will be much larger than ω_0, and this means that $A(\omega)$ is zero unless $\omega - \omega_c{}^2/\omega$ is nearly zero, *i.e.*, ω must be nearly equal to ω_c or to $-\omega_c$. Suppose first that ω equals $\omega_c + \delta$, where δ is much less than ω_c. Then:

$$k\left(\omega - \frac{\omega_c{}^2}{\omega}\right) = k\left(\omega_c + \delta - \frac{\omega_c{}^2}{\omega_c + \delta}\right) \doteqdot 2k\delta \tag{5.29}$$

Now let

$$2k = 1 \tag{5.30}$$

so that, near ω_c,

$$A(\omega) \doteqdot A_b(\delta) = A_b(\omega - \omega_c) \tag{5.31}$$

Similarly, if ω is nearly equal to $-\omega_c$, we let $\omega = -\omega_c + \delta'$, giving:

$$k\left(\omega - \frac{\omega_c{}^2}{\omega}\right) \doteqdot k(-\omega_c + \delta' + \omega_c + \delta')$$

$$= 2k\delta' = \delta' = \omega + \omega_c \tag{5.32}$$

Hence, if ω is nearly equal to $-\omega_c$

$$A(\omega) = A_b(\omega + \omega_c) \tag{5.33}$$

A comparison of equations (5.31) and (5.33) with equation (5.27) shows that the modified circuit has the desired transfer function. The introduction of the constant k, which was subsequently made equal to $\tfrac{1}{2}$, was necessary to achieve the desired bandwidth in the passband circuit.

The combination of the results in this and the previous section shows that bandpass problems can be reduced to equivalent baseband problems. The methods used in this chapter can thus be used to handle any transmission problem. One possible complication which has been overlooked in the above treatment is that the spectrum of the input signal may be so wide that it embraces both the positive and negative portions of the channel transfer function. This situation can be handled by working directly with the real time waveforms throughout. In practice the bandwidth of a modulated carrier is invariably less than the carrier frequency and the complication is therefore avoided.

5.5. The Response of Ideal Channels

We have seen in Section 5.3 that the amplitude and phase character-istics of a channel cannot be chosen independently of each other. If this is done, a non-physical result, such as the appearance of an output before the input is applied, will be obtained. Nevertheless, there are considerable mathematical simplifications when such non-physical channels are examined and the results are useful as a guide to the be-haviour of physical channels. The idealization which we shall use is that the channels have sharp cut-off frequencies, *i.e.*, the amplitude response is zero except in a finite frequency band, and that the phase characteristics are linear. In view of the conclusions of the previous section it is sufficient to consider baseband channels.

5.5.1. THE STEP RESPONSE OF A RECTANGULAR CHANNEL

The rectangular channel has the shape shown in Fig. 5.3 and defined by equation (5.6). The response to a unit step input can be obtained by the Fourier integral method, as used for the impulse re-sponse in Section 5.2.2, but the substitution of the spectrum of the step function leads to an integral which is difficult to evaluate. We therefore use the Duhamel integral method as given by equation (5.4), with a modification to allow for the use of a non-physical channel. Equation (5.4) applies only for t greater than zero, the output function $f_2(t)$ being zero for all values of t less than zero if the channel is a physical one. (The input function $f_1(t)$ is zero for t less than zero and unity for t greater than zero.) To allow for the precursor, or output before the input is applied, we add a term $f_2(0)$ to the right-hand side of equation (5.4), this term being the value of the output at time zero.

Then

$$f_2(t) = \int_0^t h(t - \tau) \cdot d\tau + f_2(0) \qquad (5.34)$$

in which $f_1(t)$ has been placed equal to unity since τ is always positive under the integral, and $h(t - \tau)$ is the impulse response obtained from equation (5.9). Equation (5.34) gives $f_2(t)$ for positive values of t. Sub-stitution for $h(t - \tau)$ gives:

$$f_2(t) = \int_0^t \frac{B \sin[\pi B(t - \tau - k)]}{\pi B(t - \tau - k)} \cdot d\tau + f_2(0) \qquad (5.35)$$

The integration variable τ is replaced by x, where

$$x = \pi B(t - \tau - k) \qquad (5.36)$$

and this leads to:

$$f_2(t) = \frac{1}{\pi} \int_{-\pi Bk}^{\pi B(t-k)} \frac{\sin x}{x} \cdot dx + f_2(0) \qquad (5.37)$$

Since the channel has unit gain in the passband, it is plausible to assume that the output $f_2(t)$ tends to equal the input $f_1(t)$ at times long after the application of the input, *i.e.*,

$$f_2(t) \to f_1(t) = 1 \quad \text{as} \quad t \to \infty \qquad (5.38)$$

and so equation (5.37) gives:

$$1 = \frac{1}{\pi} \int_{-\pi Bk}^{\infty} \frac{\sin x}{x} \cdot dx + f_2(0) \qquad (5.39)$$

A standard property of the sine integral is that:

$$\frac{1}{\pi} \int_{-\infty}^{\infty} \frac{\sin x}{x} \cdot dx = 1 \qquad (5.40)$$

and so, from equations (5.39) and (5.40) we have:

$$f_2(0) = \frac{1}{\pi} \int_{-\infty}^{-\pi Bk} \frac{\sin x}{x} \cdot dx \qquad (5.41)$$

and

$$f_2(t) = \frac{1}{\pi} \int_{-\pi Bk}^{\pi B(t-k)} \frac{\sin x}{x} \cdot dx + \frac{1}{\pi} \int_{-\infty}^{-\pi Bk} \frac{\sin x}{x} \cdot dx$$

$$= \frac{1}{\pi} \int_{-\infty}^{\pi B(t-k)} \frac{\sin x}{x} \cdot dx \qquad (5.42)$$

or

$$f_2(t) = \frac{1}{\pi} \int_{0}^{\pi B(t-k)} \frac{\sin x}{x} \cdot dx + \tfrac{1}{2} \qquad (5.43)$$

The above result has been demonstrated for positive values of t, but it may be shown by more elaborate methods, *e.g.*, the evaluation of the integral in the Fourier method, that the expression for $f_2(t)$ is in fact correct for all values of t. The sine integral occurring in this expression is available in tables and the output function, $f_2(t)$, plotted from these, is shown in Fig. 5.7. The output attains half its final value at $t = k$, so that k may be regarded as the delay. As a measure of the rise time of the output pulse, we may consider the reciprocal of the slope of the output curve evaluated at $t = k$. Now

$$\frac{\mathrm{d}f_2(t)}{\mathrm{d}t} = \frac{1}{\pi} \cdot \frac{\sin[\pi B(t - k)]}{\pi B(t - k)} \cdot \pi B$$

and if $t = k$,

$$\frac{\mathrm{d}f_2(t)}{\mathrm{d}t} = B \qquad (5.44)$$

The rise time defined in this way is, therefore, $1/B$ sec. It should be noted that the values for the decay and the rise time are the same as those obtained for the impulse response in Section 5.2.2.

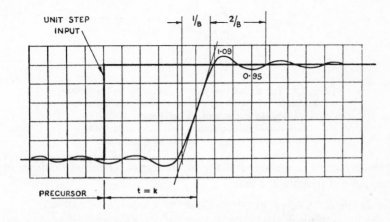

Fig. 5.7. The step response of an ideal rectangular channel.

The output shows a precursor as expected and an overshoot, followed by an oscillation. The frequency of this oscillation, which is often referred to as 'ringing' is $\frac{1}{2}B$ Hz, the filter cut-off frequency.

These results can readily be modified to apply to the situation when a unit step carrier is applied to a bandpass channel. If the channel bandwidth is B Hz (corresponding to the lowpass cut-off, $\frac{1}{2}B$), the build-up time is $1/B$ sec. This result, which remains approximately correct for practical channels of roughly rectangular amplitude characteristic, provides a useful method of assessing the performance of a channel. If a negative cancelling unit step is added at some positive time to the positive unit step, we obtain a rectangular pulse and the decay or run-down time is obviously also $1/B$ sec. We therefore see from Fig. 5.8 that the shortest pulse to give maximum output in passing through a bandpass filter of bandwidth B Hz has a duration of approximately

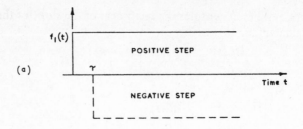

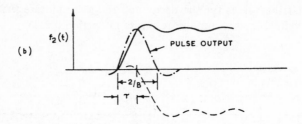

Fig. 5.8. The use of two step functions to give a rectangular pulse. The interval τ is chosen so that the output pulse just attains its maximum value
——————— positive step: ----------- negative step
—.—.— output obtained by summation.

$1/B$ sec. if this is measured at the mean height. The total duration of the pulse is roughly $2/B$ sec.

5.5.2. THE EFFECT OF A KEYED CARRIER ON A BANDPASS CHANNEL

As an example of the practical value of results, such as that obtained in the previous section, we consider a carrier wave of unit amplitude and frequency ω_c, which is suddenly switched to a filter of midband frequency ω_f and bandwidth B. We assume that ω_c is greater than ω_f and that it lies outside the passband of the filter. The spectrum of the transient and the filter amplitude characteristic are sketched in Fig. 5.9(a), only the values for positive frequencies being shown. We may ignore the negative frequencies for the reasons discussed in Section 5.4 and we further assume that the portion of the transient spectrum which gets through the filter has a uniform value over the filter band and that this value is given by the spectrum of the transient, calculated at the midband frequency, i.e.,

$$g(\omega) = \frac{1}{j(\omega_f - \omega_c)} \tag{5.45}$$

The output is the inverse transform of the product of the filter transfer function and the spectrum $g(\omega)$. Without carrying out the actual calculations, we may see that the frequency shift result of Section 5.4 shows that the output appears as a carrier of frequency equal to the filter midband frequency, ω_f, modulated by an envelope whose form will depend on the filter transfer function. If the transfer function is rectangular, the envelope will be of the (sin x)/x form, and the time interval

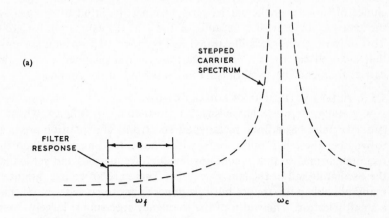

Fig. 5.9. (a) The effect of a keyed carrier on an adjacent bandpass filter.

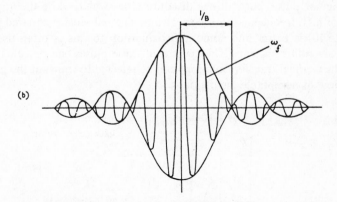

Fig. 5.9. (b) Output waveform for the situation shown in (a): the frequency of the oscillation is the filter mid-band frequency and the envelope is of (sin x)/x shape.

between the two envelope zeros on either side of the maximum will be $2/B$ sec. This statement regarding the time interval follows from a consideration of the output pulse length when a unit impulse is applied to a low pass filter of bandwidth $\tfrac{1}{2}B$. The strength of the transient is governed by the input spectrum strength, $1/(\omega_f - \omega_c)$, and, collecting

these pieces of information together, we may say that the output waveform is:

$$f_2(t) = \frac{1}{(\omega_f - \omega_c)} \cdot \frac{\sin[\pi B(t - \tau)]}{\pi(t - \tau)} \cdot \sin \omega_f t \qquad (5.46)$$

where τ is a delay which depends on the phase slope of the filter. The
choice of $\sin \omega_f t$ for the carrier term assumes the input to be $\cos \omega_c t$,
the phase change arising because of the j in the expression for $g(\omega)$.
This output waveform is illustrated in Fig. 5.9(b). This example shows
that a keyed carrier can cause interference in a channel, even if the
carrier frequency lies well outside the passband of the channel.

5.5.3. FILTERS WITH MORE GRADUAL CUT-OFFS

A rectangular pulse has a spectrum extending to infinity, whilst a
$(\sin x)/x$ pulse has a finite rectangular spectrum. The question arises as
to whether there is a pulse shape which has a less sharp rise than the
rectangular one, so that it contains few high frequencies and yet avoids
the oscillating tail of the $(\sin x)/x$ pulse. There are, of course, infinitely
many possible pulse shapes but it is unnecessary to consider more than
a small selection. The width of the frequency spectrum is largely determined by the occurrence of sharp discontinuities and these should be
avoided if possible. The half-cosine pulse is an improvement over the
rectangular pulse, but still has discontinuities which cause the appearance of high frequencies in the spectrum. The half-cosine squared pulse
(Fig. 5.10) is better and some approximation to this is often used in
pulse signalling systems. The spectra of these pulses can be calculated
and the channel transfer functions can be selected to transmit the pulses
with low or acceptable distortion.

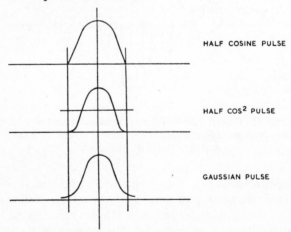

HALF COSINE PULSE

HALF COS^2 PULSE

GAUSSIAN PULSE

Fig. 5.10. Various rounded pulse shapes.

An interesting example is provided by considering the transmission of a Gaussian pulse through a network which has a Gaussian amplitude characteristic. We will first examine the possible form of the phase characteristic and also discuss a method by which such a network can be realised in practice. Suppose the amplitude response is:

$$|A(\omega)| = \exp(-c\omega^2) \tag{5.47}$$

where c is a constant, related to the half-power band width.
If $|A(\omega)|$ equals $1/\sqrt{2}\,|A(0)|$, when ω equals $\pm\pi B$, then:

$$\exp(-c\pi^2 B^2) = 1/\sqrt{2}$$

giving

$$c = \frac{1}{2\pi^2 B^2} \log_e 2 \tag{5.48}$$

We now calculate the phase characteristic by using equation (5.16) and since $N(\omega)$ equals $c\omega^2$, we have:

$$\phi(\omega) = \frac{2}{\pi} \int_0^\infty \frac{c\omega\zeta^2 \cdot d\zeta}{\zeta^2 - \omega^2} \tag{5.49}$$

This integral is awkward to handle and has to be treated as:

$$\phi(\omega) = \frac{2}{\pi} \operatorname*{Lim}_{\omega_m \to \infty} \cdot \operatorname*{Lim}_{\delta \to 0} \left[\int_0^{\omega-\delta} + \int_{\omega+\delta}^{\omega_m} \right] \frac{c\omega\zeta^2}{\zeta^2 - \omega^2} \cdot d\zeta$$

Carrying out the integration and letting δ tend to zero gives:

$$\phi(\omega) = \frac{2c\omega}{\pi} \operatorname*{Lim}_{\omega_m \to \infty} \left[\omega_m + \frac{\omega}{2} \log_e \left(\frac{\omega_m - \omega}{\omega_m + \omega} \right) \right] \tag{5.50}$$

When ω_m tends to infinity, the first term tends to infinity and the second to zero. In this limiting case, the phase can be written as:

$$\phi(\omega) = \omega\tau \tag{5.51}$$

which shows that the frequency characteristic is linear. The constant τ is therefore the group delay for the network and, since it is proportional to ω_m, it tends to infinity. A filter with a Gaussian amplitude characteristic thus involves an infinite delay.

A filter with the above characteristics can be obtained as the limiting form of the cascaded RC stages shown in Fig. 5.11. The transfer function of an individual stage is:

$$A(\omega) = \frac{1}{1 + j\omega\tau_0} \tag{5.52}$$

where $\tau_0 = $ RC, the time constant of the stage.

Equation (5.52) assumes that the network connected across the capacitor has an infinite input impedance, and this is the reason for the introduction of the buffer or isolating stages between successive RC

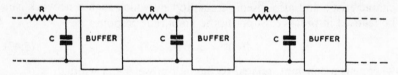

Fig. 5.11. Circuit to give a Gaussian amplitude characteristic.

sections. If each buffer stage has unity gain, the transfer function for n stages in cascade is:

$$A(\omega) = \frac{1}{(1 + j\omega\tau_0)^n} \tag{5.53}$$

from which we have:

$$|A(\omega)| = (1 + \omega^2\tau_0^2)^{-\frac{1}{2}n} \tag{5.54}$$

and

$$\phi(\omega) = n \tan^{-1}(\omega\tau_0) \tag{5.55}$$

Suppose now that n is increased indefinitely and that for each n, τ_0 is selected so that:

$$\tau_0^2 = T^2/n \tag{5.56}$$

In the limit as n tends to infinity,

$$|A(\omega)| = \lim_{n\to\infty} \left(1 + \frac{\omega^2 T^2}{n}\right)^{-\frac{1}{2}n} = \exp(-\tfrac{1}{2}\omega^2 T^2) \tag{5.57}$$

by using the result that the exponential function is the limiting form of the binomial expansion. Equation (5.57) is identical to equation (5.47) if $\frac{1}{2}T^2$ is equal to c. The corresponding value of the phase shift is:

$$\phi(\omega) = \lim_{n\to\infty} n \tan^{-1}(\omega T/\sqrt{n})$$
$$\doteqdot \lim_{n\to\infty} n\omega T/\sqrt{n}, \text{ provided } \omega T \ll \sqrt{n}$$
$$= \omega \lim_{n\to\infty} \sqrt{n} \cdot T \tag{5.58}$$

The phase characteristic again tends to be linear and the slope tends to infinity, confirming the infinite group delay deduced above. The Gaussian filter can thus be obtained in practice by taking the limit of a very large number of suitably selected isolated RC networks.

We now examine the transmission of a Gaussian pulse through a channel which has the amplitude characteristic defined by equation

(5.47). Since the phase characteristic is linear with slope τ, the output pulse is delayed by time τ. Since τ tends to infinity, this type of channel is similar to the rectangular channel in that it involves infinite delay. If the Gaussian pulse has the time waveform,

$$f_1(t) = \exp(-k_1 t^2) \tag{5.59}$$

its spectrum is given by equation (A.40) as:

$$g_1(\omega) = (\pi/k_1)^{\frac{1}{2}} \exp(-\omega^2/4k_1) \tag{5.60}$$

The output pulse has a Fourier spectrum equal to $g_1(\omega) \cdot A(\omega)$ and since the phase term of $A(\omega)$ contributes only to the delay, it can be omitted for the time being. The spectrum of the output pulse is therefore:

$$g_2(\omega) = (\pi/k_1)^{\frac{1}{2}} \exp\left(-\frac{\omega^2}{4k_1} - c\omega^2\right) \tag{5.61}$$

This is of the same form as equation (5.60) and shows that the output pulse is:

$$f_2(t) = (k_2/k_1)^{\frac{1}{2}} \exp[-k_2(t-\tau)^2] \tag{5.62}$$

the delay τ being introduced to account for the phase of $A(\omega)$. The constant k_2 can be expressed in terms of k_1 and c by noting the form of the exponent in equation (5.61):

$$\frac{1}{4k_2} = \frac{1}{4k_1} + c \tag{5.63}$$

Since c is positive, equation (5.63) shows that k_2 is less than k_1 and this implies that the amplitude of the output pulse is less than that of the input pulse and also that the output pulse is wider than the input pulse. If τ_1 is the time interval between the half-power points of the input pulse, then:

$$\exp[-k_1(\tau_1/2)^2] = 1/\sqrt{2}$$

i.e.,

$$k_1 = \frac{2}{\tau_1^2} \log_e 2 \tag{5.64}$$

The pulse length of the output pulse, τ_2, is similarly related to k_2, and equation (5.63) can thus be written as:

$$\frac{\tau_2^2}{8 \log_e 2} = \frac{\tau_1^2}{8 \log_e 2} + \frac{\log_e 2}{2\pi^2 B^2} \tag{5.65}$$

on using equation (5.48). This may also be written as:

$$\left(\frac{\tau_2}{\tau_1}\right)^2 = 1 + \left(\frac{2 \log_e 2}{\pi B \tau_1}\right)^2 \tag{5.66}$$

The increase in the pulse length, expressed as the ratio τ_2/τ_1, thus depends only on the product $B\tau_1$, i.e., channel bandwidth times input pulse length. If this product exceeds unity, the ratio τ_2/τ_1, is less than 1·09 and the increase in the pulse length is tolerable. The ratio of the amplitudes of the output and input pulses is $(k_2/k_1)^{\frac{1}{2}}$, i.e., (τ_1/τ_2) and so the reduction in amplitude is also tolerable if $B\tau_1$ exceeds unity. This shows, once again, that a reasonable criterion for channel bandwidth is that it should be at least as great as the reciprocal of the length of the pulses to be transmitted.*

5.5.4. CHOICE OF PULSE SHAPE

The comments on pulse shapes made earlier in Section 2.4.4 may now be extended in the light of our discussion of the effects of filtering. We shall consider the need to minimize the interaction between successive pulses transmitted in a telegraphy or similar digital system. It was noted in Section 2.4.4 that the time waveforms and spectra of pulses may be interchanged. In the previous section we have been looking at pulses which are restricted to finite time intervals and so have spectra extending to infinity. In the present section, we shall restrict the bandwidth of the pulses to finite values.

We begin with a pulse which has a uniform spectrum extending from $-\frac{1}{2}\omega_0$ to $\frac{1}{2}\omega_0$, i.e.

$$g(\omega) = 2\pi/\omega_0 \quad \text{if } |\omega| \leq \omega_0/2$$
$$= 0 \qquad \text{if } |\omega| > \omega_0/2 \qquad (5.67)$$

the amplitude $2\pi/\omega_0$ being selected for convenience. The corresponding waveform is

$$f(t) = \frac{1}{\omega_0} \int_{-\frac{1}{2}\omega_0}^{\frac{1}{2}\omega_0} \exp(j\omega t)d\omega$$

essentially identical to that already evaluated in Section 5.2.2. We then find from equation (5.9) with minor changes:

$$f(t) = \frac{\sin(\omega_0 t/2)}{(\omega_0 t/2)} \qquad (5.68)$$

The results of Section 5.2.2 show that we may generate such a waveform by applying an impulse to the input of an ideal low-pass filter of cut-off frequency $\omega_0/2$.

The waveform corresponding to $f(t)$ is that shown in Fig. 5.4 (with $k = 0$) and has zeros at the times

$$t_n = 2n\pi/\omega_0 \qquad (5.69)$$

---

* Channel bandwidth is here taken to be the total value B Hz, extending from $-\frac{1}{2}B$ to $\frac{1}{2}B$ Hz.

Suppose we use waveforms as defined by equation (5.68) to transmit telegraphy pulses, and select the interval between pulses to be $\tau = 2\pi/\omega_0$. The result given by equation (5.69) shows that the signal at any instant $t = n\tau$ is only influenced by the presence or absence of a pulse with a maximum at this instant. All other pulse waveforms are zero whenever t is a multiple of τ. If the telegraph receiver is set to make a decision at $t = n\tau$, it will be able to do so without any risk of contamination of the wanted pulse by the tails of previous or future pulses. The $\sin(\tfrac{1}{2}\omega_0 t)/(\tfrac{1}{2}\omega_0 t)$ waveform is thus free from the possibility of intersymbol interference.

The desirable feature of this waveform is very intimately related to the sampling theorem examined in Section 2.5 and it may be recalled that the same waveform arose in equation (2.21) in demonstrating the reconstruction of sampled waveforms. Nyquist, the author of the sampling theorem, also studied the question of intersymbol interference and demonstrated that other waveforms were also free from this interference. We shall examine this possibility for a waveform with a spectrum defined as follows:

$$g_1(-\omega) = g_1(\omega)$$

$$g_1(\omega) = 2\pi/\omega_0 \text{ if } \omega < \tfrac{1}{2}\omega_1$$

$$= \frac{\pi}{\omega_0}[1 + g_2(\omega - \tfrac{1}{2}\omega_0)] \text{ if } \tfrac{1}{2}\omega_1 < \omega < \tfrac{1}{2}\omega_2 (\omega_2 = 2\omega_0 - \omega_1)$$

$$= 0 : \omega > \tfrac{1}{2}\omega_2 \tag{5.70}$$

where $g_2(\ )$ is an odd function, *i.e.*

$$g_2(x) = -g_2(-x)$$

The result, which we shall prove, is that the waveform $f_1(t)$ corresponding to $g_1(\omega)$, also vanishes when $t = n\tau$, n being any integer excluding 0, and $\tau = 2\pi/\omega_0$.

$$f_1(t) = \frac{1}{2\pi}\int_{-\frac{1}{2}\omega_2}^{\frac{1}{2}\omega_2} g_1(\omega) \exp(j\omega t)d\omega = \frac{1}{\pi}\int_0^{\frac{1}{2}\omega_2} g_1(\omega) \cos(\omega t)d\omega$$

$$= \frac{2}{\omega_0}\int_0^{\frac{1}{2}\omega_1} \cos(\omega t)d\omega + \frac{1}{\omega_0}\int_{\frac{1}{2}\omega_1}^{\frac{1}{2}\omega_2} [1 + g_2(\omega - \tfrac{1}{2}\omega_0)] \cos(\omega t)d\omega$$

When $t = n\tau$, we know from equations (5.67) and (5.68) that

$$f(n\tau) = \frac{2}{\omega_0}\int_0^{\frac{1}{2}\omega_0} \cos(n\omega\tau)d\omega = 0$$

and we therefore obtain

$$f_1(n\tau) = \frac{1}{\omega_0}\int_{\frac{1}{2}\omega_1}^{\frac{1}{2}\omega_2} [1 + g_2(\omega - \tfrac{1}{2}\omega_0)] \cos(n\omega\tau)d\omega - \frac{2}{\omega_0}\int_{\frac{1}{2}\omega_1}^{\frac{1}{2}\omega_0} \cos(n\omega\tau)d\omega$$

Let

$$\omega' = \omega - \tfrac{1}{2}\omega_0$$

$$\cos(n\omega\tau) = \cos(n\omega'\tau + \tfrac{1}{2}n\omega_0\tau) = \cos(n\omega'\tau + n\pi)$$

$$= (-)^n \cos(n\omega'\tau)$$

Then,

$$f_1(n\tau) = \frac{(-)^n}{\omega_0} \int_{-\frac{1}{2}(\omega_0-\omega_1)}^{\frac{1}{2}(\omega_2-\omega_0)} [1+g_2(\omega')]\cos(n\omega'\tau)d\omega'$$
$$- \frac{2(-)^n}{\omega_0} \int_{-\frac{1}{2}(\omega_0-\omega_1)}^{0} \cos(n\omega'\tau)d\omega'$$

Since $\tfrac{1}{2}(\omega_2-\omega_0) = \tfrac{1}{2}(\omega_0-\omega_1)$, and $g_2(\omega')$ is an odd function, the integral of $g_2(\omega')\cos(n\omega'\tau)$ vanishes and

$$f_1(n\tau) = \frac{2(-)^n}{\omega_0} \int_0^{\frac{1}{2}(\omega_2-\omega_0)} \cos(n\omega'\tau)d\omega' - \frac{2(-)^n}{\omega_0} \int_0^{\frac{1}{2}(\omega_2-\omega_0)} \cos(n\omega'\tau)d\omega'$$

$$= 0$$

the result we set out to prove.

Any pulse with a spectrum defined by equation (5.70) thus has the property that no interference arises between pulses transmitted at intervals τ.

Each of the spectra defined by equation (5.70) satisfies the condition $g_1(\omega_0/2) = \tfrac{1}{2}g_1(0)$ and it is appropriate to regard the effective bandwidth as extending from $-\tfrac{1}{2}\omega_0$ to $\tfrac{1}{2}\omega_0$. The number of pulses which can be transmitted per second is $1/\tau$, i.e. $\omega_0/2\pi$ or B, the bandwidth as defined elsewhere in this chapter. An advantage of the spectra defined by equation (5.70) compared with the rectangular spectrum of equation (5.71) is a reduction in the size of the side-lobes of the time waveform. For example,

$$g_1(\omega) = \frac{\pi}{\omega_0}\left[1+\cos\left(\frac{\pi\omega}{\omega_0}\right)\right] \text{ if } |\omega| < \omega_0 \qquad (5.71)$$

a raised-cosine spectrum (equivalent to cosine-squared), satisfies equation (5.70), with $\omega_1 = 0$, and its time waveform is

$$f_1(t) = \frac{\pi^2}{\pi^2 - \omega_0^2 t^2} \times \frac{\sin(\omega_0 t)}{\omega_0 t} \qquad (5.72)$$

The amplitudes of the side-lobes fall off much more rapidly than do those of $f(t)$ as defined in equation (5.68), as may be seen by arguments identical to those used in Section 2.4.4.

5.6. The Integration of a Signal Waveform

There are a number of situations in communications in which the average value of a signal waveform over some specified period is desired. It will be shown in this section that the operation of integration can be effected by passing the signal through a channel with a suitably chosen transfer function and so it is appropriate to consider this problem in this chapter. Suppose that the waveform is $f(t)$ and that the integrated value over an interval T is required.

Thus we require to have an output of the form

$$F(t) = \int_{t-T}^{t} f(t')\,dt' \tag{5.73}$$

Let $g(\omega)$ be the Fourier transform of $f(t')$ so that

$$f(t') = \frac{1}{2\pi} \int_{-\infty}^{\infty} g(\omega) \cdot \exp(j\omega t')\,d\omega$$

Substituting in equation (5.67) and reversing the order of integration we obtain,

$$F(t) = \frac{1}{2\pi} \int_{-\infty}^{\infty} g(\omega) \left[\int_{t-T}^{t} \exp(j\omega t') \cdot dt' \right] d\omega$$

$$= \frac{1}{2\pi} \int_{-\infty}^{\infty} g(\omega) \left[T \frac{\sin(\omega T/2)}{\omega T/2} \exp(-j\omega T/2) \right] \exp(j\omega t)\,d\omega \tag{5.74}$$

The last equation shows that the output $F(t)$ can be obtained by transmitting the input signal through a channel which has a transfer function given by:

$$A(\omega) = T \cdot \frac{\sin(\omega T/2)}{\omega T/2} \cdot \exp(-j\omega T/2) \tag{5.75}$$

The required transfer function has an amplitude characteristic which is of the now familiar $(\sin x)/x$ shape, a gain T and a delay $T/2$. This amplitude characteristic has an infinite number of peaks which diminish rapidly as the frequency increases. An adequate approximation to the required characteristic for most purposes is obtained by a low pass filter which simulates the main lobe of the above filter. The first zero of the amplitude characteristic occurs when $\omega T = 2\pi$ and the low pass filter therefore should have a cut off frequency of about $1/T$ Hz.

The $(\sin x)/x$ filter has interesting properties. If a unit step of amplitude A is applied to a unit gain filter the output spectrum is given by the step spectrum $1/j\omega$ multiplied by the filter transfer function or,

$$\frac{A}{j\omega} \cdot \frac{\sin(\omega T/2)}{\omega T/2} \cdot \exp(-j\omega T/2)$$

The inverse transform is a sloping step of build up time T as shown in Fig. 5.12. This can be deduced by the method of equations (2.10) and (2.11). If the sloping step is represented by $f_2(t)$ then its differential $f_2'(t)$ is a rectangular pulse of length T and height A/T. Thus the spectrum of $f_2'(t)$ is from equation (2.4)

$$A \cdot \frac{\sin(\omega T/2)}{\omega T/2} \cdot \exp(-j\omega T/2)$$

The phase factor occurs because the pulse is displaced from the origin

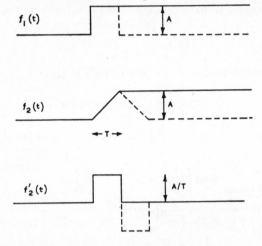

Fig. 5.12. The step response of a $(\sin x)/x$ filter.

of time. By integration with respect to t of the inverse Fourier transform (equation 2.2B) it follows that the spectrum of $f_2(t)$ is given by

$$\text{Spectrum of } f_2(t) = \frac{1}{j\omega} \times \text{Spectrum of } f_2'(t)$$

$$= \frac{A}{j\omega} \cdot \frac{\sin(\omega T/2)}{\omega T/2} \cdot \exp(-j\omega T/2) \qquad (5.76)$$

agreeing with the result above. Thus the response to a unit step is the sloping step of build-up time T.

If the unit step is cut off at time T, i.e., a rectangular pulse of duration T is applied, then the output is a triangular pulse of base $2T$ or length at the mean height of T as shown by the dotted lines in Fig. 5.12. The equivalent ideal rectangular filter with a build-up time T would have a low pass cut-off frequency of $1/2T$, that is, one half of the first zero frequency $1/T$ of the $(\sin x)/x$ filter. Thus the penalty in terms of bandwidth to obtain sharp corners to the sloping step output is quite severe. The $(\sin x)/x$ integrating filter is important in sampling theory as already referred to in Chapter 2.

5.7. Aperture or Scanning Effect

An effect which is closely related to the results of the previous section arises in the facsimile systems described in Chapter 1. If the light spot which scans the picture has a finite size, the intensity of the reflected light and hence the electrical output from the photo cell is determined by the average picture density under the light spot. Any detail smaller than the light spot cannot be reproduced faithfully. Similar problems occur in connection with television camera tubes and with magnetic recording devices, in which the finite width of the magnetic gap used to record information on the magnetic storage tape plays the same part

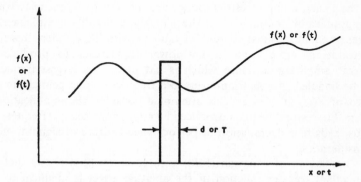

Fig. 5.13. Illustration of the aperture effect.

as the width of the light spot in the facsimile system. The effect is known generally as the aperture or scanning effect and a discussion of the facsimile case is applicable to the other cases mentioned.

Consider a spatial function $f(x)$ representing the light intensity along one scanning line (Fig. 5.13) and let d be the length of the spot measured in the direction of this line. For simplicity, the light spot may be assumed rectangular such that the effective width of the scanning line is much less than d. If the scanning velocity is v and the spot starts from $x = 0$ at time, $t = 0$, then $x = vt$ and $d = vT$, where T is the equivalent time length of the aperture. Apart from the scale factor v, we may therefore work in time and the scanning spot is represented by the rectangular pulse of duration, T, as shown in Fig. 5.13. The photo-cell output at any instant will be proportional to the average value of the function $f(t)$ over the duration of the pulse and we may therefore write the photo-cell output as $F(t)$, where

$$F(t) = \int_{t-\frac{1}{2}T}^{t+\frac{1}{2}T} f(t') \, . \, dt' \qquad (5.77)$$

This equation is identical to equation (5.73) if the integration interval is symmetrical about time t. The argument developed in the previous section therefore shows that the effect of the finite spot size is exactly the same as that observed when the signal $f(t)$ is passed through a channel with a transfer function:

$$A(\omega) = \frac{2 \sin(\omega T/2)}{\omega} \tag{5.78}$$

the exponential term in equation (5.75) reducing to unity. The form of $A(\omega)$ shows that the frequencies in the spectrum of $f(t)$ which are just less than or larger than the value $1/T$ are considerably attenuated by the finite size of the light spot. To avoid serious frequency distortion of the signal, we must restrict the spectrum of $f(t)$ to frequencies which are appreciably less than $1/T$. Alternatively, a suitable complementary filter or equalizer must be used in tandem with the scanner, this filter introducing attenuation at the lower frequencies to produce an overall amplitude response which is flat up to the largest frequency to be handled. If equalization up to a frequency approaching the first zero of $A(\omega)$ at $\omega = 2\pi/T$ is attempted considerable amplification is needed to restore the loss introduced by attenuating the low frequencies. This leads to a degradation of performance because of signal-to-noise considerations.

Comparison of equation (5.78) with equation (2.4) shows that the equivalent transfer function of the aperture effect is identical to the Fourier transform of the time waveform corresponding to the aperture pulse. This result is true for any form of scanning aperture; for example, if the aperture is graded so that the light density in the spot is tapered towards the edges, then the equivalent time waveform is a rounded pulse. The frequency characteristic of the channel, which has the same effect as the aperture, then has the same form as the Fourier transform of this pulse. This statement can be confirmed by using the convolution theorem given in Section A.3.5.

Similar arguments apply to pulse sampling systems in which time waveforms are sampled by pulses of finite width. The theory developed in Chapter 2 for sampling systems assumed the use of infinitely narrow pulses, or at least of pulses whose duration was very short compared with the sampling period. In practice, there are many reasons why the use of such short pulses is undesirable. One reason is that the energy content of a pulse amplitude modulated signal is then very small so that the final recovered signal will be at a very low level and will have a poor signal-to-noise ratio. Also, much of the energy available from the source is wasted if excessively narrow pulses are used. The need to make the pulses in a time division multiplex system as long as possible has been explained in Section 4.9. The sampling switch shown in Fig. 4.18

averages the input function over the time interval for which the switch contact remains closed and again leads to an aperture effect identical to that discussed above. The transfer function corresponding to the switch which generates the succession of pulses is multiplied by the transfer function of the reconstituting filter to give the overall characteristic of the complete system.

Chapter 6

Noise and its Limiting Effect on Communication

6.1. Introduction

The limit to the range of air-borne sound waves which can be detected and recognized by the human ear is determined by extraneous noises in the neighbourhood of the recipient. In the limit we say that the talker is 'drowned' by the local noise. Even supposing this extraneous noise did not exist and that the human ear was rather more sensitive than it is, there would be a limit set by the noise produced by the Brownian or random motion of the air molecules. When the sounds become attenuated such that the wave motion of the air particles is comparable in magnitude to this random motion, they are to all intents and purposes lost.

A precisely analogous state of affairs exists with the electromagnetic waves of line and radio communication. There are extraneous electromagnetic waves produced by other communications systems, by many electrical devices, and by atmospheric electricity. These extraneous waves find their way into a channel to a greater or lesser extent in almost all systems. Their effect will usually be a random signal superimposed on the wanted signal. In the case of telephony, a random noise will be heard along with the wanted signal. For this reason these extraneous waves are usually known as noise or interference, although they may occur at frequencies much higher than audio. If the interference is mainly from another communication channel it is called crosstalk in the telephony case, and crossfire in the telegraphy case. The crosstalk may be intelligible or unintelligible speech, the former being much more annoying when reception conditions are limiting.

Even supposing that a channel can be screened or protected in some way against all extraneous interference, there remains a limiting type of noise known as fluctuation or thermal noise. This is caused by the random motion of the electrons in conductors and bodies associated with, or in proximity to, the channel. These motions cause bodies to radiate electromagnetic energy very weakly, the radiated power being proportional to the absolute temperature T. Any body in thermal equilibrium with its surroundings absorbs from other bodies a power equal to that radiated by itself, so that there is no net loss or gain of energy.

In line communication, the random motions of atoms and electrons in the conductors and bodies surrounding them radiate electromagnetic energy, some of which is guided by the conductors to the terminals of the receiving equipment, where it appears as a noise voltage. It can be shown from thermodynamic arguments that the mean of this voltage squared is proportional to the resistive part R of the impedance, looking back into the line. In radio communication a similar situation exists. Bodies surrounding a receiving aerial radiate energy, some of which is picked up by the aerial, according to its directivity pattern, causing a noise voltage to appear at the aerial terminals. It can be shown that the mean of this voltage squared is proportional to the radiation resistance of the aerial plus a small allowance for the ohmic resistance of the aerial.

Similar noise voltages appear at the terminals of any resistor, the mean square voltage being proportional to the value R of the resistance. If these noise voltages were amplified by a hypothetically perfect amplifier, it would be found that the mean square noise voltage in a given small bandwidth is constant and independent of the mid-frequency of the band selected. Thus the mean square noise voltage is proportional to the total bandwidth B. This result is important and it shows that the bandwidth employed for transmitting a signal should always be made as small as possible. Ideally the bandwidth should only be sufficient to accommodate the frequencies in the signal, but in practice it may be somewhat larger. If the noise voltage is transposed to the audible frequency band it sounds rather like rushing water.

6.2. Mathematical Description of Noise

6.2.1. STATISTICS OF NOISE WAVEFORMS

Any noise waveform is caused by the superposition of a large number of events occurring in a random manner. For example, thermal noise, discussed in the previous section, arises from the random motion of electrons in conductors. It is impossible to make precise statements about the magnitude of the effect at any particular time instant, but since the number of events concerned is very large the average behaviour is well defined and a satisfactory description of the noise waveform can be made in statistical terms. A typical noise waveform is shown in Fig. 6.1(a), and shows fluctuation about an average value, denoted by $\overline{x(t)}$. In all the cases with which we will be concerned, this average value can be obtained by considering the noise to be absent, $e.g.$ if a direct current is flowing through a resistor this average value will simply be the direct current as calculated from ordinary circuit theory. We can therefore regard the noise waveform as being the fluctuation about this average value and consider the enlarged diagram in Fig. 6.1(b). An obvious measure of the magnitude of the noise is the r.m.s. value of this waveform: in practice it is usually simple to work directly with the mean square value and this is defined by the equation:

$$\overline{x^2(t)} = \underset{T\to\infty}{Lt} \frac{1}{T} \int_{-T/2}^{T/2} x^2(t)\,dt \tag{6.1}$$

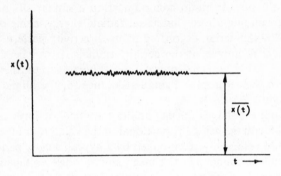

Fig. 6.1. (a) Typical noise waveform.

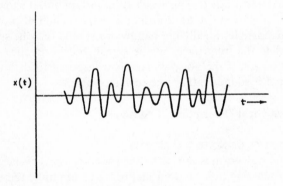

Fig. 6.1. (b) Enlarged diagram to show fluctuation about
mean value.

The limiting process is necessary since both the integral and the observation period are infinitely large. The notation of a horizontal bar implies that the average value of the quantity under the bar is to be taken. It should be noted that $\overline{x(t)}^2$ *i.e.* the square of the average of $x(t)$ (in this case zero) is clearly not equal to $\overline{x^2(t)}$.

We also need to know the likelihood of the value of the noise waveform attaining any particular magnitude, and this information is provided by the probability distribution. To determine this we divide the time of observation T into a large number of small intervals and the waveform magnitude into a large number of steps. The probability of the magnitude having a value in a particular step is then the ratio of the number of intervals for which the waveform magnitude is in the step selected, to the total number of intervals considered. The lengths of the intervals and the sizes of the steps are made infinitesimally small and the probability distribution is obtained: it is written in the form

$$dp = F(x)dx \qquad (6.2)$$

which implies that the probability of x lying between x and $x + dx$ (*i.e.* a small step of size dx) is $F(x)dx$. A general statistical theorem, the Central Limit theorem, states that the probability distribution of any quantity which arises as the sum of the effects of a large number of separate contributions is the Gaussian distribution for which

$$F(x) = \frac{1}{\sigma(2\pi)^{\frac{1}{2}}} \exp(-x^2/2\sigma^2) \qquad (6.3)$$

where

$$\sigma^2 = \overline{x^2} \qquad (6.4)$$

This theorem is appropriate for all noise phenomena of interest: for example, if the noise waveform is that caused by thermal noise in a resistor, it arises from the combined effects of a very large number of electrons moving within the resistor. The quantity σ is referred to in

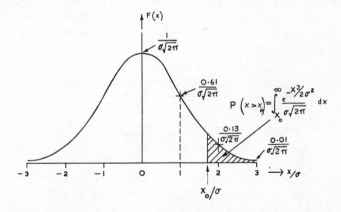

Fig. 6.2. The Gaussian probability distribution.

statistics as the standard deviation and equation (6.4) shows that it equals the r.m.s. value of the waveform.

The shape of the Gaussian probability distribution is shown in Fig. 6.2, where $F(x)$ is plotted against x/σ.

The probability that x exceeds x_0 is found by integrating equation (6.2) over the range x_0 to infinity, *i.e.*

$$P(x > x_0) = \int_{x_0}^{\infty} \frac{1}{\sigma(2\pi)^{\frac{1}{2}}} \exp\left(-\frac{x^2}{2\sigma^2}\right) dx \tag{6.5}$$

The symbol $P(x > x_0)$ means the probability that x exceeds x_0. This can be found from Fig. 6.2 by evaluating the area under the curve to the right of x_0: this is shown by the shading. The skirts of the curve extend to plus and minus infinity but rapidly become smaller.

From the way in which probability was defined above, the probability that x has some finite value, *i.e.* that x exceeds minus infinity, must be unity. It follows that:

$$P(x > -\infty) = \int_{-\infty}^{\infty} \frac{1}{\sigma(2\pi)^{\frac{1}{2}}} \exp\left(-\frac{x^2}{2\sigma^2}\right) dx = 1 \tag{6.6}$$

The validity of this result can be established from the calculations carried out in Section A.3.6 in connection with the Fourier transform of a Gaussian pulse.

The important properties of the Gaussian distribution are shown by Fig. 6.2. The distribution is symmetrical with respect to x equal to zero and decreases steadily as the magnitude of x increases. Some further numerical results are given in Table 6.1: the probabilities of a certain value of x_0 being exceeded are given for x_0 expressed as a multiple of σ and also in decibels above the reference level taken at $x = \sigma$. Since the distribution curve is symmetrical, $P(x < -x_0) = P(x > x_0)$ and so $P(|x| > x_0) = 2 P(x > x_0)$. $P(|x| > x_0)$ gives the probability that the magnitude of x exceeds x_0 irrespective of the sign of x_0.

TABLE 6.1

Probabilities calculated from the Gaussian distribution.

Level x_0 x_0/σ	dB (relative to $x_0 = \sigma$)	$P(x > x_0)$	$P(|x| > x_0)$
1	0	0·158	0·317
1·28	2·2	0·1	0·2
2·33	7·3	10^{-2}	2×10^{-2}
3·09	9·8	10^{-3}	2×10^{-3}
3·72	11·4	10^{-4}	2×10^{-4}
4·25	12·6	10^{-5}	2×10^{-5}
4·75	13·5	10^{-6}	2×10^{-6}

It is a feature of the Gaussian distribution that there is always a finite probability, however small, that any specified high level will be exceeded. In practice, systems are not linear up to high levels, and so there is a small probability that a noise peak will overload the system. Within the limits of linear operation, *i.e.* at levels comparable to the signal, the Gaussian law gives a method of determining the probability that noise peaks will interfere with the operation of the system.

In many communications problems we are faced with the problem of what happens when two noise signals are combined. The answer depends very considerably on whether or not the two signals are independent, *i.e.* whether or not the probability of a particular value of one noise signal is affected by the presence of the other. Noise signals generated by separate sources can be regarded as independent in this

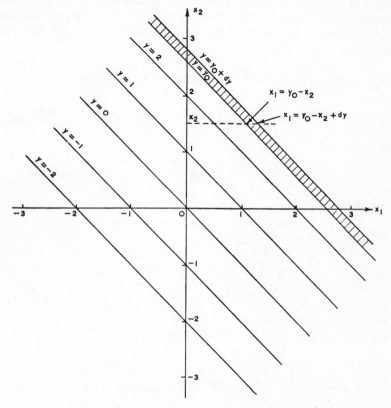

Fig. 6.3. Summation of two noise signals: $y = x_1 + x_2$

sense and the probability that two such signals have values in specified ranges is then equal to the product of the separate probabilities that the individual signals are each in their specified range. Suppose then that two noise signals x_1, x_2 have the Gaussian distributions of r.m.s. values σ_1, σ_2 respectively, and that they are added to give a noise waveform, y. At every instant:

$$y = x_1 + x_2 \tag{6.7}$$

The value of y can thus be read off a chart such as that in Fig. 6.3 and this chart will be used to find the probability that y lies between the values y_0 and $y_0 + dy$, *i.e.* in the shaded area. This means that we must sum the probabilities of x_1, x_2 which lead to values of y within this range. If the value of x_2 is first fixed, then the corresponding permissible range of x_1 is between $y_0 - x_2$ and $y_0 - x_2 + dy$ and the probability of this is $F_1(y - x_2)dy$ where F_1 is the Gaussian function with r.m.s. value σ_1. We now allow x_2 to have any value and add the probabilities of the possible values: this means we multiply by $F_2(x_2)dx_2$ and integrate over the permissible range of x_2. This step is only permissible since x_1 and x_2 are independent. The probability that y lies between y_0 and $y_0 + dy$ is therefore:

$$P(y_0 < y < y_0 + dy) = \int_{-\infty}^{\infty} F_2(x_2)F_1(y - x_2)\,dy\,dx_2 =$$

$$= \int_{-\infty}^{\infty} \frac{1}{\sigma_2(2\pi)^{\frac{1}{2}}} \exp\left(\frac{-x_2^2}{2\sigma_2^2}\right) \frac{1}{\sigma_1(2\pi)^{\frac{1}{2}}} \exp\left(-\frac{(y - x_2)^2}{2\sigma_1^2}\right) dy\,dx_2$$

$$= \frac{1}{2\pi\sigma_1\sigma_2} \int_{-\infty}^{\infty} \exp\left[-\frac{y^2}{2\sigma_1^2} + \frac{2x_2 y}{2\sigma_1^2} - \frac{x_2^2}{2}\left(\frac{1}{\sigma_1^2} + \frac{1}{\sigma_2^2}\right)\right] dy\,dx_2$$

Since

$$\frac{x_2^2}{2}\left(\frac{1}{\sigma_1^2} + \frac{1}{\sigma_2^2}\right) - \frac{2x_2 y}{2\sigma_1^2} =$$

$$= \frac{1}{2}\left[\left(\frac{1}{\sigma_1^2} + \frac{1}{\sigma_2^2}\right)^{\frac{1}{2}} x_2 - \frac{y/\sigma_1^2}{\left(\frac{1}{\sigma_1^2} + \frac{1}{\sigma_2^2}\right)^{\frac{1}{2}}}\right]^2 - \frac{y^2}{2\left(1 + \frac{\sigma_1^2}{\sigma_2^2}\right)\sigma_1^2}$$

we may simplify the integral by defining

$$z = \left(\frac{1}{\sigma_1^2} + \frac{1}{\sigma_2^2}\right)^{\frac{1}{2}} x_2 - \frac{y/\sigma_1^2}{\left(\frac{1}{\sigma_1^2} + \frac{1}{\sigma_2^2}\right)^{\frac{1}{2}}}$$

and then:

$$P(y_0 < y < y_0 + dy) =$$

$$= \frac{1}{2\pi\sigma_1\sigma_2} \int_{-\infty}^{\infty} \exp\left[\frac{-y^2}{2\sigma_1^2} + \frac{y^2}{2\sigma_1^2\left(1 + \frac{\sigma_1^2}{\sigma_2^2}\right)} - \frac{z^2}{2}\right] \frac{dy\,dz}{\left(\frac{1}{\sigma_1^2} + \frac{1}{\sigma_2^2}\right)^{\frac{1}{2}}}$$

$$= \frac{\exp\left[-\frac{y^2}{2} \cdot \frac{1}{(\sigma_1^2 + \sigma_2^2)}\right] dy}{2\pi(\sigma_1^2 + \sigma_2^2)^{\frac{1}{2}}} \int_{-\infty}^{\infty} \exp\left(-\frac{z^2}{2}\right) dz$$

The integral is a special case of equation (6.6) and equals $(2\pi)^{\frac{1}{2}}$ giving

$$P(y_0 < y < y_0 + \mathrm{d}y) = \frac{1}{(\sigma_1{}^2 + \sigma_2{}^2)^{\frac{1}{2}}} \cdot \frac{1}{(2\pi)^{\frac{1}{2}}} \cdot \exp\left[\frac{-y^2}{2(\sigma_1{}^2 + \sigma_2{}^2)}\right] \mathrm{d}y \tag{6.8}$$

This shows that the probability distribution of y is also Gaussian and that the mean square value of y is $(\sigma_1{}^2 + \sigma_2{}^2)$. The mean square value of a noise waveform obtained by adding two independent noise waveforms is therefore the sum of the mean square values of these two waveforms. This result can be extended to any number of independent waveforms. Since the mean square value of a voltage or current waveform gives the power dissipated, apart from a constant, we see that the above result is equivalent to the statement that noise powers add. It must be stressed that this result is only valid if the noise waveforms are statistically independent.

6.2.2. FREQUENCY ANALYSIS OF NOISE WAVEFORMS

We have seen in previous chapters that the behaviour of communication circuits can be analysed very simply by regarding time waveforms as composed of spectra of sinusoidal signals, and it is therefore natural to seek a corresponding method of analysis for noise waveforms. In principle, the spectrum of a noise waveform could be obtained by taking the Fourier transform of the time waveform but a practical difficulty arises in that the condition that the transform can be evaluated in this way is not satisfied. In any case, this transform would contain all the information relating to the waveform, such as the particular instants at which peaks occurred. Since we are only concerned with average properties, this suggests we should look for an alternative approach. This is indicated by two points, the first being that we are interested in mean square values as shown by the previous section, and the second being that most noise waveforms can be regarded as formed by the superposition of a large number of pulses occurring at random instants. This second point is clearly true for thermal noise since the passage of an electron corresponds to a pulse of current. The Fourier transforms of pulses considered in Appendix A all show that the amplitude of the transforms do not depend on the instants at which the pulses occur. This suggests that if we consider the amplitude of the spectrum only, we will obtain a function which gives the required average properties by removing the randomness associated with the times at which the individual pulses occur. Further the first point indicates the choice of the square of the amplitude rather than the amplitude itself. The difficulty associated with the evaluation of the integral remains but is dealt with by a limiting process similar to that used in equation (6.1) where the mean square value is defined.

The ideas outlined in the previous paragraph lead to the following method of handling noise signals. Suppose the time waveform is $f(t)$ and that the Fourier transform is evaluated for a time interval $-T/2$ to $T/2$ *i.e.* we evaluate

$$g_T(\omega) = \int_{-T/2}^{T/2} f(t) \exp(-j\omega t) \, dt \qquad (6.9)$$

We now define a power spectrum $G(\omega)$ by

$$G(\omega) = \operatorname*{Lt}_{T \to \infty} \frac{|g_T(\omega)|^2}{T} \qquad (6.10)$$

The term 'power' is appropriate since, if $f(t)$ is either a voltage or a current, $|g_T(\omega)|^2 d\omega$ will be proportional to the power associated with the frequencies in the interval ω to $\omega + d\omega$. The limiting process obtains an average for this 'power' in a sense similar to that used in equation (6.1). Strictly, the average will be the mean square voltage or current in the frequency interval depending on whether $f(t)$ is a voltage or current waveform.

In order to make use of this definition we must establish some further results. It will be noticed that $G(\omega)$ is defined in terms of the Fourier transform of the time waveform, and we begin by finding the time function of which $G(\omega)$ is the transform. To do this we use the Convolution Theorem derived in Section A.3.5. If $g_1(\omega)$ and $g_2(\omega)$ are the transforms of the time functions $f_1(t)$ and $f_2(t)$ respectively, then the product $g_1(\omega)g_2(\omega)$ is the transform of the time function, $\int_{-\infty}^{\infty} f_1(x)f_2(t - x)\,dx$. We take $g_1(\omega)$ as the transform $g_T(\omega)$ defined by equation (6.9) for a time function $f(t)$: the nature of this definition implies that $f(t)$ vanishes outside the range $-T/2$ to $T/2$. The second transform $g_2(\omega)$ is defined in the same way but with $f(-t)$ in place of $f(t)$ *i.e.*

$$g_2(\omega) = \int_{-T/2}^{T/2} f(-t) \exp(-j\omega t) \, dt = \int_{-T/2}^{T/2} f(t) \exp(j\omega t) \, dt$$
$$= g_T{}^*(\omega) \quad \text{since } f(t) \text{ is a real function.}$$

We may write the convolution theorem as:

$$g_1(\omega)g_2(\omega) = \int_{-\infty}^{\infty} \left[\int_{-\infty}^{\infty} f_1(x)f_2(t - x)\,dx \right] \exp(-j\omega t) \, dt$$

and the choice of $g_1(\omega)$, $g_2(\omega)$ above gives:

$$g_T(\omega)g_T{}^*(\omega) = \int_{-\infty}^{\infty} \left[\int_{-T/2}^{T/2} f(x)f(x - t)\,dx \right] \exp(-j\omega t) \, dt$$

The limits on the inner integral can be taken as $-T/2$ to $T/2$ since $f(x)$ vanishes outside this range. Hence,

$$G(\omega) = \operatorname*{Lt}_{T \to \infty} \frac{|g_T(\omega)|^2}{T} = \operatorname*{Lt}_{T \to \infty} \frac{g_T(\omega)g_T{}^*(\omega)}{T}$$

$$= \operatorname*{Lt}_{T \to \infty} \frac{1}{T} \int_{-\infty}^{\infty} \left[\int_{-T/2}^{T/2} f(x)f(x-t)\,dx \right] \exp(-j\omega t)\,dt$$

$$= \int_{-\infty}^{\infty} \left[\operatorname*{Lt}_{T \to \infty} \frac{1}{T} \int_{-T/2}^{T/2} f(x)f(x-t)\,dx \right] \exp(-j\omega t)\,dt$$

which shows that $G(\omega)$ is the Fourier transform of the function $R(t)$ defined by the equation:

$$R(t) = \operatorname*{Lt}_{T \to \infty} \frac{1}{T} \int_{-T/2}^{T/2} f(x)f(x-t)\,dx \qquad (6.11)$$

and so

$$G(\omega) = \int_{-\infty}^{\infty} R(t) \exp(-j\omega t)\,dt \qquad (6.12)$$

The inverse Fourier transform gives the corresponding result:

$$R(t) = \frac{1}{2\pi} \int_{-\infty}^{\infty} G(\omega) \exp(j\omega t)\,d\omega \qquad (6.13)$$

The function $R(t)$ is called the auto-correlation function (A.C.F.) of the noise waveform. A method by which this function may be measured will be discussed below. Some simple results can be easily obtained. If t is made zero in equations (6.11) and (6.13), we find:

$$R(0) = \operatorname*{Lt}_{T \to \infty} \frac{1}{T} \int_{-T/2}^{T/2} f^2(x)\,dx = \frac{1}{2\pi} \int_{-\infty}^{\infty} G(\omega)\,d\omega$$

which becomes, on using equation (6.1),

$$\overline{f^2(t)} = \int_{-\infty}^{\infty} G(\omega)\,\frac{d\omega}{2\pi} \qquad (6.14)$$

Suppose $f(t)$ is a voltage developed across a 1 ohm resistor. Then $\overline{f^2(t)}$ is the average power dissipated in the resistor and the integral on the right is the integrated power of the various frequencies represented: $G(\omega)$ can thus be regarded as a power density per unit frequency interval as discussed earlier. The presence of the 2π in association with the $d\omega$ means we take the frequency unit as the Hz.

The definition of $G(\omega)$ in equation (6.10) shows that it must be a real positive quantity and it follows from equation (6.13) that

$$R(t) \leqslant \frac{1}{2\pi} \int_{-\infty}^{\infty} G(\omega) \, d\omega$$

since $|\exp(j\omega t)|$ equals unity. Hence:

$$R(t) \leqslant R(0) \tag{6.15}$$

The auto-correlation function is useful in noise analysis since it can often be more easily calculated than the power spectrum: when this is the case, the power spectrum is obtained from equation (6.12). Examples of this are given later.

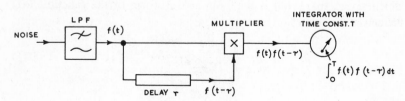

Fig. 6.4. Circuit designed to measure the auto-correlation function of a noise signal.

The auto-correlation function can be measured by the circuit shown in Fig. 6.4. This circuit is designed to measure the properties of noise in the base-band. The noise signal $f(t)$ is split into two parts, one of which is passed through a delay network giving a time delay τ. The two parts are then fed to a multiplier unit, which can be of any of the types discussed in Chapter 3, and the output of this multiplier is therefore $f(t)f(t - \tau)$. The multiplier is followed by a low pass filter with a time constant T and this acts as an integrator giving as its output $\int_0^T f(t)f(t - \tau) \, dt$. Noise signals of the type with which we are concerned have properties which are independent of the choice of time origin and it is easily shown that this integral is therefore proportional to $R(\tau)$. A series of readings can be taken for different values of τ and the auto-correlation function can then be plotted.

6.2.3. THE SHOT NOISE FORMULA

A simple case of noise arises in a temperature limited diode. In such a valve, the anode current is determined only by the rate at which electrons are emitted from the cathode, the anode potential being sufficiently great to attract all electrons to the anode. The anode current can be regarded as the sum of a succession of current pulses, each pulse being caused by the transit of one electron from the cathode to the anode. Suppose an electron leaves the cathode at time t: the current pulse can be written as $i(t)$. This function $i(t)$ is zero if t is less than zero and it is also zero when t exceeds τ, the time taken by the electron to travel from the cathode to the anode. The charge transferred by each electron, is the electronic charge, e and so:

$$\int_0^\tau i(t)\,dt = e \qquad (6.16)$$

If the number of electrons emitted per second is N, the total charge transferred per second is Ne and this is clearly the d.c. anode current I_0:

$$I_0 = Ne \qquad (6.17)$$

The electrons are emitted at random instants and we may use the statistical result that mean square values can be added: this is permissible since any individual electron is not affected by the others, the individual current pulses thus being independent in the statistical sense. A simplification is made by assuming that the transit time, τ, is negligibly short. This means that $i(t)$ can be represented by an impulse function of strength given by equation (6.16): *i.e.*

$$i(t) = e \,.\, \delta(t) \qquad (6.18)$$

The frequency spectrum of the single pulse $i(t - t_n)$ which occurs at time t_n is

$$g_n(\omega) = e \exp(-j\omega t_n)$$

giving

$$|g_n(\omega)|^2 = e^2 \qquad (6.19)$$

The magnitude $|g_n(\omega)|$ is independent of the time at which the pulse occurs. The value of $|g_n(\omega)|^2$ for the NT pulses which occur in a period T is given by adding the values of $|g_n(\omega)|^2$ for the individual pulses and the power spectrum $G(\omega)$ is therefore:

$$G(\omega) = \mathop{Lt}_{T\to\infty} \frac{NT|g_n(\omega)|^2}{T} = Ne^2$$
$$= eI_0 \qquad (6.20)$$

This power spectrum is the value of the mean square current associated with a bandwidth of 1 Hz, centred on the angular frequency, ω. The spectrum is defined for both positive and negative frequencies and there are equal contributions from the bands centred on $+\omega$ and $-\omega$. These two contributions are usually added together and the result is expressed in the form:

$$\overline{i^2} = 2\,eI_0\,df \qquad (6.21)$$

which is equivalent to saying that the mean square current in the frequency f to $f + df$ is $2\,eI_0\,df$. This formula is often referred to as the Schottky formula, Schottky being the first to make a theoretical study of this problem. It has been shown experimentally to be valid to a high degree of accuracy and temperature-limited diodes are widely used as standard noise generators. The mean square noise current per unit bandwidth is seen from the formula to be independent of the mid-band frequency. Noise which is independent of frequency in this way is called white noise, the adjective white being used, since white light contains equal amounts of all frequencies within the visible band of electromagnetic radiation. This independence of frequency only holds strictly when the electron transit time is assumed negligibly short. In practice, equation (6.21) is valid for frequencies appreciably less than the reciprocal of the transit time.

The argument in this section relies on the assumption that the individual pulses are statistically independent. This is justified for temperature-limited operation but most valves operate under space-charge limited conditions. The motion of an electron from cathode to anode is then influenced by the space charge, *i.e.* by the other electrons in the region between the cathode and the anode. In this case, the electrons are not statistically independent and the analysis has to be modified to take this into account. Details of this will be found in the references which are listed in the Bibliography.

6.2.4. THERMAL NOISE

The occurrence of thermal noise has been mentioned in the introduction to this chapter and the close connection with black body radiation has been pointed out. Various methods, all based on thermodynamics, have been used to calculate the mean square noise voltage or current associated with a resistor at absolute temperature, T. One approach is to consider the random thermal motion of the electrons within a resistor. The statistical distribution of the electron velocities can be obtained by a method similar to that used in the kinetic theory of gases. The random motion of each electron corresponds to a current within the resistor and the summation of the effects of all the electrons, making allowance for the statistical distribution of the velocities, gives the following expression for the mean square current:

$$\overline{i^2} = 4kTG\,\mathrm{d}f \tag{6.22}$$

where
k is Boltzmann's constant (1.37×10^{-23} joules per degree Kelvin)
T is the absolute temperature (degrees Kelvin)
G is the conductance of the resistor.
This equation gives the mean square value of the current delivered in the frequency band, $\mathrm{d}f$, (interpreted as in equation (6.21)) by a current

generator acting in parallel with the resistor as in Fig. 6.5(a). An alternative representation can be found by using Thevenin's theorem. The instantaneous voltage across the resistor is

$$v = Ri$$

if i is the instantaneous current delivered by the current generator. Squaring and averaging gives

$$\overline{v^2} = \overline{(Ri)^2} = R^2\,\overline{i^2}$$

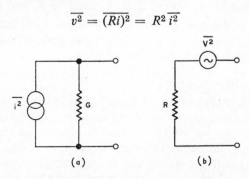

(a)   (b)

Fig. 6.5. Equivalent circuits to represent the thermal noise in a resistor. (a) Current generator: $\overline{i^2}$ = 4kTGdf : G = 1/R (b) Voltage generator $\overline{v^2}$ = 4kTRdf :

since R is a constant. The mean square e.m.f. of a voltage generator in series with R equals $\overline{v^2}$ by Thevenin's theorem and substitution for $\overline{i^2}$ gives:

$$\overline{v^2} = 4kTR\,\mathrm{d}f \tag{6.23}$$

The corresponding circuit representation is shown in Fig. 6.5(b). Equations (6.22) and (6.23) show that thermal noise is white: this is strictly true only for frequencies at which quantum effects can be neglected (i.e. of black body radiation as given by the Rayleigh-Jeans Law). Quantum effects can in general be neglected in the frequency range considered in communications studies, the only significant exception being the use of Maser amplifiers.

Equation (6.23) applies to any kind of resistor, including the radiation resistance of an aerial. In the case of an aerial, the temperature is that of the body at which the aerial is pointing, i.e. the body on which radiation from the aerial would fall. For many communication aerials, most of the radiation from the aerial is ultimately incident on the earth's surface and the temperature of the earth's surface (about 300°K) is used. In some special application, e.g. radio astronomy and satellite communication, very directive aerials are used and all the radiation from the aerial is in effect transmitted into outer space: the effective temperature of the aerial radiation resistance is then only a few degrees Kelvin.

If a resistance R_1 is connected across R, noise power will be transferred from R to R_1. If R_1 is at the same temperature as R, an equal amount of noise power must be transferred from R_1 to R, otherwise the second law of thermodynamics is violated. It may be verified from equation (6.23) that the two noise powers are equal. Maximum power transfer from R to R_1 occurs if R_1 is equal to R, and the power so transferred is $\overline{v^2}/4R = kT\mathrm{d}f$. This amount of power, kT per Hz, is called the available noise power. It may be noted that the available noise power is independent of the resistance value.

6.2.5. BAND-LIMITED NOISE

In the previous sections we have been considering white noise, which extends throughout the whole of the frequency spectrum, at least up to limiting frequencies which are usually outside the bands used for communication. However, in practical communication systems, the noise, together with the signal, will be transmitted by frequency-selective networks and only those frequencies within the passband of the system will appear at the output. We will therefore be interested in the nature of a noise waveform restricted to a particular band of frequencies, extending from, say, $f_0 - \tfrac{1}{2}B$ to $f_0 + \tfrac{1}{2}B$. This is the situation for a system of mid-band frequency, f_0, and band-width B Hz. If only the noise is present, the output appears similar to a carrier wave of frequency f_0 modulated by a noise signal containing only frequencies between 0 and $\tfrac{1}{2}B$. We may express such a signal in the form:

$$f(t) = x(t)\cos(\omega_0 t) + y(t)\sin(\omega_0 t) \tag{6.24}$$

$x(t)$, $y(t)$, being noise signals with identical independent Gaussian distributions. x,y can be regarded as the amplitudes of the in-phase and quadrature components respectively. The probability that x lies between x and $x + \mathrm{d}x$ is

$$\frac{1}{\sigma\sqrt{2\pi}} \exp\left[-\frac{x^2}{2\sigma^2}\right] \mathrm{d}x$$

σ being the r.m.s. value of x. The distribution of y is identical in form.

The probability that x lies between x and $x + \mathrm{d}x$ and that y simultaneously lies between y and $y + \mathrm{d}y$ is

$$\frac{1}{2\pi\sigma^2} \exp\left[-\frac{x^2 + y^2}{2\sigma^2}\right] \mathrm{d}x\,\mathrm{d}y$$

since the two quantities are statistically independent. This expression gives the probability that the pair of values x,y lies within the shaded area of Fig. 6.6(a).

An alternative expression for $f(t)$ is:

$$f(t) = r \cos(\omega_0 t + \phi) \qquad (6.25)$$

in which r, ϕ are respectively the amplitude and phase of the modulated carrier. Since:

$$x^2 + y^2 = r^2 \quad \text{and} \quad y/x = \tan\phi$$

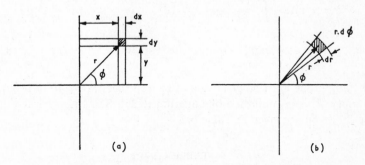

Fig. 6.6. Representation of band-limited noise: (a) in terms of in-phase and quadrature components. (b) in terms of magnitude and phase.

the probability of a pair of values r, ϕ can be predicted from the expression for the probability of the pair x, y. The product $dx\, dy$ equals $r\, dr\, d\phi$ and so the joint probability can be written as $(1/2\pi\sigma^2)\exp(-r^2/2\sigma^2)r\, dr\, d\phi$. This is now interpreted as the probability that the pair of values r, ϕ lies in the shaded region of Fig. 6.6(b). The angle ϕ only appears in the differential, which means that all values of ϕ are equally likely. The possible range of ϕ is from 0 to 2π so that the probability of a value between ϕ and $\phi + d\phi$ must be $d\phi/2\pi$. The joint probability is the product of $d\phi/2\pi$ and $r/\sigma^2 \exp(-r^2/2\sigma^2)dr$, so that the probability of a value of r between r and $r + dr$ must be $r/\sigma^2 . \exp(-r^2/2\sigma^2)dr$. This is called a Rayleigh distribution and is plotted in Fig. 6.7. Band-limited

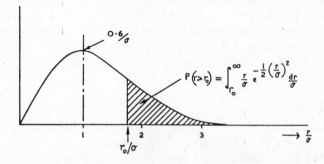

$$P(r > r_0) = \int_{r_0}^{\infty} \frac{r}{\sigma} e^{-\frac{1}{2}\left(\frac{r}{\sigma}\right)^2} \frac{dr}{\sigma}$$

Fig. 6.7. The Rayleigh probability distribution.

noise can thus be regarded as a modulated carrier of random phase, all values being equally likely, and with an amplitude which has a Rayleigh probability distribution. The carrier frequency is the mid-band frequency, f_0, and it follows that the average number of zeros per second is $2f_0$.

From the Rayleigh distribution, we have:

$$P(r > r_0) = \int_{r_0}^{\infty} \frac{r}{\sigma^2} \exp\left(\frac{-r^2}{2\sigma^2}\right) dr = \int_{r_0/\sigma}^{\infty} x \exp\left(-\tfrac{1}{2}x^2\right) dx$$

$$= \exp\left(-\frac{r_0^2}{2\sigma^2}\right) \tag{6.26}$$

The amplitude must be positive and so we expect $P(r > 0)$ to be equal to unity. This is confirmed by equation (6.26). Some numerical values for $P(r > r_0)$ are given in Table 6.2.

TABLE 6.2.

Probabilities calculated from the Rayleigh distribution

Level r_0/σ	dB (relative to $r_0 = \sigma$)	$P(r > r_0)$
1	0	0·607
2·14	6·6	0·1
3·05	9·7	10^{-2}
3·72	11·4	10^{-3}
4·29	12·6	10^{-4}
4·80	13·6	10^{-5}
5·26	14·4	10^{-6}

6.3. Signal to Noise Ratios

If a communication channel is to function satisfactorily the output signal cannot be allowed to fall below some specified level above the output noise. The tolerable difference in level between the signal and the noise depends upon the nature of the signal being transmitted and on the type of system used. The ratio of the signal power to the noise power is called the signal-to-noise ratio and is usually expressed in decibels, thus:

$$\text{Signal to noise ratio} = 10 \log_{10}\left(\frac{S}{N}\right) \tag{6.27}$$

where S is the signal power, taken as a mean value for the signal being transmitted and N is the average noise power.

In most practical cases, the signal-to-noise ratio is sufficiently large for S to be measured as the mean power when both signal and noise are present. The noise power, N, can, of course, be measured when the signal is disconnected from the sending end of the channel.

Noise power can be regarded as being of two kinds: first, there are the fundamental sources of noise, such as thermal noise in resistors and shot noise in electronic devices, which are unavoidable. However, in addition, other forms of noise can enter a communication channel via power supplies to valves and transistors, or by stray electric and magnetic field coupling. These forms can be minimized by proper design of the equipment used. The fundamental sources are inevitably present and it follows that as a signal travels through a channel, extra noise is added to it and the signal-to-noise ratio is degraded. It is this degradation of the signal-to-noise ratio with which we will be concerned and we now consider various ways in which it may be expressed.

Consider a channel with an input signal power S_1 and a power gain G. The output signal power S_2, is therefore GS_1. The input noise power, N_1, is also amplified and contributes an output noise power GN_1. In addition, noise will be introduced within the channel and so the total noise output, N_2, will exceed GN_1. The signal to noise ratio at the output, GS_1/N_2, is therefore less than the signal to noise ratio at the input, S_1/N_1, since N_2 exceeds GN_1. A measure of the degradation caused by the channel is the reduction in signal to noise ratio and we define the noise factor of the channel to be:

$$\text{Noise factor } F = \frac{\text{Signal to noise ratio at input}}{\text{Signal to noise ratio at output}}$$

$$= \frac{S_1/N_1}{GS_1/N_2} = \frac{N_2}{GN_1} \tag{6.28}$$

The advantage of defining a noise factor in this way is that the signal power cancels from the expression, leaving a measure which depends only on the channel and not on the signal level at which the channel is being operated. An ideal channel introduces no extra noise and has N_2 equal to GN_1. Its noise factor is therefore unity. The more noise the channel introduces, the larger the noise factor becomes. The noise factor is often expressed in decibels (*i.e.* as $10 \log_{10} F$) since the output signal to noise ratio in dB can then be found by subtracting the noise factor in dB from the input signal to noise ratio in dB.

The definition of noise factor is only valid if the output signal is of the same form as the input signal. For example, it may be used if the input and output signal modulation is the same, even though there may be a change in carrier frequency within the channel.

It will be shown in Chapter 8 that by using signal transformations, *i.e.* changes in the form of the signal, it is possible to transmit a baseband signal with a high signal-to-noise ratio by means of another signal or carrier wave with a lower signal-to-noise ratio. The noise factor definition is inadequate to describe this situation and an improvement allowance must be made. If this is expressed in decibels it will be of opposite sign to the noise factor.

The noise factor of a channel may often be allowed for in calculations by supposing that all the extra noise is introduced at the channel input. Since the channel gain is G and the total output noise power is N_2, the effective input noise power on this supposition is N_2/G. The actual noise input power to the channel is N_1 and so the extra to be attributed to the channel is $N_2/G - N_1$. This may be written as $N_1[(N_2/GN_1) - 1]$, *i.e.* $N_1(F - 1)$. The noisiness of a channel of noise factor, F, can thus be allowed for by introducing the extra noise power, $N_1(F - 1)$, at the input. The use of this approach is illustrated by the calculation of the noise factor of two channels in cascade. Suppose the two channels have power gains G_1, G_2 and noise factors F_1, F_2 and are coupled as in Fig. 6.8. It is assumed that the networks are matched and that the noise factor F_2 of the second network is defined on the assumption of an input noise power, N_1. The output noise power from network 1 is $F_1 N_1 G_1$ and

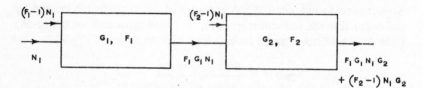

Fig. 6.8. The noise factor of two channels in cascade.

added to this at the input of network 2 is the equivalent extra power $(F_2 - 1)N_1$. The output noise power from network 2 is therefore $G_2[F_1 N_1 G_1 + (F_2 - 1)N_1]$. The definition of the noise factor (equation (6.28)) can be interpreted as:

$$F = \frac{\text{Actual noise output power}}{\text{Noise output power if networks are noiseless}} \qquad (6.29)$$

For the cascade connection, the numerator of this expression is the value calculated above and the denominator is simply the input noise power, N_1, multiplied by the overall gain, $G_1 G_2$. Hence,

$$F = \frac{G_2 N_1[F_1 G_1 + (F_2 - 1)]}{N_1 G_1 G_2}$$

$$= F_1 + \frac{(F_2 - 1)}{G_1} \qquad (6.30)$$

If the first network has a large gain, F is equal to F_1, independently of the noise factor of the second stage. The result may be extended to any number of networks, with gains G_1, G_2, G_3 and noise factors F_1, F_2, F_3, and connected in the order 1, 2, 3 from the input, to give:

$$F = F_1 + \frac{F_2 - 1}{G_1} + \frac{F_3 - 1}{G_1 G_2} + \frac{F_4 - 1}{G_1 G_2 G_3} \qquad (6.31)$$

The noise factor is widely used in communication work but it has one drawback, namely, its dependence on the input noise power, N_1. Suppose, for example, that we consider the behaviour of a receiver fed from an aerial of radiation resistance, R. If the aerial is matched to the receiver, N_1 is the available noise power of the radiation resistance and has the value $kT_1 B$, where T_1 is the effective temperature of the aerial and B is the receiver bandwidth in Hz. As pointed out in the previous section, the effective temperature of the aerial can be dependent on which direction it is pointing: N_1 is therefore not a constant and the noise factor will vary as N_1 varies. The basic property of the receiver in this case is the equivalent input for the noise power generated within the receiver. This noise input can be expressed as $kT_e B$, where T_e is defined as the effective noise input temperature of the amplifier. T_e can be regarded as the increase in the temperature of the aerial to account for noise generated in the amplifier. T_e is a constant for the receiver, independent of the effective temperature of the aerial.

If the temperature of the aerial is T_1 and the effective noise temperature of the receiver is T_e the total input noise power is $k(T_1 + T_e)B$. A noiseless receiver has an input noise power $kT_1 B$, so that the noise factor is dependent on the aerial temperature according to the equation:

$$F = (T_1 + T_e)/T_1 = 1 + T_e/T_1 \qquad (6.32)$$

This shows the dependence of the noise factor on the aerial temperature.

In many situations, the noise temperature of the input circuit is ordinary room temperature and the value usually taken for this is 290°K. If T_e is 60°K, then the corresponding noise factor is $1 + 60/290$, i.e. 1·21 or 0·8 db. In practice, the noise factor is adequate for all situations except those involving low noise temperature aerials and low noise amplifiers, such as occur in satellite ground stations.

6.4. The Detection of Pulses in the presence of Noise

As an elementary application of the idea of signal-to-noise ratio consider its application to the detection of a baseband pulse in noise, a problem which arises in both telegraph and radar systems. In Fig. 6.9(a), a rectangular pulse to which noise has been added is to be identified by a decision circuit. Such a decision circuit sets a threshold level, shown by the dotted line; when noise, or signal plus noise, are below the line the circuit transmits 0, and when they are above the line it transmits 1. In synchronous telegraphy the circuit may only transmit its decisions at the middle of telegraph elements, such as the middle of the rectangular pulse in Fig. 6.9(a). In radar a similar decision circuit is formed by setting a bias level on a cathode-ray tube such that signal plus noise below the bias does not illuminate the tube, whereas signal plus noise above the level illuminates or 'paints' on the tube. In the latter case the decision circuit can operate at all times.

A decision circuit can make errors due to noise peaks. At the left of Fig. 6.9(a), a noise peak can convey the *mark* signal when *space* is intended. In radar such a noise peak above the threshold would be termed a 'false alarm'. At the right of Fig. 6.9(a) a signal is present, but a negative peak of sufficient magnitude can cause the circuit to convey *space* instead of *mark*. In radar, if the noise peak depressed the entire top of the signal pulse, it would be termed a 'missed target'.

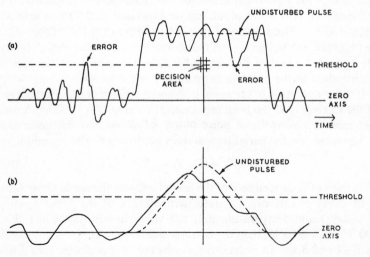

Fig. 6.9. The detection of pulses in the presence of noise: (a) rectangular pulse: (b) half sine pulse.

The probability of either kind of error can be expressed in terms of the probability distribution for the noise. Suppose the noise has a Gaussian distribution with an r.m.s. value, σ, and the rectangular pulse has an amplitude E. Let the threshold be set at kE. The probability that a mark is obtained in the absence of a pulse is $P(x > kE)$, where the noise waveform is denoted by x. A space is obtained if the noise depresses the pulse level below the threshold, *i.e.* if $x + E$ is less than kE

or $x < -(1 - k)E$. The probability of this is $P(x < -(1 - k)E)$. In telegraphy systems it is usual to equalize the probabilities of the two types of error, giving:

$$P(x > kE) = P(x < -\overline{1 - k} E) \tag{6.33}$$

Since the Gaussian distribution is symmetrical, this equation can only be satisfied if k equals 0·5, showing that the threshold must be set at half the pulse level. The probability of either error is then:

$$P(x > \tfrac{1}{2}E) = \frac{1}{\sigma\sqrt{2\pi}} \int_{\frac{1}{2}E}^{\infty} \exp\left(-\frac{x^2}{2\sigma^2}\right) dx$$

$$= \frac{1}{\sqrt{2\pi}} \int_{\frac{1}{2}\cdot E/\sigma}^{\infty} \exp(-y^2/2)\,dy \tag{6.34}$$

The probability depends only on the ratio E/σ and this ratio is simply the square root of the signal to noise ratio. Table 6.1 can therefore be used to construct a curve showing error probability against signal-to-noise ratio. For example, an error probability of 10^{-3} corresponds to the x/σ ratio of 3·09: this means that $E/2\sigma$ must equal 3·09, giving $(E/\sigma)^2 = (6·18)^2 = 38·2$. The signal-to-noise ratio needed to give an error probability of 10^{-3} is therefore 15·8 dB. This is 6 dB greater than the dB figure in Table 6.1 because we are interested in the ratio of σ to $\tfrac{1}{2}E$. Other values can rapidly be obtained by adding 6 dB to the results in Table 6.1 and the curve in Fig. 6.10 is then obtained.

In synchronous telegraphy it may be noted that the square corners do not contribute, since the decision circuit operates at the middle of the pulse, and that signal power may therefore be saved by sending a more rounded pulse. A half sine pulse in fact saves 3 dB transmitted power and a more favourable curve would be produced, as shown in Fig. 6.10. This introduces an important point, of a type which frequently occurs in communications. In order to keep the noise power as low as possible the bandwidth must be reduced to the minimum compatible with the type of signal to be detected. The rectangular pulse is, therefore, not practical and some type of smooth pulse is to be preferred. The optimum shape will be considered in Chapter 7, but Fig. 6.9(b) shows a more practical representation of signal plus noise as observed at the output of a receiving filter. The noise has been filtered down to a bandwidth comparable with that required to pass a more optimum shape of pulse. It should be noted, however, that this has caused the sloping edges of the pulses to be shifted in time by the noise to a greater extent than in the previous figure. This is unimportant in synchronous telegraphy but it can be objectionable where the leading edge of the pulse is used for time measurement as in radar. The magnitude of this time displacement by noise is treated in Chapter 10.

In radar it is usual to set the threshold at a level such that the probability of false alarms is very low, say 1 in 10^6 or less depending upon the system. When the signal amplitude is just equal to the threshold, the probability of detection is about 50 per cent, because noise may depress or augment the signal. However, this refers to a single pulse at long range. As the target comes to shorter ranges more pulses are received, the probability of missed target falls and the probability of detection (1—probability of missed target) increases. It should be noted that the baseband signal argument is not strictly applicable to radar systems with carrier detection. It does, however, provide a useful understanding of the problems involved.

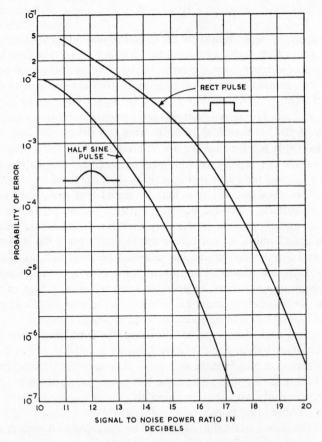

Fig. 6.10. Error probabilities in a telegraphy system with rectangular and half sine pulses. (Decision level set at half amplitude).

An important deduction can be made from the binary telegraphy curves of Fig. 6.10. A small increase of signal-to-noise ratio results in a large reduction of error rate. For example, in the half sine pulse case, improving the ratio from 14·3 dB to 15·6 dB makes a tenfold reduction in error rate.

6.5. Demodulation in the presence of Noise

The detection of signals in the presence of noise is the central problem in communications and depends on the signal-to-noise ratio which can be achieved in the communication system being studied. The changes in signal-to-noise ratio caused by transmission through a linear network can be discussed in terms of the noise factor introduced in Section 6.3, but a different approach is required when non-linear networks are involved. The most important non-linear networks in this respect are those used for the demodulation process. The simplest example of demodulation is the extraction of a d.c. signal proportional to the amplitude of an incoming carrier wave. We will examine the changes in signal-to-noise ratio associated with this process. We begin in the next section by considering an ideal process and then examine the closely analogous problem of frequency changing. Finally, the types of detectors used in practice are considered.

6.5.1. COHERENT DETECTION

Suppose a channel has midband frequency, f_c, and bandwidth B, and that the channel output contains a carrier wave at the mid-band frequency and white noise within the bandwidth B. The problem is to produce a d.c. output proportional to the carrier amplitude with the minimum deterioration of signal-to-noise ratio. It will be assumed that the baseband extends from 0 to $\frac{1}{2}B$ Hz, this being consistent with the channel bandwidth B. The results derived can thus be applied to a carrier telegraphy system in which the channel transmits pulses at the mid-band frequency.

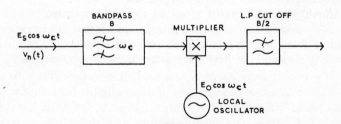

Fig. 6.11. Coherent detector.

Suppose that the carrier frequency is highly stable and that a local oscillator providing a constant amplitude output of identical frequency, and having the same phase as the carrier, is available at the receiving end of the channel. The coherent detector being considered is shown in Fig. 6.11 and consists essentially of an ideal multiplier into which the incoming signal, plus associated noise, and the local oscillator output are fed. The input signal is $E_c \cos(\omega_c t)$, the noise waveform is $v_n(t)$ and the local oscillator output is $E_0 \cos(\omega_c t)$. The product obtained from the multiplier is therefore:

$$f(t) = E_0 \cos(\omega_c t)[E_c \cos(\omega_c t) + v_n(t)]$$

$$= \tfrac{1}{2}E_0 E_c[1 + \cos 2\omega_c t] + E_0 v_n(t) \cos(\omega_c t) \qquad (6.35)$$

The first term is the required d.c. output, proportional to E_c, the carrier amplitude, while the second term is rejected by the baseband filter. The final term results in baseband noise. The noise waveform $v_n(t)$ corresponds to band limited noise of the type discussed in Section 6.2.5 and can be expressed in the form

$$x(t) \cos(\omega_c t) + y(t) \sin(\omega_c t) \quad \text{(c.f. equation 6.24)}$$

Hence:

$$E_0 v_n(t) \cos(\omega_c t) = \tfrac{1}{2}E_0[x(t)\{1 + \cos(2\omega_c t)\} + y(t) \sin(2\omega_c t)] \qquad (6.36)$$

Since $x(t)$, $y(t)$ contain frequencies within the range 0 to $\tfrac{1}{2}B$ Hz only, all terms in the expression on the right, except the first, are rejected by the baseband filter. The final output from this filter is therefore the signal term $\tfrac{1}{2}E_0 E_c$ and noise term $\tfrac{1}{2}E_0 x(t)$. The signal to noise ratio of this output is $(\tfrac{1}{2}E_0 E_c)^2$ divided by $(\tfrac{1}{2}E_0)^2 \overline{x^2(t)}$, i.e. $E_c^2/\overline{x^2(t)}$. The input signal power is $\tfrac{1}{2}E_c^2$ and the input noise power is $\tfrac{1}{2}[\overline{x^2(t)} + \overline{y^2(t)}]$ i.e. $\overline{x^2(t)}$ since $x(t)$ and $y(t)$ have identical statistical properties. The input signal to noise ratio is therefore $\tfrac{1}{2}E_c^2/\overline{x^2(t)}$ and we see that the output signal to noise ratio is twice this value. The coherent detector gives a 3 dB improvement in signal-to-noise ratio. The reason for this improvement is that the combination of the multiplier and low pass filter eliminates the quadrature noise waveform, $y(t) \sin(\omega_c t)$.

It should be noted that if a second system is operated in the same channel by using a carrier in quadrature with the first, it will experience the same 3 dB improvement after coherent detection. The coherent detector is thus an ideal device for determining the presence and amplitude of a carrier in noise. It is, however, frequently difficult to apply in practice because of the problem of obtaining the local carrier in the correct phase with sufficient stability, and because of phase instabilities of the signal carrier due to propagation. It depends upon a suitable multiplying device which behaves ideally. Suitable electronic devices such as multigrid valves and semiconductor diode switches can in fact be used to achieve something very close to the above performance, as discussed in Chapter 3.

6.5.2. FREQUENCY CHANGING

The arguments of the previous section can be adapted to show that given a suitable multiplier device it is possible to change the frequency of a carrier signal in noise, with no perceptible change in signal-to-noise ratio. The local oscillator applied to the multiplier must now have a frequency, ω_0, different from the carrier frequency, ω_c. This local frequency is often called the heterodyne frequency. The new frequency of the signal will be the sum or difference of the two frequencies, ω_0 and ω_c. The new frequencies will have attendant noise bands. It is assumed that there is no overlapping of these bands with one another. If the frequencies are suitably chosen the effect of a multiplier can be realized by a square law detector, *i.e.* a non-linear device giving an output proportional to the square of the instantaneous applied voltage. A similar application of a square law device has been examined in Chapter 3. The output of the square law device is proportional to

$$[E_0 \cos(\omega_0 t) + E_c \cos(\omega_c t) + v_n(t)]^2 =$$
$$= E_0{}^2 \cos^2(\omega_0 t) + [E_c \cos(\omega_c t) + v_n(t)]^2$$
$$+ 2E_0 \cos(\omega_0 t)[E_c \cos(\omega_c t) + v_n(t)] \qquad (6.37)$$

With a suitable choice of the heterodyne frequency, the first two terms produce products which can be ignored and interest centres upon the third term which is equal to the output produced by an ideal multiplier. Considering the signal and noise portions we obtain:

$$2E_0 E_c \cos(\omega_0 t) \cos(\omega_c t)$$

and

$$2E_0 x(t) \cos(\omega_0 t) \cos(\omega_c t) + 2E_0 y(t) \cos(\omega_0 t) \sin(\omega_c t)$$

respectively.

By expanding these into sum and difference products, and assuming that each of them is filtered by a filter of bandwidth equal to the channel supplying the carrier, it can be seen that the signal-to-noise ratio of each product is simply equal to the ratio at the input to the multiplier or detector.

6.5.3. THE RECTIFIER DETECTOR

We now consider the more usual forms of detector which rely on the use of non-linear rectifiers, in place of the ideal coherent detector discussed in Section 6.5.1. As before, the input to the detector will be taken as the combination of a carrier $E_c \cos(\omega_c t)$ and a noise signal, expressed in the form $x(t) \cos(\omega_c t) + y(t) \sin(\omega_c t)$. The noise is assumed to be white, covering the frequency range $f_c - \frac{1}{2}B$ to $f_c + \frac{1}{2}B$, so that the waveforms $x(t)$, $y(t)$, which are real functions of time contain frequencies in the range 0 to $\frac{1}{2}B$. The bandwidth B is taken to be much less than the carrier frequency. The noise is assumed to have a Gaussian probability distribution. The input signal-to-noise ratio, calculated as in Section 6.5.1 is $E_c^2/2\overline{x^2(t)}$ and will be denoted by S_{in}. The output from the detector is delivered as a baseband which will cover the possible modulation frequencies, if a modulated carrier is used. The present treatment will be restricted to unmodulated carriers for simplicity but can be extended to cover modulated carriers. The results obtained will be applied in the next section to carrier telegraphy and it may therefore be assumed that the bandwidth B of the input channel to the detector has been selected to deal with the required telegraphy rate. The baseband is not specifically limited at this stage except that it is regarded as passing only audio frequencies, *i.e.* frequencies very much less than the carrier frequency f_c.

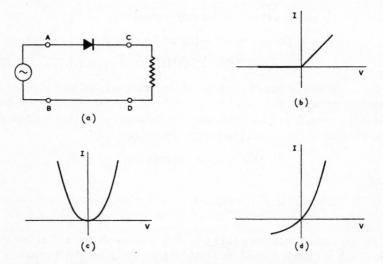

Fig. 6.12. (a) Simple detector circuit. Typical non-linear characteristics are shown in (b) linear detector, (c) parabolic detector, (d) square law detector

An ideal rectifier has zero resistance in the forward direction and infinite resistance in the reverse direction but for the present purpose an element with a voltage-current characteristic as shown in Fig. 6.12(*b*) is equally good. A detector circuit, using such an element is shown in Fig.

6.12(*a*), the detected baseband output being developed across CD. The detector characteristic shown in Fig. 6.12(*b*) is approximated by two linear portions and is said to be piecewise linear. If the input to this detector is an unmodulated carrier, the output is a d.c. signal directly proportional to the carrier amplitude. There is therefore a linear relation between the detector output and the input in the absence of noise and so the term 'linear detector' is often applied to this arrangement. There is, however, no simple relation between the output noise and the input noise and an analysis of the effect of this type of detector on signal to noise ratio is rather involved.

A further possible detector characteristic is shown in Fig. 6.12(*d*). This type of characteristic is readily achieved from valves or transistors and its current-voltage characteristic can be approximated by the equation:

$$i = av + bv^2 \tag{6.38}$$

The linear term, av, makes no contribution to the rectification and so can be ignored. The performance of the element as a detector is therefore the same as that of an element with the characteristic:

$$i = bv^2 \tag{6.39}$$

shown in Fig. 6.12(*c*). The form of the relation in equation (6.39) leads to the names 'square-law detector' or 'parabolic detector'. The signal-noise performance of a square-law detector can be assessed by the arguments given below.

The input to the detector is the combination of an unmodulated carrier and noise, as described earlier in this section, *i.e.*:

$$v = E_c \cos(\omega_c t) + x(t) \cos(\omega_c t) + y(t) \sin(\omega_c t) \tag{6.40}$$

The detector output is directly proportional to the current passing through the rectifier element and we may therefore regard i, given by equation (6.39) as the detector output. When v is defined by equation (6.40) we have

$$i = b[E_c^2 \cos^2(\omega_c t) + 2E_c x(t) \cos^2(\omega_c t)$$
$$+ 2E_c y(t) \cos(\omega_c t) \sin(\omega_c t) + x^2(t) \cos^2(\omega_c t)$$
$$+ y^2(t) \sin^2(\omega_c t) + 2x(t)y(t) \cos(\omega_c t) \sin(\omega_c t)] \tag{6.41}$$

The detector output is filtered by a baseband channel which will suppress all frequencies except those which are much less than the carrier frequency. The expression in equation (6.41) can be rewritten as:

$$i = \tfrac{1}{2}b[E_c{}^2 + 2E_c x(t) + x^2(t) + y^2(t)]$$
$$+ \tfrac{1}{2}b[E_c{}^2 + 2E_c x(t) + x^2(t) - y^2(t)]\cos(2\omega_c t)$$
$$+ b[E_c y(t) + x(t)y(t)]\sin(2\omega_c t) \qquad (6.42)$$

Since the noise waveforms $x(t)$, $y(t)$ are slowly varying time functions, containing frequencies in the range 0 to $\tfrac{1}{2}B$, the first of the three terms on the right of the above equation will also contain only relatively low frequencies and will be accepted by the baseband channel. The second and third terms, however, correspond to a modulated carrier of frequency $2f_c$ and are eliminated by the baseband channel. We therefore conclude that the output from the detector circuit is:

$$i = \tfrac{1}{2}b[E_c{}^2 + 2E_c x(t) + x^2(t) + y^2(t)] \qquad (6.43)$$

This detector output can be regarded as composed of three terms. The first term, $\tfrac{1}{2}bE_c{}^2$, is the d.c. output corresponding to the input carrier and is the desired output signal. The output signal power developed in a 1 ohm resistor can thus be calculated:

$$\text{Output signal power} = (\tfrac{1}{2}bE_c{}^2)^2 \qquad (6.44)$$

It should be noted that the results which will be obtained for the output signal to noise ratio are independent of the value of the load resistor and 1 ohm has been selected as the most convenient choice. The second term $bE_c x(t)$ is a noise signal and since it consists simply of a constant multiplying the noise waveform $x(t)$, its statistical properties and frequency spectrum will be the same as those of $x(t)$. This term arises because of a beating mechanism between the carrier and the noise frequency components in the input to the detector. It will therefore be referred to as the carrier-noise beat contribution and we have:

$$\text{Output noise power (carrier-noise beat contribution)}$$
$$= \overline{[bE_c x(t)]^2} = b^2 E_c{}^2 \overline{x^2(t)} \qquad (6.45)$$

The third term, $\tfrac{1}{2}b[x^2(t) + y^2(t)]$ is also a noise term but it is less simple in form and raises some interesting problems. This term does not depend on the carrier amplitude and exists even when the carrier is zero. A detailed examination in terms of frequency components shows that it arises because of beats between noise components and we will therefore refer to it as the noise self-beat contribution.

The average value of this term is

$$\tfrac{1}{2}b\overline{[x^2(t) + y^2(t)]} = b\,\overline{x^2(t)} \qquad (6.46)$$

since $\overline{x^2(t)}$ and $\overline{y^2(t)}$ are equal. The detected output thus contains a d.c. term given by:

$$\text{d.c. current from noise self-beat contribution} = \overline{bx^2(t)} \qquad (6.47)$$

The total power delivered by this term to a 1 ohm resistor is:

Total power (noise self-beat contribution)

$$= \overline{\{\tfrac{1}{2}b[x^2(t) + y^2(t)]\}^2}$$
$$= \frac{b^2}{4} [\overline{x^4(t)} + \overline{2x^2(t)y^2(t)} + \overline{y^4(t)}]$$
$$= \frac{b^2}{2} [\overline{x^4(t)} + \{\overline{x^2(t)}\}^2]$$

since $x(t)$, $y(t)$ have identical statistical distributions and are independent.

The function $x(t)$ has a Gaussian probability distribution and the method of finding averages shows that:

$$\overline{x^4(t)} = 3\{\overline{x^2(t)}\}^2$$

leading to the result:

Total power (noise self-beat contribution)

$$= \frac{b^2}{2} [3\{\overline{x^2(t)}\}^2 + \{\overline{x^2(t)}\}^2]$$
$$= 2b^2\{\overline{x^2(t)}\}^2 \qquad (6.48)$$

This total power includes that due to the d.c. term evaluated in equation (6.46), the remainder being accounted for by the low frequency noise components. These low frequency components are responsible for the fluctuations in the output and we are concerned with the power associated with them. This is given by:

Fluctuation noise power (noise self-beat contribution)
$$= \text{Total power} - \text{d.c. power}$$
$$= 2b^2\{\overline{x^2(t)}\}^2 - b^2\{\overline{x^2(t)}\}^2 = b^2\{\overline{x^2(t)}\}^2 \qquad (6.49)$$

The noise self-beat contribution to the output thus involves equal amounts of power in the d.c. and fluctuation terms.

We may now use these results to determine the output signal-to-noise ratio. In calculating this ratio, we are primarily concerned with the effect of the noise in causing fluctuations in the output—either because such fluctuations appear as a random movement of an indicating instrument such as a meter, or because of their audibility if the output is delivered to a loudspeaker. It is therefore reasonable in calculating this signal-to-noise ratio to ignore the d.c. power arising from the noise self-beat contribution. The total fluctuation noise power is therefore obtained from equations (6.45) and (6.49) and is:

Total fluctuation noise power $= b^2 E_c{}^2 \overline{x^2(t)} + b^2 \{\overline{x^2(t)}\}^2$ (6.50)

giving:

Output signal to noise ratio $S_{out} = \dfrac{(\frac{1}{2}bE_c{}^2)^2}{b^2 E_c{}^2 \overline{x^2(t)} + b^2 \{\overline{x^2(t)}\}^2}$

Since the input signal noise ratio, S_{in}, equals $E_c{}^2/2\overline{x^2(t)}$, this equation can be rewritten as:

$$S_{out} = \frac{S_{in}{}^2}{2S_{in} + 1} \qquad (6.51)$$

If the input signal-to-noise ratio is much larger than unity, the output signal to noise ratio is approximately equal to $\frac{1}{2}S_{in}$, the square law detector, thus causing a 3 dB reduction in signal-to-noise ratio. This

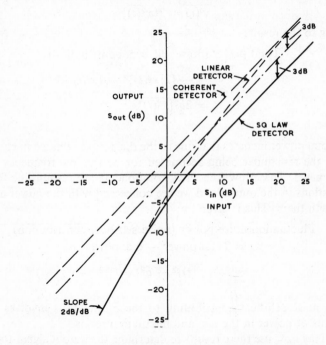

Fig. 6.13. The relationship between output and input signal to noise ratios for coherent, linear and square law detectors.

approximation is equivalent to ignoring the noise self-beat contribution. The fact that the noise self-beat contribution is much less than the carrier-noise beat contribution is readily observed with any fixed gain radio receiver. The noise level from the speaker increases when the receiver is tuned to an unmodulated carrier, relative to the level when

no carrier is present, showing that the carrier-noise beat contribution is the dominant one, provided the carrier amplitude is sufficiently large.

For input signal-to-noise ratios which are much less than unity, $2\,S_{in}$ is negligible compared with unity and equation (6.51) shows that S_{out} is now equal to S_{in}^2. In this case, the square law detector causes a very serious degradation of the signal-to-noise ratio. The approximation here is equivalent to ignoring the carrier-noise beat contribution compared with the noise self-beat contribution.

The relation between S_{out} and S_{in} for the different types of detector is shown in Fig. 6.13. The square law detector results are calculated from equation (6.51) and those for the coherent detector from the analysis given in Section 6.5.1. The output signal to noise ratio from a square law detector is -4.7 dB if the input signal-to-noise ratio is unity, and, in the small signal region, the output ratio drops by 2 dB for each 1 dB drop in the input ratio.

Results for the linear detector obtained from the literature on this subject are also shown in Fig. 6.13. For large signal-to-noise ratios the linear detector has the same performance as the coherent detector, the output ratio being 3 dB greater than the input ratio. The reason for this 3 dB improvement is that the linear detector operates on the envelope or amplitude of its input. An examination of the expression for the combination of signal and noise at the detector input shows that the amplitude depends only on the in-phase noise component, $x(t)\cos(\omega_c t)$, provided the signal-to-noise ratio is large. The quadrature noise component, $y(t)\sin(\omega_c t)$, affects only the phase of the detector input and does not therefore contribute to the baseband output. Half the input noise is thus eliminated by the linear detector. For very small signal-to-noise ratios the linear detector performance is the same as for the square law detector, the output signal-to-noise ratio being in both cases equal to the square of the input signal-to-noise ratio. The transition from the linear dependence between the input and output signal to noise ratios to the square law dependence occurs for both detectors at input ratios just below unity; the transition is often referred to as the threshold of AM detection as it represents the signal-to-noise level at which the modulation is lost in the background noise.

In the above analysis, no restriction was placed on the bandwidth of the baseband channel other than its ability to exclude frequencies much larger than B. The baseband channel is in practice chosen to be sufficiently large to accept any modulation and in view of the restriction on the input band to the frequency range $f_c - \frac{1}{2}B$ to $f_c + \frac{1}{2}B$, it follows that the baseband need never cover more than the range 0 to $\frac{1}{2}B$. This obviously raises the question as to whether any significant portion of the output noise lies outside this range. It has been seen that the square law detector has two noise terms in its output. The first term, the carrier-

noise beat contribution, has the same spectral distribution as the noise waveform $x(t)$ as shown in Fig. 6.14(b). It is therefore white noise over the frequency range 0 to $\frac{1}{2}B$, the same range as proposed for the baseband channel. The second term, the noise self-beat contribution, has, however, a different spectral distribution and an examination of this by considering the beating action in detail shows that the power density per Hz decreases linearly from zero frequency to a zero value at a frequency equal to B as shown in Fig. 6.14(c). The part of this noise which

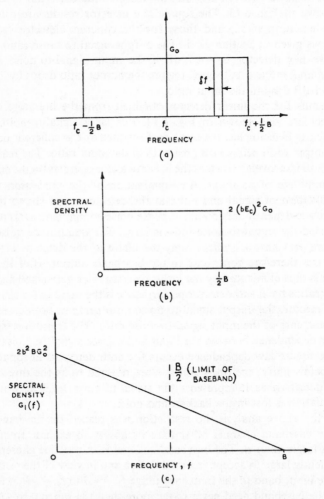

Fig. 6.14. (a) Input noise spectrum centred on the carrier frequency. (b) Output noise spectrum caused by noise beating with the carrier. (c) Output noise spectrum caused by self beats between noise components.

lies in the frequency range $\frac{1}{2}B$ to B can therefore be suppressed without any loss in modulation output. This cuts the noise self-beat contribution to three-quarters of the power calculated above. This will of course have a negligible effect on the output signal-to-noise ratio if the input signal to noise ratio is much larger than unity.

6.5.4. THE ENVELOPE DETECTOR AND ITS APPLICATION TO CARRIER TELE-GRAPHY

The ideal linear detector whose characteristic is shown in Fig. 6.12(b) produces a baseband output which is proportional to the peak positive amplitude or envelope of the carrier or carrier plus noise. Approximations to this in practice are achieved by semiconductor rectifiers with a high ratio of reverse to forward resistance when operated in a circuit, with a load resistance which is high in relation to the forward resistance but low in relation to the reverse resistance. Although the calculation of the relation between output and input signal to noise ratios for all values of the ratio presents difficulties, it is possible to make assumptions which are adequate when the ratio is moderately high, such as is necessary for a practical system.

A good example is the calculation of the error probabilities in an on-off carrier telegraphy system with envelope detection. The mark signal corresponds to a pulse of carrier and the space to the absence of carrier. The channel is assumed to have a bandwidth correctly chosen in relation to the signalling speed and we are concerned with the noise in the spaces and the noise at the maximum value of the carrier. The signal-to-noise ratio is assumed to be at least 5 to 1 in voltage at the maximum of the mark. The system may consist of an RF filter, frequency changer, and IF filter, none of which should cause any fundamental deterioration of the signal-to-noise density ratio. These are then followed by the envelope detector and a decision circuit or threshold which delivers 0 or 1 to the telegraph apparatus.

In the gaps between pulses, the noise is of the type discussed in Section 6.2.5, *i.e.* band limited noise with an envelope which follows the Rayleigh probability distribution.

Thus the noise is represented by:

$$v_n(t) = x(t) \cos \omega_c t + y(t) \sin \omega_c t$$

$$= r(t) \cos [\omega_c t + \theta(t)] \tag{6.52}$$

The noise envelopes $x(t)$ and $y(t)$ are independent of one another and each has an r.m.s. value of σ. The total noise power is

$$N = \frac{\sigma^2}{2} + \frac{\sigma^2}{2}$$

$$= \sigma^2 \tag{6.53}$$

The envelope $r(t)$ has a Rayleigh-type probability distribution. The r.m.s. value is $\sigma\sqrt{2}$ and equation (6.26) gives the probability of the envelope crossing a threshold r_0:

$$P(r > r_0) = \exp\left(-r_0^2/2\sigma^2\right) \tag{6.54}$$

For a given value of $P(r > r_0)$, say 0·001, we have by taking logarithms

$$\log_e 1000 = r_0^2/2\sigma^2$$

or

$$r_0 = 3·72\sigma \tag{6.55}$$

Thus, if the threshold r_0 is set at 3·72 times the r.m.s. value of the total noise the probability of a space being read as a mark is 0·001.

During the mark periods the signal plus noise may be represented by

$$v(t) = [E_c \cos \omega_c t + x(t) \cos \omega_c t] + y(t) \sin \omega_c t \tag{6.56}$$

This may be imagined to consist of a carrier of amplitude E_c having superimposed on it the noise component of amplitude $x(t)$, which is fluctuating according to a Gaussian distribution. Since we are considering ratios of E_c to $x(t)$ or $y(t)$ which are larger than $\sqrt{2} \times 5 \doteqdot 7$, to a first approximation the quadrature noise component $y(t) \sin\omega_c t$ may be imagined to produce a random variation of the phase of E_c as illustrated by Fig. 6.15. The envelope detector is relatively insensitive to this phase variation if it is small, and it will give an output proportional to $E_c + x(t)$ except on those occasions where $x(t)$ is negative and larger than E_c. Since an envelope detector cannot produce a negative output by definition on those occasions where the noise depresses the carrier below the zero axis, it will merely give an output equal to the magnitude of the

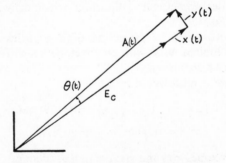

Fig. 6.15. Phasor diagram for a carrier with in-phase $x(t)$ and quadrature $y(t)$ components of noise. Envelope detector is insensitive to $y(t)$. FM limiter-discriminator is insensitive to $x(t)$.

amount by which the noise amplitude exceeds the carrier amplitude. However, since we have assumed a good signal-to-noise ratio the probability of this happening is quite small, and the distribution of signal plus in-phase noise can be taken to a be Gaussian distribution superimposed on the carrier. Thus, as the carrier is added to the noise, the envelope distribution changes from Rayleigh type to a biassed Gaussian type. Strictly speaking, the probability distribution curve starts at the origin in all cases.

To obtain the probability of errors due to mark being read as space we have to find the probability that

$$E_c + x(t) < r_0$$

This can only happen with $x(t)$ negative, *i.e.* the in-phase noise has become anti-phase.

The probability is then given by:

$$P(x > E_c - r_0) = \frac{1}{\sigma\sqrt{2\pi}} \int_{E_c - r_0}^{\infty} \exp\left(-x^2/2\sigma^2 \cdot dx\right) \tag{6.57}$$

If we take $P = 0.001$ the same as for errors during space, we obtain from Table 6.1 that $(E_c - r_0)/\sigma = 3.09$.

From equation (6.55) we obtain

$$E_c/\sigma = 3.09 + 3.72$$

$$= 6.81 \tag{6.58}$$

Since σ is r.m.s. noise voltage and E_0 is peak carrier voltage the signal-to-noise ratio in volts is

$$\sqrt{S_{out}} = \frac{E_c}{\sqrt{2}\,\sigma} = \frac{6.81}{\sqrt{2}} = 5 \quad \text{or} \quad 14 \text{ dB} \tag{6.59}$$

The threshold is set at $3.72/6.81$ or about 55% of the value of E_c. The signal-to-noise ratio deduced for an error rate of 0.001 is just about the minimum to permit the simplifying assumption that the quadrature noise may be neglected. For higher signal-to-noise ratios the assumption is fully justified and error rates would be very small. In practice, the threshold is set at about half of the peak value of the pulse and the advantage to be gained from setting it at 55% would be very marginal.

6.6. Effect of Noise on Modulation: Comparison of Amplitude and Frequency Modulation

So far in this chapter we have been mainly concerned with the detection of a carrier in the presence of noise and we have considered modulation only in the very simple form of the marks and spaces in a telegraphy system. In the majority of telecommunication systems, however, we are concerned with complex modulation signals arising from, say, speech, and we are then concerned with the effect of noise on the intelligibility of the modulation output from the detector. This means that in considering the signal-to-noise ratio at the output, we must regard the signal power as the power produced by the modulation. The d.c. output associated with the carrier is no longer of any significance and is therefore not included in calculating the output power.

The calculations for an amplitude modulated carrier proceed on very similar lines to those already discussed in earlier sections. We make the following assumptions:

(a) The input signal-to-noise ratio is large.

(b) The detector is an idealised envelope detector, giving an output directly proportional to the instantaneous amplitude of the modulated carrier.

(c) The modulation consists of a single sinusoid of angular frequency, ω_m.

With these assumptions we may write the input to the detector as:

$$v = E_c[1 + m \cos(\omega_m t)] \cos(\omega_c t) + x(t) \cos(\omega_c t) + y(t) \sin(\omega_c t)$$

$$(6.60)$$

where m is the modulation ratio and the noise is expressed in the same form as in previous sections. We may regard this signal as having an instantaneous amplitude given by:

$$A(t) = [\{E_c(1 + m \cos \omega_m t) + x(t)\}^2 + y^2(t)]^{\frac{1}{2}}$$

and since we are assuming a large value for the input signal-to-noise ratio, we have:

$$A(t) \doteqdot E_c(1 + m \cos \omega_m t) + x(t) \qquad (6.61)$$

It should be noted that this approximation breaks down if m is nearly unity for the times at which the modulation signal has its peak negative value. In practice, this represents only a small fraction of the total period of the modulation, and the error made by using the approximation will be small. Further, amplitude modulation systems are rarely operated with m greater than about 0·8, for the reasons stated in an earlier chapter.

The detector output is proportional to $A(t)$ and so:

$$v_0(t) = aA(t) = a \cdot E_c[1 + m \cos \omega_m t] + a \cdot x(t)$$

where a is the detector constant of proportionality. The modulation output is therefore $am \cdot E_c \cos \omega_m t$ and the noise output is $a \cdot x(t)$. Referring the output to a one ohm resistor, we therefore have:

$$\text{Output signal power} = \tfrac{1}{2}a^2 m^2 E_c^2$$

calculated from the modulation term only for the reasons given at the beginning of this section.

$$\text{Output noise power} = \overline{a^2 x^2(t)}$$

$$\text{Output signal-to-noise ratio} = \frac{a^2 m^2 E_c^2}{2a^2 \overline{x^2(t)}} \qquad (6.62)$$

The input carrier to noise ratio is $E_c^2/2\overline{x^2(t)}$, calculated as in Section 6.5.1. The actual signal power input to the detector exceeds the input carrier power by the power associated with the modulation term, but in practice it is more convenient to define the input signal-to-noise ratio in terms of the carrier power only, since a constant figure can be used for a given transmitter irrespective of the depth of modulation being used. Equation (6.62) can thus be written as

$$S_{AM} = m^2 S_{in} \qquad (6.63)$$

where S_{in} is the carrier-to-noise ratio at the detector input, and S_{AM} is the output to noise ratio for amplitude modulation of depth, m.

A comparable calculation can now be carried out for frequency modulation, using the same assumptions as above, except that the detector is taken as an ideal frequency discriminator. The input to the discriminator is

$$v = E_c \cos[\omega_c t + m_p \sin(\omega_m t)] + x_1(t) \cos(\omega_c t) + y_1(t) \sin(\omega_c t) \qquad (6.64)$$

where m_p is the modulation index. The noise contribution is white Gaussian noise extending over the band required to deal with the maximum values of modulation index and modulation frequency. The method by which this band is determined has been examined in Chapter 3. To simplify the calculations as far as possible, we treat the modulation and noise outputs separately. If the noise is ignored, we have for the instantaneous frequency of v,

$$\omega_i = \frac{d}{dt}[\omega_c t + m_p \sin(\omega_m t)]$$

$$= \omega_c + \omega_m m_p \cos(\omega_m t) \qquad (6.65)$$

and the output from an ideal discriminator is

$$v_0 = k(\omega_i - \omega_c) = k\omega_m m_p \cos(\omega_m t) \qquad (6.66)$$

where k is the proportionality constant of the discriminator. The corresponding output power in a one ohm resistor is therefore:

$$\text{Output signal power} = \tfrac{1}{2}k^2 \omega_m^2 m_p^2 \qquad (6.67)$$

When noise is present and the modulation is absent, equation (6.64) becomes:

$$v = E_c \cos(\omega_c t) + x_1(t) \cos(\omega_c t) + y_1(t) \sin(\omega_c t)$$

which may be expressed as:

$$v = A(t) \cos[\omega_c t + \theta(t)]$$

where

$$\tan \theta(t) = \frac{y_1(t)}{E_c + x_1(t)}$$

The detector will be preceded by a limiter which removes the time variation of the amplitude $A(t)$, leaving the noise affecting only the phase $\theta(t)$. Further, since the input signal-to-noise ratio is very large, $x_1(t)$ and $y_1(t)$ are both numerically much less than E_c, so that:

$$\theta(t) \doteqdot y_1(t)/E_c \qquad\qquad (6.68)$$

This result can also be derived by a phasor argument as shown in Fig. 6.15.

To evaluate the discriminator output corresponding to $\theta(t)$, we must obtain an expression for the instantaneous frequency, $d\theta(t)/dt$. This requires a more detailed examination of the noise waveform, $y_1(t)$, which we have stated to be white Gaussian noise over the frequency range covered by the discriminator input. If the band of the discriminator input covers the frequency range $f_c - \frac{1}{2}B$ to $f_c + \frac{1}{2}B$, then $y_1(t)$ has frequency components in the range 0 to $\frac{1}{2}B$. We divide this range into N small intervals each of width df, so that $N df$ equals $B/2$. Since $y_1(t)$ corresponds to white noise, the average mean square noise associated with each interval will be the total mean square noise divided by N, the number of intervals. We can therefore express $y_1(t)$ as:

$$y_1(t) = \sum_{r=1}^{N} a_r \sin(\omega_r t + \phi_r) \qquad\qquad (6.69)$$

where

(i) $$\omega_r = \pi df + 2\pi(r - 1) df \qquad\qquad (6.70)$$

i.e. the mid-angular frequency of the rth interval,
(ii) the mean square value of each component is

$$\tfrac{1}{2}\overline{a_r^2} = \overline{y_1^2(t)}/N \quad \text{and} \qquad\qquad (6.71)$$

(iii) ϕ_r is an arbitrary phase angle with equal probability for any value in the range 0 to 2π. We now have for the discriminator output due to noise:

$$v_{0,n} = k\frac{\mathrm{d}\theta(t)}{\mathrm{d}t} = \frac{k}{E_c}\frac{\mathrm{d}y_1(t)}{\mathrm{d}t}$$

$$= \frac{k}{E_c}\sum_{r=1}^{N}\omega_r a_r \cos(\omega_r t + \phi_r) \qquad (6.72)$$

The discriminator noise output thus consists of a set of components, the rth one having angular frequency ω_r and r.m.s. amplitude $k\omega_r a_r/\sqrt{2}\,E_c$. The r.m.s. noise voltage is thus proportional to the frequency of the component considered. The average power delivered to a one ohm resistor by the rth component is $k^2\,\overline{a_r^2}\omega_r^2/2E_c^2$ and is associated with a frequency interval of width $\mathrm{d}f$ centred on $\omega_r/2\pi$. We may therefore say that the output noise has a power density $G_0(f)$ where:

$$G_0(f)\mathrm{d}f = 4\pi^2 k^2 f^2\,\overline{a_r^2}/2E_c^2$$

$$= 4\pi^2 k^2 f^2\,\overline{y_1^2(t)}/NE_c^2$$

$$= 4\pi^2 k^2 f^2\,\overline{y_1^2(t)}\,2\mathrm{d}f/BE_c^2 \qquad (6.73)$$

The results obtained above for $\overline{a_r^2}$ and N have been used in arriving at the final answer. The power density function is parabolic with frequency.

Equation (6.73) holds for all frequencies in the range 0 to $\frac{1}{2}B$ but the bandwidth of the output channel following the discriminator can be restricted to pass only frequencies within the range required to deal with the modulation signal. If the highest modulation frequency is b, then the total output noise power is:

$$\int_0^b G_0(f)\,\mathrm{d}f = \frac{8\pi^2 k^2\,\overline{y_1^2(t)}}{BE_c^2}\int_0^b f^2\,\mathrm{d}f$$

$$= \frac{8\pi^2 k^2 b^3\,\overline{y_1^2(t)}}{3BE_c^2} \qquad (6.74)$$

Combining this result with that in equation (6.67) we obtain for the output signal-to-noise ratio:

$$\frac{S_{\mathrm{FM}}}{N} = \frac{1}{2}k^2 m_p^2\,\omega_m^2\,\frac{3BE_c^2}{8\pi^2 k^2 b^3\,\overline{y_1^2(t)}} = \frac{3m_p^2 f_m^2 BE_c^2}{4b^3\,\overline{y_1^2(t)}} \qquad (6.75)$$

$$\frac{3\beta^2 f_s^2 BE_c^2}{4b^3\,\overline{y(t)^2}}$$

The performance of the AM and FM systems can now be compared. Clearly, for a valid comparison, it is reasonable to take equal carrier powers but it is less obvious what the relative noise powers should be. Most of the noise sources we have discussed have white spectra, so that the total noise power can be taken as the product of the system bandwidth and the noise power per unit bandwidth. A valid comparison between the AM and FM systems must assume identical noise backgrounds and this requires that the noise powers per unit bandwidth be the same in the two cases. The input bandwidths are B for the FM system and $2b$ for the AM system, so that the mean square noise levels should satisfy:

$$\frac{\overline{x^2(t)}}{2b} = \frac{\overline{x_1^2(t)}}{B} = \frac{\overline{y_1^2(t)}}{B} \tag{6.76}$$

Equation (6.75) now becomes:

$$S_{FM} = \frac{3m_p^2 f_m^2 E_c^2}{2b^2 \overline{x^2(t)}} = \frac{3m_p^2 f_m^2}{b^2} \cdot \frac{S_{AM}}{m^2}$$

$$= \frac{3f_d^2}{b^2} \cdot \frac{S_{AM}}{m^2} \tag{6.77}$$

where f_d is the frequency deviation of the frequency modulated signal. In Chapter 3 we considered a typical FM system for broadcasting in which the maximum frequency deviation (f_d) is 75 kHz and the upper limit of the baseband channel (b) is 15 kHz. Comparing the performance of this system with an AM system, using the same baseband, and 100 per cent modulation ($m = 1$), we see that:

$$S_{FM}/S_{AM} = 3 \times (75)^2/(15)^2 = 75$$

The FM system thus shows a signal to noise improvement over the AM system of 19 dB.

At the beginning of this section it was stated that the carrier-to-noise ratio is assumed large and in fact it must exceed a certain value before the FM improvement can be realized. It was also stated that the discriminator should be preceded by a limiter to eliminate any AM. If we consider the unmodulated carrier wave and the total noise power in the main carrier channel (240 kHz in the example above) then, if the noise component in anti-phase with the carrier should exceed the carrier in amplitude, the noise would completely obliterate the carrier because the limiter output would be determined by the noise. The carrier-to-noise ratio should, therefore, be such that the probability of this happening is small. The actual ratio chosen is arbitrary but a figure of about 12 dB is usual, for which the probability of breakthrough of the in-phase component can be found by entering Table 6.1 (extended) at about 15 dB. The probability of break-through is seen to be of the order of 10^{-8}, which is very small.

This is known as the threshold in an FM system.* For signal-to-noise ratios below this value the output signal-to-noise ratio deteriorates

---

* See Appendix C for a fuller discussion.

badly, falling below the AM case. For better signal-to-noise ratio the FM improvement of equation (6.77) is applicable.

When the signal-to-noise ratio in the main FM channel is 12 dB the signal-to-noise ratio in the reference AM channel is considerably better, because it would have a bandwidth of $2b$ which is much less than the FM bandwidth. In the above example it would only be 30 kHz for the AM channel and 240 kHz for the FM channel, so that the signal-to-noise ratio in the AM channel would be 9 dB better or $12+9 = 21$ dB actual signal-to-noise ratio.

Thus we could represent the FM input-output signal-to-noise ratio characteristic on Fig. 6.16 by the curve shown which is drawn for $b = 15$ kHz, $f_d = 75$ kHz and an overall bandwidth of 240 kHz. The threshold is clearly shown below which the performance deteriorates rapidly. Above the threshold the FM improvement applies.

The triangular shape of the noise output voltage against frequency suggests that the input signal at the transmitter modulator should be

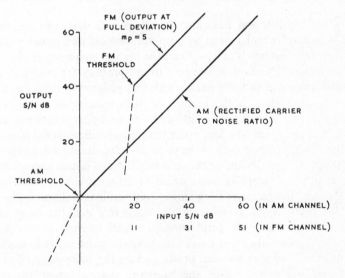

Fig. 6.16. Signal to noise ratio improvement in an FM system of modulation index = 5.

emphasized for increasing modulating frequencies, a technique known as pre-emphasis. After the discriminator there should be a de-emphasis network to restore a flat overall voltage-frequency characteristic. The slope of the pre-emphasis and de-emphasis networks should be such as to provide a noise output uniform with frequency. Since a triangular noise voltage-frequency characteristic is equivalent to noise power increasing at 6 dB per octave the emphasis should be of this order.

If the input modulating signal has a uniform frequency spectrum the lower frequencies must be attenuated and the higher frequencies emphasized, otherwise overloading would occur. However, with speech and music, the information has in general a spectrum decreasing above 1 or 2 kHz so that it is usual to apply emphasis above 2 kHz and attenuation of the lower frequencies is not necessary. At the receiving end de-emphasis by a network which provides an overall flat system reduces the high frequencies and the high frequency noise. The net signal-to-noise improvement can be calculated by making input signal assumptions, but for sound systems subjective testing is usually employed and an extra improvement of about 4.5 dB is usually obtained. FM systems have other advantages, such as transmitters working continuously at full power, so that further comparison with AM becomes a question of comparing detailed practical systems and is therefore beyond the present scope.

6.7. The Effect of Sampling on Signal to Noise Ratio

The sampling theorem described in Chapter 2 demonstrates that a waveform can be reproduced by amplitude-modulated pulses provided that the sampling rate is at least twice the highest frequency in the waveform. Certain distortions may occur in the process, for example, aperture distortion due to finite pulse width and phase distortion due to the re-constituting filter. If the input waveform consists of a signal plus noise, and the noise is restricted to the same upper frequency as the signal, it is clear that to a first order the output will be signal plus noise and the signal-to-noise ratio at the re-constituted output will be identical with the input signal-to-noise ratio. In practice any circuit imperfections, such as jitter of sampling pulse width or inherent circuit noise due to low levels at the reconstituting stage, may result in a deterioration.

If the input signal is substantially noise-free and the noise enters the system via the wide band channel which conveys the amplitude modulated pulses, then this noise will cause the pulse peaks to fluctuate, and noise will appear between pulses and on the pulse edges. For simplicity we will only consider the baseband case, although the results are identical for the carrier frequency case if detection is ideal. If the pulses are sampled at the middle (or the noise between pulses is gated out) a certain value of signal-to-noise ratio will be implicit in the values of the samples. If the peak signal voltage E is constant, and the pulse length is τ and the recurrence period T, then the low pass cut-off frequency of the filter accepting the pulses is assumed to be $1/2\tau$. The pulses at the output of the filter will be roughly triangular in shape, but they will still have the amplitude E because the build-up time of such a filter is τ seconds. If the noise power density per Hz of the channel is

G_0, the signal-to-noise power ratio in a 1 ohm load for the samples at the middle of the pulses would be

$$\frac{S}{N} = \frac{E^2}{G_0/2\tau} \tag{6.78}$$

If the signal is reconstituted by passing the samples through a filter of low pass cut-off frequency equal to half the pulse recurrence frequency, the output signal-to-noise ratio is substantially the same as the signal-to-noise ratio measured at the middle of the pulses. The amplitude of the sampling pulse will be the signal value plus or minus the sampled value of the noise. The signal portion will have a line spectrum with an envelope of $(\sin x)/x$ form, depending upon the width of the sampling pulse. The amplitudes of the noise pulses will be independent, but each pulse will have a continuous spectrum with the same form of envelope. The proportion of the signal spectrum passing through the reconstituting filter will be the same as the proportion of the noise spectrum passing the filter. Thus the output signal-to-noise ratio of the input pulse is given by equation (6.78) above.

Simple reconstitution by filtering of the wide band signal and noise, without gating-out of the noise between signal pulses, results in a gross deterioration of signal-to-noise ratio. In this case the d.c. component of the signal would be given by:

$$E \cdot \tau/T$$

The noise passing the re-constituting filter of bandwidth $\frac{1}{2}f_r = 1/2T$ would be given by

$$G_0/2T$$

The signal-to-noise power ratio in this case is then

$$\frac{(E \cdot \tau/T)^2}{G_0/2T} = \frac{\tau}{T} \cdot \frac{E^2}{G_0/2\tau} \tag{6.79}$$

Thus the signal-to-noise ratio is reduced by the factor τ/T or the duty cycle. Sampling is therefore an essential process in the re-constitution of a pulse modulated signal.

It is important to understand that transmitting a signal in pulsed form as compared with continuous form does not provide any change in signal-to-noise ratio, provided the same average power is used in each case and the bandwidths are correctly chosen. Thus a steady signal having the same mean power as the above signal would have a power of:

$$E^2 \cdot \tau/T$$

To accommodate the same information capacity as the pulse signal the steady signal would need to be passed through a filter of cut-off frequency $1/2T$. The signal-to-noise power ratio would be

$$\frac{S}{N} = \frac{E^2\tau/T}{G_0 \cdot 1/2T} = \frac{E^2}{G_0/2\tau} \tag{6.80}$$

or the same as equation (6.78) above. Thus the signal-to-noise ratio as defined by the peak pulse power to the total noise power in the band necessary to pass the pulse, is the same as the signal-to-noise ratio provided by the same mean power in a bandwidth $f_r/2$. Such a band would transmit information frequencies up to $f_r/2$ exactly as for the pulse case.

6.8. The Tandem Connection of Channels with Noise

6.8.1. THE TANDEM CONNECTION OF CONTINUOUS CHANNELS

In complex communications systems, such as those used for telephony or telegraphy, it is usual for a link between two distant points to be established by the connection in tandem of a number of individual channels. The operational flexibility of such an arrangement is greatest if each channel is designed to have zero insertion loss and to have constant values of input and output impedance. Under these conditions, the total insertion loss remains zero, irrespective of the number of channels connected in tandem. Signal-to-noise performance will, however, be affected by the number of the channels in use and we will examine the effect of this. The problem is simplified by assuming that there are n identical channels in tandem and that each channel generates an identical amount of noise. The overall noise performance can thus be calculated with the help of equation (6.31): if the noise factor of each channel is F_1, then the noise factor of n channels in tandem is:

$$F = F_1 + \frac{F_1 - 1}{G_1} + \frac{F_1 - 1}{G_1 G_2} + \cdots \frac{F_1 - 1}{G_1 G_2 \ldots G_{n-1}} \tag{6.81}$$

Each power gain is unity since each channel has zero insertion loss and equation (6.81) can therefore be rewritten as

$$(F - 1) = n(F_1 - 1) \tag{6.82}$$

If the signal-to-noise ratio at the input is S_{in} then the signal-to-noise ratio at the output of the n channels is:

$$S_n = S_{in}/F \tag{6.83}$$

For systems of the type considered, the noise is almost entirely that generated in the channels, so that large values of F_1 are involved. This means that equation (6.82) can be approximated by

$$F = nF_1 \qquad (6.84)$$

Also, the effect of operating channels in tandem is more easily appreciated by comparing the output from n channels with the output if only a single channel is used. The output signal-to-noise ratio for a single channel is:

$$S_1 = S_{in}/F_1 \qquad (6.85)$$

Equations (6.83)–(6.85) therefore give:

$$\frac{S_n}{S_1} = \frac{F_1}{F} = \frac{1}{n}$$

i.e. $$S_n = S_1/n \qquad (6.86)$$

This means that if signal-to-noise ratios are expressed in dB, there is a degradation of $10 \log_{10} n$ dB when n channels are operated in tandem. For example, if the number of channels is doubled, the signal-to-noise ratio is reduced by 3 dB.

A further degradation that must not be overlooked is the reduction of overload point that occurs in the overload characteristic of n channels in tandem, as compared with the overload point of one channel. If the amplitudes of a sinusoidal input to a continuous channel are plotted as abscissae and the corresponding fundamental output amplitudes as ordinates, the typical overload curve will be obtained. The point at which the output departs from linearity by some specified figure, often 1 db, is nominally taken to be the overload point. For several links in tandem the nominal overload point at the receiving end will be reached with smaller input signals, thus decreasing again the dynamic range available between overload point and noise level.

6.8.2. THE TANDEM CONNECTION OF DIGITAL LINKS

A binary digital link of baseband type will be either a single or double current telegraph circuit, and its voice frequency equivalent will be a single-tone or two-tone telegraph circuit. The radio frequency equivalent will generally be either interrupted CW or frequency shift keying. In all cases there will be some form of threshold or decision circuit at the receiving end to deliver the two condition mark-space signal to the next link. If the link is synchronous these decisions are made at the time centre of each telegraph element so that the signal sent forward is completely regenerated. It is identical to the signal at the sending end, excepting for errors due to noise peaks (as in Section 6.4). If there are n

such links in tandem the errors are cumulative, and the error rate overall, will be *n* times that for one link. However, from Fig. 6.10, we have seen that small increases in signal-to-noise ratio result in large reductions in error rate. Binary digital links are thus eminently suitable for tandem connection, provided some form of reconstitution process takes place at each switching point.

Chapter 7

Digital Communication

7.1. Introduction

The need to transmit streams of digits arises naturally in relation to telegraphy and data communication. In recent years, there has been a number of developments which point to the desirability of using digital methods for other systems such as telephony and an account of these will be given in this chapter. We begin by providing some of the reasons why such methods have been developed. The most important is that the possible signals to be handled in digital communication can be uniquely defined. For example, in the simplest and most widely used case we will be handling two-level or binary signals and only two possibilities exist, 0 or 1. Synchronous operation as described in Section 2.3 is possible so that the function of a receiver is reduced to the very basic one of deciding whether a 0 or a 1 is present. It has already been apparent in previous chapters that signalling waveforms, channel bandwidths, etc., can be selected to minimize the probability that a wrong decision is made. In summary, it is much simpler to optimize the design of a communication channel if only digital signals are being handled.

The next reason in favour of digital operation arises from a consideration of the tandem connection of a number of channels such as may arise in any trunk telephone call. If analogue signals are used, the signal is progressively degraded as the number of channels connected together increases. This has been demonstrated in Section 6.8 by considering the signal-to-noise ratio. A fundamental difference arises for digital signals in that amplification by a repeater inserted in a channel is replaced by regeneration. The input to the repeater, now a regenerator, is a train of pulses modified by distortion and noise. A decision circuit determines whether a 0 or 1 is present and the succession of such decisions is used to regenerate the original pulse train free from distortion and noise. The regenerated signal may differ from the original as a result of decision errors but as we have already noted in Section 6.4, the signal-to-noise ratio required to achieve low error rates is quite modest. Further discussion of such errors will be provided later.

The next advantage following directly from the ability to regenerate and to achieve low error rates with small signal-to-noise ratios is that much poorer quality channels can be used. The first widespread use of digital methods in telephony has arisen as a result of this possibility.

Existing junction cable pairs suitable for only one or two telephone channels when analogue signals are employed may be converted by the insertion of regenerative repeaters to digital operation with 24 channel capacity.

The last point which will be mentioned is perhaps obvious but nevertheless important. There is no essential difference between digital telephone signals and telegraphy or data signals, so that conversion to digital telephony makes possible a single integrated network capable of handling telephony, telegraphy or data.

The starting point for a discussion of digital communication is the sampling theorem examined in Section 2.5. A typical telephone signal within the frequency band 0·3-3·4 kHz can be sampled at an 8 kHz rate. This exceeds the Nyquist rate of 6·8 kHz by an adequate amount to allow suitable filters to be designed. The next step is to convert each sample value into a series of binary digits and two processes are involved in this. The first is quantization by which each sample is represented by a discrete value chosen from a fixed number of possible levels. This is examined in the next section. The second is coding by which each level is represented by a series of binary digits. The final result of sampling, quantizing and coding is to replace the analogue signal by a series of pulses—the whole process is therefore called pulse code modulation (PCM).

The basic idea of pulse code modulation was described by A. H. Reeves in 1937. The equipment required to implement PCM is however very much more complex than that needed for analogue operation and was too expensive when valves were the only available active devices. The position was transformed when solid state devices became available, together with low-cost integrated circuitry. From 1962 onwards, there has been a steady increase in the use of PCM for telephony and it is reasonable to speculate that ultimately PCM will predominate.

7.2. Quantization

7.2.1. PRINCIPLES

Each sample of a sampled waveform is represented by a level as indicated in Fig. 7.1. The number of levels is usually chosen as some power of 2 in order to facilitate the coding operation to be described in Section 7.3. Fig. 7.1 has been drawn for the equally spaced levels shared between positive and negative values. The set of samples shown may thus be represented by the series of positive and negative numbers at the bottom of the diagram.

This result may be illustrated in a different form by showing the levels as a function of the signal amplitude as in Fig. 7.2. The signal amplitude is restricted to lie within the range $-s_m$ to s_m and the number of levels

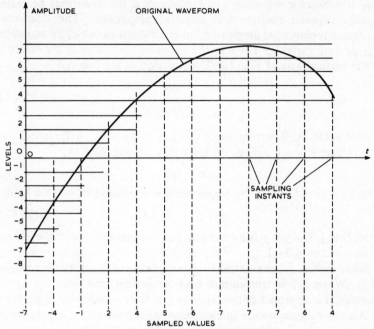

Fig. 7.1. Quantization of sampled signal.

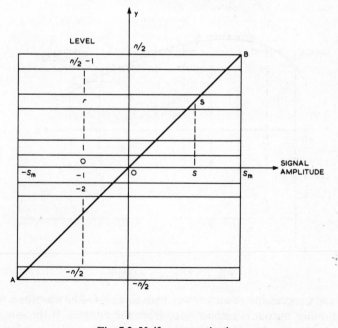

Fig. 7.2. Uniform quantization.

used is n. Since n will always be a power of 2, the levels may be shared equally between positive and negative amplitudes. The quantizing operation replaces the amplitude s (a continuous variable) by an integer r (a discrete variable). Let y be the continuous variable in the direction of the vertical axis of Fig. 7.2. The equation of the line AB is

$$y = \frac{ns}{2s_m} \tag{7.1}$$

so that points A, B correspond to $(-s_m, -n/2)$ and $(s_m, n/2)$ respectively. The integer r which defines the level may be expressed as

$$r = [y] \tag{7.2}$$

where $[y]$ is defined as the largest integer contained in y. This implies

$$[y] \leqq y < [y+1] \tag{7.3}$$

Note that $[y] = y$ if y is an integer. The integers defining the levels run from $-n/2$ to $n/2-1$.

Each integer $[y]$ may arise from any value of y satisfying equation (7.3). When $[y]$ is transmitted over a channel and observed at the receiver, it is appropriate to take $[y]+\frac{1}{2}$ as the best estimate of the input y. An error ε thus occurs in each received sample and ε will lie within the range $-\frac{1}{2}$ to $\frac{1}{2}$ as shown in Fig. 7.3. This error is called the quantizing

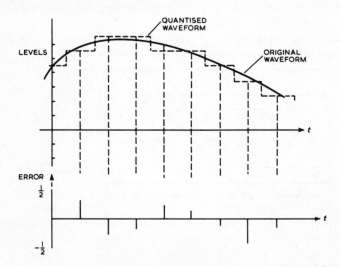

Fig. 7.3. Quantizing errors.

error and a succession of such errors leads to a noise-like waveform when a continuous output is reconstituted from the samples. If the sampling rate satisfies the Nyquist condition, successive quantizing errors are statistically independent and the mean square noise arising from quantization is therefore the mean square value of ε. The probability distribution for ε is uniform within the range $-\frac{1}{2}$ to $\frac{1}{2}$:

$$p(\varepsilon)\mathrm{d}\varepsilon = \mathrm{d}\varepsilon \text{ for } -\tfrac{1}{2} \leq \varepsilon \leq \tfrac{1}{2} \tag{7.4}$$

and so

$$\overline{\varepsilon^2} = \int_{-\frac{1}{2}}^{\frac{1}{2}} \varepsilon^2 \mathrm{d}\varepsilon = 1/12 \tag{7.5}$$

We must now make some assumption relating to the signal statistics in order to arrive at the mean square signal level. The simplest assumption is that s is uniformly distributed in the range $-s_m$ to s_m and from equation (7.1) it follows that y is uniformly distributed in the range $-n/2$ to $n/2$:

$$p(y)\mathrm{d}y = \mathrm{d}y/n \text{ for } -n/2 \leq y \leq n/2 \tag{7.6}$$

giving

$$\overline{y^2} = \frac{1}{n}\int_{-n/2}^{n/2} y^2 \mathrm{d}y = \frac{n^2}{12} \tag{7.7}$$

The signal-to-noise ratio arising from quantization is therefore.

$$\left(\frac{S}{N}\right)_Q = \frac{\overline{y^2}}{\overline{\varepsilon^2}} = n^2 \tag{7.8}$$

a very simple result. Values of this ratio for 64, 128 and 256 levels are 36, 42 and 48 dB respectively.

This calculation is adequate to indicate the approximate magnitude of the quantization signal-to-noise ratio but needs refinement before being applied to practical cases. The major weakness lies in the assumption relating to the signal probability which makes no allowance for such factors as the difference in loudness of speakers using a telephone. Uniform quantization as discussed above is inefficient because of such differences, and an improvement can be achieved by non-uniform quantization. This is discussed in the next section.

7.2.2. NON-UNIFORM QUANTIZATION

The result given by equation (7.8) may be interpreted as implying that the quantization signal-to-noise ratio equals the square of the number of levels over which the signal varies. The signal-to-noise ratio thus falls as the mean-square signal level is reduced. This disadvantage can be mitigated by using levels which are closely spaced for low amplitude signals but with progressively wider spacing as the amplitude increases. How much improvement can be gained in this way will now be investigated.

Non-uniform quantization may be discussed in terms of a diagram corresponding to Fig. 7.2 but with a non-linear relation between y and s

as shown in Fig. 7.4. The relation is assumed to be symmetrical for positive and negative signals and it suffices to draw the diagram for positive values of s. We now have, in place of equation (7.1), a functional relation:

$$y = y(s) \tag{7.9}$$

such that

$$n/2 = y(s_m) \tag{7.10}$$

to maintain the number of quantizing levels at n.

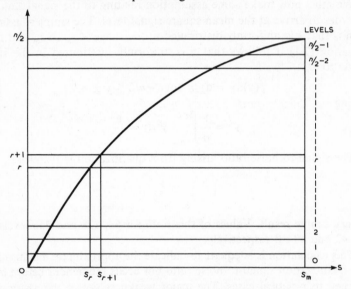

Fig. 7.4. Non-uniform quantization.

A curve such as in Fig. 7.4 implies that y is a compressed version of s in the sense that the rate of change of y with s decreases as s increases. Before we examine quantization, we deal with the question of recovering the original signal from y. This may be achieved by deriving an output signal s_o equal to a function $f(y)$ such that

$$f\{y(s)\} = s.$$

Then, s_o equals s as required. The process involving the conversion from s to y and y to s is similar to that which occurs in the compander described in Section 4.7.

We now consider quantization. Suppose

$$y(s) = r \text{ when } s = s_r \tag{7.11}$$

The level corresponding to a signal s within the range $s_r \leq s \leq s_{r+1}$ is then r, which may be expressed as $[y(s)]$. The output signal $s_{o, r}$ corresponding to level r is chosen as $\frac{1}{2}(s_r + s_{r+1})$, *i.e.* the mid-point of the range of s corresponding to level r. The mean square quantizing error in s_o for input signals in this range is found by the argument used in deriving equation (7.5) to be $(s_{r+1}-s_r)^2/12$. When s_r and s_{r+1} are reasonably close:

$$y(s_{r+1}) - y(s_r) = r+1-r$$

$$\doteq \left(\frac{dy}{ds}\right)_{s_r} (s_{r+1} - s_r)$$

whence

$$s_{r+1} - s_r \doteq 1/(dy/ds)_{s_r} \tag{7.12}$$

We conclude that the mean square quantizing error for signals of amplitude, s, is $1/12(dy/ds)^2$. The probability distribution for signal amplitude s is taken to be $p(s)$ and so:

$$\text{Mean square quantizing error} = \int_{-s_m}^{s_m} \frac{p(s)ds}{12(dy/ds)^2} \tag{7.13}$$

Also,

$$\text{Mean square signal level} = \int_{-s_m}^{s_m} s^2 p(s)ds \tag{7.14}$$

and so:

$$\left(\frac{S}{N}\right)_Q = \frac{12 \int_{-s_m}^{s_m} s^2 p(s)ds}{\int_{-s_m}^{s_m} \frac{p(s)ds}{(dy/ds)^2}} \tag{7.15}$$

An interesting conclusion may be drawn from equation (7.15) for the special case in which

$$\left(\frac{dy}{ds}\right)^2 = \frac{d^2}{s^2} \tag{7.16}$$

where d is some constant.

The signal-to-noise ratio then takes the value

$$\left(\frac{S}{N}\right)_Q = 12d^2 \tag{7.17}$$

irrespective of the nature of the probability distribution, $p(s)$. Since $p(s)$ determines the mean square signal level, the quantization signal-to-noise ratio is independent of the signal level, the result we hoped to be able to achieve.

Unfortunately, the relation between y and s is not one which can be used. On integrating equation (7.16), subject to the condition in equation (7.10) we obtain:

$$y = d \log_e (s/s_m) + n/2 \qquad \text{for } s > 0$$

and

$$y = -d \log_e (-s/s_m) - n/2 \text{ for } s < 0 \qquad (7.18)$$

The plot of y against s is shown in Fig. 7.5; the signal range $-s_1$ to s_1 where s_1 is obtained from

$$\log_e (s_1/s_m) = -n/2d$$

is excluded from the signal range which can be quantized.

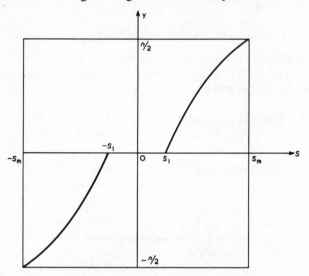

Fig. 7.5. Logarithmic quantizing relation.

Although the logarithmic relation given by equation (7.18) is unsuitable, it serves to indicate what relations are appropriate in practice. A large number of possibilities has been examined all with the basic property that the plot of y against s is linear for small values of s and logarithmic for large values of s. A typical case is defined by

$$y = \frac{n}{2} \frac{\log_e (1+cs)}{\log_e (1+cs_m)} \text{ if } s \geq 0 \qquad (7.19)$$

and satisfies both $y = 0$ for $s = 0$ and $y = n/2$ for $s = s_m$.

The quantizing signal-to-noise ratio is now dependent on the mean square signal level but much less strongly than is the case for linear quantization. This is illustrated in Fig. 7.6 which has been calculated using equation (7.19) with $cs_m = 255$ and $n = 128$. The signal amplitude probability distribution is assumed to be Gaussian. For comparison, results are also shown for linear quantization with 128 levels and signals with Gaussian probability distribution. The improvement resulting from non-uniform quantization is shown by the very much slower

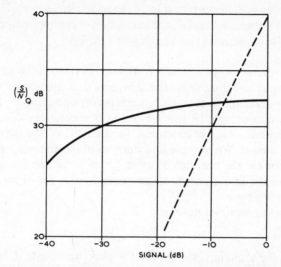

Fig. 7.6. Dependence of quantization signal-to-noise ratio on signal level. ——— non-uniform quantization, based on equation (7.19). ...... uniform quantization (0 dB signal level corresponds to a 10^{-3} probability that peak clipping occurs).

rate at which $(S/N)_Q$ changes with signal level. The maximum mean square signal level has been chosen to keep the probability that signal peaks exceed s_m to the tolerably low level of 10^{-3}, and has the value $0.185\ s_m^2$. If this level is exceeded, the quantizing action will cause appreciable clipping of the signal peaks thus introducing distortion.

A further change in the quantizing relation, by approximating the continuous curve defined by equation (7.19) as a series of straight lines,

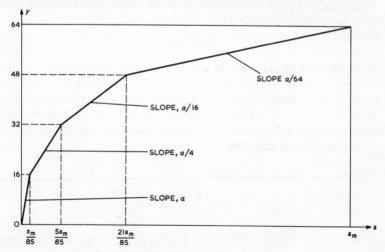

Fig. 7.7. Piecewise-linear quantization curve.

facilitates the coding operation to be described in Section 7.3. A typical example is shown in Fig. 7.7, the curve shown corresponding closely to the signal-to-noise ratio results given in Fig. 7.6.

7.2.3. TRANSMISSION REQUIREMENTS FOR QUANTIZED SIGNALS

The output resulting from the sampling and quantizing operations will be a series of levels at regular intervals corresponding to the Nyquist rate. This output may be transmitted as a succession of pulses, each pulse amplitude being proportional to the level of the corresponding quantized sample. We now use an extension of the analysis for the detection of pulses in the presence of noise, given in Section 6.4, to examine the bandwidth and signal-to-noise requirements for transmission of the multi-level pulses.

A typical transmitted pulse is

$$y(t) = A_r, \text{ if } -\tau/2 < t < \tau/2 \tag{7.20}$$

where τ is the sampling interval and the pulse amplitude A_r is restricted to a value $(r+\frac{1}{2})a$, where r is an integer in the range $-n/2$ to $(n/2)-1$. This choice of amplitudes gives n levels symmetrically positioned about zero. The value of $y^2(t)$ for a single pulse is A_r^2 and if all pulse amplitudes are equally likely, the average value of $y^2(t)$ over a long series of pulses is

$$\overline{y^2} = \frac{1}{n} \sum_{-\frac{1}{2}n}^{\frac{1}{2}n-1} (r+\tfrac{1}{2})^2 a^2 = \frac{2}{n} \sum_{0}^{\frac{1}{2}n-1} (r+\tfrac{1}{2})^2 a^2$$

$$= (n^2-1)a^2/12 \tag{7.21}$$

(Note that n will always be even, so that $\frac{1}{2}n$ is an integer.)

The received pulse corresponding to $y(t)$ will be contaminated by the noise added within the channel:

$$y_0(t) = y(t) + x(t) \tag{7.22}$$

where $x(t)$ will be assumed, as usual, to be a Gaussian noise waveform with r.m.s. value σ. The receiver will select the level to which $y_0(t)$ approximates most closely, i.e. it selects level r' such that $y_0(t) - r'$ is a minimum. When $x(t)$ is zero, r' equals r as required. An error will be made whenever $x(t)$ exceeds $a/2$ or is less than $-a/2$ and the probability of such an error is therefore:

$$P(|x| > a/2) = 2P(x > a/2)$$

$$= \left(\frac{2}{\pi}\right)^{\frac{1}{2}} \int_{a/2}^{\infty} \exp\left(\frac{-x^2}{2\sigma^2}\right) \frac{dx}{\sigma}$$

$$= \left(\frac{2}{\pi}\right)^{\frac{1}{2}} \int_{a/2\sigma}^{\infty} \exp\left(\frac{-x_1^2}{2}\right) dx_1 \tag{7.23}$$

The expression for this error is similar to that already discussed (equation 6.34) and the error probability depends only on the ratio a/σ. We may therefore specify a maximum permissible error rate and this defines a minimum value, k_e, for the ratio a/σ. The minimum signal-to-noise ratio required at the output from the transmission channel is then:

$$\frac{S}{N} = \frac{\overline{y^2}}{\sigma^2} = \frac{(n^2-1)a^2}{12\sigma^2} = \frac{(n^2-1)}{12} k_e^2 \qquad (7.24)$$

The channel bandwidth is determined by the number of pulses to be transmitted per second, *i.e.* the Nyquist sampling rate W_s. The discussion in Chapter 5 has led to the conclusion that pulses can be transmitted over a baseband channel at a rate $2B$ where B is the channel bandwidth. The bandwidth required to handle W_s pulses per second is therefore $W_s/2$.

It is appropriate at this stage to recapitulate the results so far obtained. We find that a signal limited to a bandwidth B may be sampled at a rate $2B$ and that the samples may be transmitted over a channel of bandwidth B. If n quantizing levels are used the quantization signal-to-noise ratio is n^2. We also see from equation (7.24) that the channel signal-to-noise ratio required is $(n^2-1)k_e^2/12$. Since the bandwidth of the channel is the same as that of the original signal, we might equally well have transmitted the signal directly and then the channel signal-to-noise ratio would be applicable to the output signal. We would in practice be better off, since k_e equals 8·8 for a 10^{-5} probability of making an error in selecting the correct received level at the receiver. The channel signal-to-noise ratio is therefore better by a factor of $(8·8)^2/12$, *i.e.* approximately 8 dB, than the quantization signal-to-noise error, if n is large compared with unity. The only result of sampling and quantization is to degrade the signal-to-noise performance by 8 dB.

What then, if anything, can be gained by sampling and quantization? The answer lies in the additional process of coding. The sampled values are not transmitted directly as pulses with n possible levels but in some different form. The choice of form is based on the same kind of argument such as that which leads to the representation of numbers in decimal or binary or some other scale. We would not contemplate using 100 separate symbols to represent the integers from 1 to 100 but we are effectively doing just this if we transmit a separate amplitude for each of the n levels required for quantization. The use of the idea of number scales leads to the suggestion that each sample is coded as a number in a suitable scale. Two changes in the channel requirements now arise. Firstly, we shall require to transmit a pulse for each digit in the coded sample and the bandwidth has to be increased to accommodate the increased number of pulses per second. Secondly, the number of levels required in each pulse reduces to the scale of the number used, and the signal-to-noise ratio needed is correspondingly reduced according to equation (7.24). We thus trade an increase in bandwidth for a reduction in signal-to-noise ratio. This trade-off is examined in detail in Table 7.1

TABLE 7.1

Signal-to-Noise Ratio and Bandwidth Requirements for 256 Level Quantization: $k_e = 8 \cdot 8$ $(10^{-5}$ *error probability)*

Number scale (m)	Number of digits (d)	Channel bandwidth (Hz) $(dW_s/2)$	Channel signal-to-noise ratio (dB) $10 \log_{10} [(m^2-1)k_e^2/12]$
256	1	$W_s/2$	56·3
16	2	W_s	32·2
4	4	$2W_s$	19·9
2	8	$4W_s$	12·9

for $n = 256$, this providing a number of simple possibilities. A plot in Fig. 7.8 shows the rapidity with which the signal-to-noise ratio drops as the channel bandwidth increases to accommodate more digits per

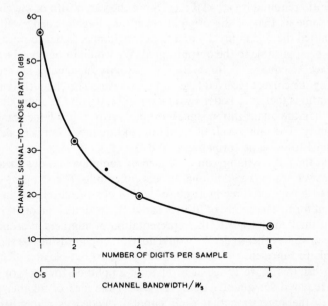

Fig. 7.8. Channel bandwidth and channel signal-to-noise ratio for 256 quantization levels available per sample. The results are calculated for an error probability of 10^{-5}.

sample. The equation of the continuous curve shown in Fig. 7.8 is easily obtained. Since

$$m^d = 256$$

$$d \log_2 m = 8$$

and so

$$W = 4W_s/\log_2 m$$

Also

$$\frac{S}{N} = \frac{(m^2 - 1)k_e^2}{12}$$

giving

$$m^2 = 1 + \frac{12S}{k_e^2 N}$$

$$2 \log_2 m = \log_2 \left(1 + \frac{12S}{k_e^2 N}\right)$$

The curve in Fig. 7.8 therefore satisfies the equation

$$W \log_2 \left(1 + \frac{12S}{k_e^2 N}\right) = 8W_s = \text{number of binary digits per second} \quad (7.25)$$

Although the argument given above demonstrates this relation only for the restricted choice of W corresponding to the entries in Table 7.1, it is possible to extend its validity to other values by considering groups of samples. Further discussion of the significance of the relation in equation (7.25) will be found in Chapter 8.

For our immediate purpose, we restrict attention to binary coding and note that the signal-to-noise ratio required for a binary channel is very modest. A relatively poor quality channel may therefore be used to transmit binary coded samples and we have thus confirmed one of the advantageous features of PCM noted in Section 7.1.

7.2.4. MULTIPLEXING

The most convenient method for multiplexing PCM is time-division (TDM) as described in Section 4.9. If the number of channels is N, the bits corresponding to each channel may be interleaved and the transmission bandwidth required becomes N times that of a single PCM channel, i.e. $N \times (W_s/2) \times \log_2 n$ from the results of the previous section.

A typical case occurs for 24-channel telephony. The sampling rate is 8 kHz and 128 levels are used, requiring 7 bits per sample. An extra bit is included to permit signalling operations. The minimum bandwidth required for the 24-channel group is thus $24 \times 4 \times 8 = 768$ kHz. In practice, the bandwidth must be appreciably greater than this to provide adequate discrimination between pulses. It is becoming increasingly common to specify channels by the bit rate they can handle ($1 \cdot 536$ M bits/s in this example) rather than by bandwidth.

7.3. Coding

7.3.1. CODING FOR PCM

The necessity for coding the quantized samples has been demonstrated in Section 7.2.3. from signal-to-noise considerations and it was assumed that a simple binary number code would be suitable. There are however many other possibilities and the most effective choice depends on the technique used for coding and decoding. A detailed account of these techniques would require more extensive treatment of digital circuitry than is appropriate in this volume. Simple examples of coding and decoding will be helpful in appreciating the problems which may arise, and these will now be considered.

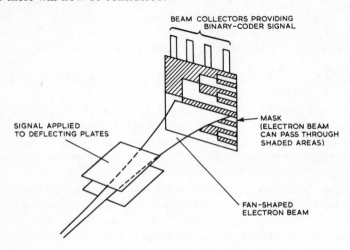

Fig. 7.9. Electron beam encoder.

An early form of encoder, shown in Fig. 7.9, is similar in construction to a cathode-ray tube but with a mask and a series of collecting strips in place of a screen. The beam is fan shaped and is deflected by the signal to be coded, outputs being obtained from the collectors if the beam passes through a hole in the mask. The pattern of the holes determines the code: for example, the mask in Fig. 7.10(a) provides binary code, the digits being produced in parallel. A simple modification, the introduction of a second set of deflecting plates which drive a pencil beam by a linear time base with period equal to the sampling interval, provides binary code in serial form. A disadvantage of the arrangement in Fig. 7.10(a) is that errors may arise if the beam overlaps two adjacent levels, for example as a result of imperfect focusing. A particularly bad case arises if levels 7 and 8 are overlapped, the encoder output being 1 1 1 1 (corresponding to level 15) rather than 0 1 1 1 (level 7) or 1 0 0 0 (level 8).

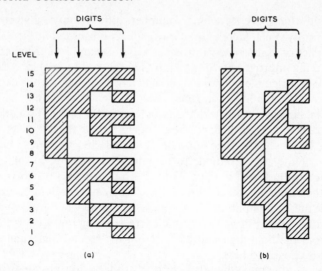

Fig. 7.10. Masks for beam encoder: (a) Binary code; (b) Gray code.

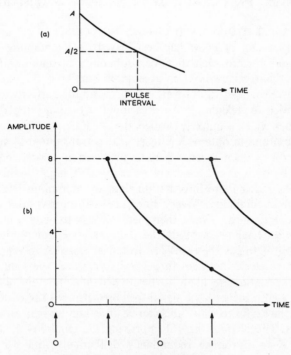

Fig. 7.11. Shannon-Rack decoder.

A code in which successive binary numbers differ in only 1 digit removes this difficulty. Such a code, the Gray code, is illustrated in Fig. 7.10(b).

The electron beam encoder provides a useful example of the possibility of obtaining improved performance by suitable code selection. A second example in which the operation is completely dependent on the choice of code occurs in the Shannon-Rack decoder. The sample must be coded as a serial binary number with digits arriving in the order of significance, the least significant being first. The decoder relies on the decay behaviour of an RC circuit, the time constant being selected to correspond to a halving of the signal output in the interval between two digits. The operation is illustrated in Fig. 7.11 for the bit stream 0 1 0 1 corresponding to the decimal number 10 since the least significant digit occurs first. A single pulse of amplitude A applied to the CR network gives an output $2^{-n} A$ after a delay of n. The output corresponding to 0 1 0 1 is the sum of A from the final digit and $2^{-2} A$ from the second digit, and equals 10 if A is chosen to be 8.

Low-cost integrated digital circuitry is available to translate from one code to another and it is therefore feasible to use different codes at different stages in a complete PCM system. The examples of a code and decoder given here are used to illustrate the principles. Practical coders and decoders currently used make use of integrated digital circuits.

7.3.2. ERROR DETECTING AND CORRECTING CODES

The discussion of PCM has been based on maintaining a channel signal-to-noise ratio such that the probability of making an error in selecting the received level is acceptably small. The figure of 10^{-5} which has been used to illustrate the discussion is perfectly adequate if PCM is used for telephony. An occasional error in a sample will cause a noise click which is unlikely to affect the intelligibility of a conversation. A similar situation obtains if English text is transmitted by telegraphy, for an error in a single letter may be observed and corrected from the context. In both cases, the message received is highly redundant, *i.e.* it contains much more information than the minimum level required to achieve intelligibility. A very different situation may arise when data is being communicated since then there is likely to be very little, if any, redundancy. Each element of the data may be independent of the others, and if this is the case it is no longer possible to remove errors from the received message. In general, therefore, very much smaller error probabilities are likely to be needed in data transmission. One possibility is to increase the signal-to-noise ratio but there are practical limits to the extent to which this can be done and we seek an alternative approach. The obvious way of improving the reliability of data transmission is to introduce redundancy deliberately and we will now examine a number of ways in which this can be done. As a preliminary we shall investigate the occurrence of errors in blocks of binary digits.

Suppose the probability of an error in an individual digit is p. We assume that the receiver decision circuitry is set so that the probabilities of reading an 0 as 1 and of reading a 1 as 0 are equal. We further assume that successive digits are statistically independent in the sense that the error probability is identical for each digit. A basic result from probability theory shows that the probability of observing r errors in a sequence of m digits is:

$$P(r:m) = \frac{m!}{r!(m-r)!} p^r(1-p)^{m-r} \qquad (7.26)$$

We are interested primarily in the situation where p is very small and where r will be much less than m. An adequate approximation to equation (7.26) is then

$$P(r:m) = (mp)^r/r! \text{ for } r>0 \qquad (7.27)$$

When mp is less than 10^{-2}, it is reasonable to ignore all possibilities except the occurrence of no errors, or of a single error.

A simple way of introducing redundancy is to repeat each message block, and to compare the received blocks. The probability that the same digit will be in error in both blocks is the probability of one error in the first block times the probability of an error in the same digit of the second block, *i.e.* $mp \times p = mp^2$. This gives the probability that an error will remain undetected. It is relatively easy to make this probability very small, *e.g.* it is 10^{-8} if $m = 100$ and $p = 10^{-5}$, but the channel bandwidth must be doubled to cope with twice the digit rate and further it is impossible to determine which message is correct. Straight repetition is thus very inefficient.

A procedure which is less demanding on bandwidth than repeating blocks is to use a simple parity check, an extra digit being added to the block of m digits to maintain parity, *i.e.* to keep the number of 1's an even number. If only one error occurs in the message block, the parity check will be violated and an error indicated. It may be that the error is in the check bit but since the detection of an error requires a request that the message be repeated no harm has resulted. The parity check will fail to detect an error if two digits in the block are incorrect, the probability of this being $(m+1)^2p^2/2$. For $m = 100$ and $p = 10^{-5}$, the probability of failure is approximately 0.5×10^{-6}. Reducing the block size decreases this probability; for example the probability of missing an error is 10^{-7} when $m = 44$ and $p = 10^{-5}$. Since the parity check requires only one extra digit per block the increase in channel bandwidth needed to accommodate this is relatively small.

Further discussion of the error behaviour of codes is facilitated by using the concept of Hamming distance, Hamming being the name of the originator of this concept. The distance between any two characters in a code (formed by a block of m binary digits) is the number of digits in which the two characters differ and the minimum or Hamming distance is the smallest value obtained when all possible pairs of characters are considered. The Hamming distance for a simple binary code is obviously 1. It is impossible to detect errors if the Hamming distance is unity for an error in one digit converts one character of the code to another. If this distance is two or greater, however, a single error cannot change a character into another member of the code set and the error may therefore be detected. A further possibility exists of both detecting and correcting a single error if the Hamming distance is 3 or greater. An example will illustrate this. We form a code with Hamming distance 3 from three digit numbers by using 0 0 0 and 1 1 1 as the characters. A single error in 0 0 0 produces a block say 0 1 0 which has a smaller distance from 0 0 0 than it does from 1 1 1. The single error in this case the 1 in the second digit, may thus be detected and removed. This simple example is a very extravagant one, since three digits are required to transmit each of the two possible characters.

Hamming and later workers have constructed a large number of error correcting and detecting codes and have established a very sophisticated mathematical theory on the properties of such codes. An early example proposed by Hamming will serve to illustrate the technique by which errors are detected and corrected. This code contains 7 digits of which 4 are used for the message and 3 are used as check digits, thus providing 3 possible parity checks. The parity checks locate the position of one digit in error by providing a three bit binary number defining the digit position. The bits of this number are 0 for a successful parity check and 1 otherwise. Suppose one of the check bits in the code is in error, giving two successful parity checks and one unsuccessful. The number locating the error must therefore be 0 0 1, 0 1 0, or 1 0 0, *i.e.* decimal 1, 2 or 4. It follows that the check bits should have locations 1, 2 and 4 in the 7 digit code character.

The way in which parity checks are made is derived by considering the binary representation of each digit position. Suppose digit 6 is in error—we require the check to produce 1 1 0 to locate this digit and we deduce that digit 6 should appear in the checks associated with parity bits 2 and 4 but not with that including bit 1. Consideration of the other message digits 3, 5, 7 shows that parity checks should be applied to

   (i) digits 1, 3, 5, 7 to provide a_1
   (ii) digits 2, 3, 6, 7 to provide a_2
   (iii) digits 4, 5, 6, 7 to provide a_3

In each case, $a = 0$ for a successful check and 1 for an unsuccessful check. The digit in error is located by the binary number $a_3\ a_2\ a_1$.

The complete code is constructed by allotting simple 4 digit binary numbers to digits 3, 5, 6, 7 and selecting the check digits to satisfy the parity rules given above. The list of characters is given in Table 7.2.

TABLE 7.2

Hamming Error Correcting Code

Character	Digit number						
	1	2	3	4	5	6	7
0	0	0	0	0	0	0	0
1	1	1	0	1	0	0	1
2	0	1	0	1	0	1	0
3	1	0	0	0	0	1	1
4	1	0	0	1	1	0	0
5	0	1	0	0	1	0	1
6	1	1	0	0	1	1	0
7	0	0	0	1	1	1	1
8	1	1	1	0	0	0	0
9	0	0	1	1	0	0	1
10	1	0	1	1	0	1	0
11	0	1	1	0	0	1	1
12	0	1	1	1	1	0	0
13	1	0	1	0	1	0	1
14	0	0	1	0	1	1	0
15	1	1	1	1	1	1	1

To correct a single error

(i) $a_1 = 0$ if digits 1, 3, 5, 7 provide parity check
    $= 1$ if not.

(ii) $a_2 = 0$ if digits 2, 3, 6, 7 provide parity check
    $= 1$ if not.

(iii) $a_3 = 0$ if digits 4, 5, 6, 7 provide parity check
    $= 1$ if not.

Digit in error is $a_3\ a_2\ a_1$.

If two digits are wrong, Hamming's code detects an error but provides incorrect information regarding correction. However, it is much more efficient than straight repetition of 4 digit blocks for single errors in that it not only detects the errors but corrects them, and the increase in the number of digits to be transmitted is less.

The application of the results of coding theory to practical situations requires careful consideration of two factors. Firstly, the need to increase the rate at which digits are transmitted may lead to an increase in the error probability for each digit and an assessment must be made of the effectiveness of the expanded code at this increased error rate. Secondly, the assumption that the digit error probabilities are independent may not be valid: a channel subject to fading will produce significantly greater error rates at the times when fading is most severe and whole blocks of digits may then be in error. Some form of automatic request for repeat transmission of wrong data is usually the most effective method of dealing with this situation.

7.4. Choice of Waveforms in Digital Systems

The factors which influence the choice of pulse shapes to be used for digital systems have already been discussed and Section 5.4.4 in particular contains an account of some of the relevant factors in relation to baseband signals. The results for baseband signals are immediately applicable to AM but some further comments are needed to cover PM and FM possibilities, which are widely used in practice.

The use of FM to transmit a binary coded signal means that two separate frequencies f_1 and f_2 are used for 1 and 0 respectively: the term frequency shift keying, FSK, is normally used to describe this arrangement. Signals with more than two levels may be transmitted either by using a separate frequency to represent each level, or more usually by combinations of frequencies. A typical arrangement is to transmit 2 frequencies selected from 5 to provide $5!/2!3!$, *i.e.* 10 levels: this is obviously convenient for transmitting numerical data in decimal notation.

If PM is used, the carrier phase is switched to one of a number of possible values, *e.g.* for a binary system 0° or 180° or for a four-level system 0°, 90°, 180°, 270°. In this case the name phase shift keying, PSK, is used. The receiver can detect which level has been transmitted by comparing the phase of the incoming signal with that of a reference. The stability of this reference must be very high, and an alternative arrangement, differential PSK (DPSK), is often used to avoid the need for a reference carrier. In DPSK, the phase difference between successive intervals is used to convey the information, this difference being measured at the receiver by comparing the received waveform with the same waveform delayed by one interval.

In an AM system, the pulses are shaped to achieve economy of channel bandwidth and to minimize intersymbol interference (Sections 5.4.3 and 5.4.4). In FSK and PSK systems, the transitions from one level to another are correspondingly shaped but the analysis becomes much more complicated since both FM and PM are inherently non-linear operations. A detailed examination of the effects of distortion arising within the transmission channel must be made in all cases. Such distortion, particularly that arising from differential group delay, will modify the pulse shapes sufficiently to prevent the ideal behaviour discussed in the sections quoted from holding in practical systems. Further investigation, often experimental or based on detailed computer calculations, is then needed to establish the margin available to differentiate between different levels, and to determine the optimum time within the pulse interval at which the decision on level should be made. The results of such investigations are conveniently displayed as eye patterns, the description 'eye' being applied in view of the shape these patterns take. An example is shown in Fig. 7.12, for a binary system, and indicates

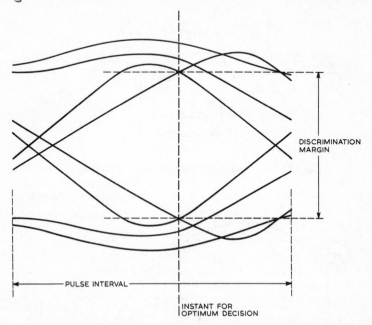

Fig. 7.12. 'Eye' pattern for binary system. Each curve in the diagram corresponds to a different sequence of 0's and 1's.

both the margin available for discrimination of 1 and 0 and the optimum time necessary to make this decision. Each individual curve corresponds to a different succession of 1's and 0's. Eye patterns have similar shapes whatever form of modulation is used.

7.5. Delta Modulation

The basis of pulse code modulation discussed in earlier sections has been to sample the input waveform and to transmit coded forms of the sample values. An alternative is to transmit the difference between successive samples. In principle these differences can be represented to any desired degree of accuracy by quantizing and coding exactly as for the original samples.

In the particular case of delta modulation, a single binary digit is used to describe the change from one sample value to the next. This digit is derived by comparing the signal amplitude at a sampling instant with the value obtained by reconstructing the signal from the previous digits. Since each digit is in effect a crude measure of the time differential of the signal waveform, the reconstruction process must involve integration. The block diagram in Fig. 7.13 indicates a simple form of delta-modulator and Fig. 7.14 indicates the waveforms at various stages.

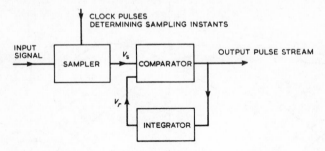

Fig. 7.13. Block diagram for system using delta modulation.

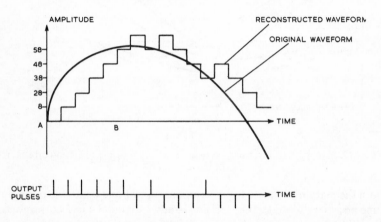

Fig. 7.14. Waveforms for delta modulation.

The input waveform, V_s, which will be a continuous function of time is compared with the reconstructed signal V_r, the difference $V_s - V_r$ being examined by the comparator circuit at a series of sampling instants. This circuit produces a pulse at each sampling instant, the pulse value being 1 if $V_s \geq V_r$ and -1 if $V_s < V_r$. The output pulse stream is thus a bipolar digital signal which may easily be converted to a unipolar signal (*i.e.* 1's and 0's) for transmission if desired. A digit value 1 causes the integrated output to increase by a step δ and the value -1 causes a change $-\delta$. The reconstructed waveform, V_r, is thus quantized. Note from Fig. 7.14 that the effective quantizing level is 2δ, alternate samples being restricted to sets of levels which are separated by δ. The demodulator at the receiver consists of an integrator, producing an output identical to V_r, followed by a low pass filter with a cut-off frequency below the sampling frequency. This filter thus removes the effect of sampling exactly as in the case of PCM.

The equipment needed either at the transmitter or at the receiver is much simpler when delta modulation is used than in the case of PCM. The attractive features of PCM, in particular the use of regenerators rather than repeaters, may thus be achieved relatively easily. A complete analysis of delta modulation is much more complicated than that for PCM but it would show that the performance of a simple delta-modulation system is less effective than can be achieved with PCM for comparable channel bandwidths and signal-to-noise ratios. A variety of modifications which make delta modulation more competitive with PCM have been proposed and include the use of double integration in reconstituting the signal, and the use of quanta of variable size. It appears that systems for telephony at present using PCM may be improved with delta modulation and that the circuitry needed would not be prohibitively complex.

A very crude estimate of the bandwidth needed for delta modulation can be obtained by noting that the limiting factor is the possibility of 'slope clipping' as illustrated by the waveforms of Fig. 7.14 during the period A to B. The maximum slope $df(t)/dt$, which can be represented faithfully is $f_s\delta$ where f_s is the sampling frequency. The maximum slope which will occur in waveform $f(t)$ depends on the r.m.s. value of the waveform, σ, and the highest frequency B which the waveform contains. In the extreme case when $f(t) = \sqrt{2}\sigma\sin(2\pi Bt)$, the maximum slope is $2\pi\sqrt{2}B\sigma$. This extreme is very unlikely to occur in any practical baseband waveform and a fraction, typically 0·25, of the above value is taken as the largest slope likely to arise. We therefore require:

$$f_s\delta \geqq \pi\sqrt{2}B\sigma/2 \qquad (7.28)$$

A second relation between δ and f_s may be found by considering the quantizing noise. An argument similar to that given in Section 7.2.1 shows that the mean square quantizing noise is $\delta^2/12$. The spectrum of this noise is much wider than that of the baseband signal and a fraction of the noise will be eliminated when the reconstructed signal is passed through a low-pass filter. If the spectrum of this noise is assumed to be uniform over the frequency band 0 to $\frac{1}{2}f_s$ required to transmit the pulses, the effective mean square noise in the baseband is $\delta^2 B/6f_s$. The signal-to-noise ratio of the output signal after reconstruction and filtering is therefore

$$\frac{S}{N} = \left(\frac{6\sigma^2 f_s}{\delta^2 B}\right)$$

which becomes

$$\frac{S}{N} = \frac{12}{\pi^2}\left(\frac{f_s}{B}\right)^3 \tag{7.29}$$

after substituting for δ from equation (7.28), using the equality condition in this equation. Values for f_s and δ/σ may thus be estimated from equation (7.28) and (7.29) if (S/N) and B are specified. For example, if $S/N = 30$ dB and $B = 3\cdot4$ kHz, we find $f_s = 32$ kHz and $\delta/\sigma = 0\cdot24$.

Chapter 8

Communication Theory

8.1. A Measure of Information

The objective of any communications system is to convey information from a source to a recipient and any study of communications must attempt to provide a basis to assess how effectively this objective is achieved. The word information can be used in a variety of senses and a distinction must be drawn between information in the sense of 'volume of data' and information in the sense of the meaning to be attached to the data available. The term 'semantics' is applied in the latter case. The communications engineer is concerned exclusively with volume of data and not with semantics, and throughout the remainder of this chapter the word information will be used only in this restricted sense, *i.e.* as a measure of the amount of data to be communicated. The theory which will be presented is a simplified account of a very general treatment developed by C. E. Shannon of the Bell Telephone Laboratories and first published in 1948. The theory is important because it provides an absolute standard of performance which cannot be exceeded by any communications channel and also highlights the factors which limit the performance of a channel. An appreciation of the nature of these factors has led directly to the introduction of several new techniques.

The key idea in arriving at a suitable method of quantifying information is that the receipt of information in the form of a message implies uncertainty in the mind of the recipient before the message arrives. A measure of the information content of a message can thus be based on the amount of uncertainty it has removed. The input to a communication channel is a message source, and in all situations of interest, the recipient will have knowledge of the possible messages which might be received from this source. The quantity of most interest is therefore some form of average of the information content of the set of possible messages. Some simple examples will be given to clarify these ideas.

The most elementary form of message source is one which has only two possible messages. We take a specific example by supposing that these messages refer to the results of a coin-tossing experiment. Each toss can provide either heads or tails. It is reasonable to take the communication of the result of such an experiment as a basis for defining the unit of information. We notice that this message can be communicated in binary digital form, *e.g.* 1 if a head occurs, 0 if a tail occurs. Each message thus requires 1 binary digit or bit and this suggests that the unit of information be taken as the bit. There are two factors which must be examined further before this idea can be extended to provide a definition for the information content of more complicated messages. Firstly, we expect in a coin-tossing experiment that heads and tails are equally likely and secondly that the result of each experiment is not affected by any previous experiments. Bearing these two factors in mind, we therefore take as our starting point the definition that the information content of a message is one bit if this message is one of two equi-probable possibilities and if the message probability is not affected by the receipt of previous messages.

We are now able to extend our definition and begin by considering a source which provides two messages with probabilities which are not equal. As an example, we consider messages relating to the weather, *e.g.* 'rainy' or 'fine'. If such a message is sent from a region which is known to have predominantly rainy weather, the recipient expects to receive the message 'rainy', and his state of uncertainty is not greatly reduced if this message is received. The message 'fine' is unexpected and may therefore be regarded as carrying more information than 'rainy'. The numerical measure of the information content of these two messages should therefore satisfy the following conditions: the message which is least expected, *i.e.* having the smaller probability, should have the higher information content and a message which is certain, *i.e.* it is always transmitted, carries no information. The information content of a message with unit probability should thus be zero. Finally, if the message probability is $\frac{1}{2}$ we have defined the information content to be 1 bit. Shannon proposed that the information content of a message having probability, p, should be taken as $-\log_2 p$. It is easily verified that this expression satisfies all the conditions listed above.

If a message source provides two messages with probabilities p_1, p_2 (such that $p_1 + p_2 = 1$), the average information content per message is obtained by considering a long series of say N messages. The number of messages of the two types will be Np_1 and Np_2 respectively, and the information contents of the two types are $-\log p_1$ and $-\log p_2$ respectively. The average information content per message is then:

$$H = \frac{-Np_1 \log_2 p_1 - Np_2 \log_2 p_2}{N} = -p_1 \log_2 p_1 - p_2 \log_2 p_2 \qquad (8.1)$$

The quantity H is conveniently called the information content of the source. The expression for H is identical in form to that for the negative entropy of a two-state physical system in which the state probabilities are p_1, p_2. The term 'negentropy' is therefore sometimes used for H.

The dependence of H on p_1 is of interest. When p_1 is zero or unity, H is zero corresponding to situations in which only one message occurs. The maximum value of H is 1 bit when $p_1 = p_2 = \frac{1}{2}$.

Equation (8.1) may be extended to cover a message source providing a set of m messages with probabilities p_1, $p_2...p_m$, these probabilities being unaffected by previous messages. The result for the average information content per message is

$$H = - \sum_{i=1}^{m} p_i \log_2 p_i \qquad (8.2)$$

The maximum value for H occurs when the messages all have probability $1/m$ and is $\log_2 m$. Suppose for example the messages are the 26 letters of the English alphabet. The value of H would be $\log_2 26 = 4\cdot7$ bits if all letters were equally likely. In practice this is not so and H becomes $4\cdot2$ bits if the letter probabilities are used.

Further extensions of the basic definition of information content may be made to take account of the dependence between successive messages. The need for this is evident from English text where the probabilities of letters are very dependent on what has gone before, *e.g.* the probability that a 'q' is followed by a 'u' is almost unity. The concept of conditional probability is needed. Suppose the messages are a_1, $a_2...a_m$. Then the conditional probability of message a_i, given that the previous message was a_j, is denoted by $P(a_i | a_j)$. An averaging process similar to that described earlier leads to the expression:

$$H = - \sum_{i,j} P(a_j)P(a_i | a_j) \log_2 P(a_i | a_j) \qquad (8.3)$$

for the average information content when conditional probabilities are considered. We shall not pursue this extension further except to note that H is reduced relative to its value in equation (8.2). The reduction can serve as one measure of the 'redundancy' of a message source, redundancy implying that more information is being transmitted than needed. Such redundancy is often valuable in practice since it allows the detection and correction of errors, as already discussed in Section 7.3.2.

8.2. Source Coding

The source information content defined in the previous section is useful in determining the choice of a code to be associated with each message during its transmission. Our discussion of this topic will be restricted to the simplest case in which the message source provides m messages with probabilities which are independent of any previous message. We denote the messages by a_1, a_2, $...a_m$ and their probabilities by $p_1, p_2, ...p_m$. Each message will be represented by a code word consisting in general of d digits each digit having b possible values. The available number of code words is thus b^d, and must be at least as great as the number of messages required to allow each code word received at the communication terminal to be translated uniquely into the corresponding message.

The Murray code in Fig. 2.6 provides an example: the 32 possible messages are represented by 32 5-digit binary words. In this example, the average information content per message, H, would be $\log_2 32 = 5$ bits if the messages were equally likely, and this information content equals the number of binary digits per codeword. In practice, the messages are not equally likely and H is less than 5 bits. Can some more economical coding be found?

In principle, the answer is yes. One of the basic results of information theory is that the output from a message source can be coded so that the average number of bits per codeword is equal to H. To achieve the optimum code, it may be necessary to use groups of L messages and to provide codewords of N bits for each such group. The theoretical result is that N/L tends to H as L becomes very large. The larger the value of L required, the greater will be the delay involved in coding and decoding the groups of messages so that in practice it may be preferable to accept a coding system which is less efficient than the theoretical possible optimum.

There are some special cases in which the optimum code can be achieved without requiring the use of long message groups. The most obvious is in fact the case when the messages are equally likely, as indicated by the the 32 message example considered above.

In other cases it is possible to select variable-length code words, using the shortest codes for the most probable messages, in order to obtain an average number of bits per codeword equal to the average information content per message. A systematic procedure has been devised by Huffman to derive codes which are optimum in the sense that the average number of bits per codeword is as close to H as possible. We will illustrate this procedure by an example.

Suppose four messages, a_1 to a_4, have probabilities $\frac{1}{2}$, $\frac{1}{4}$, $\frac{1}{8}$, $\frac{1}{8}$. Then

$$H = \tfrac{1}{2} \log_2 2 + \tfrac{1}{4} \log_2 4 + 2 \times \tfrac{1}{8} \log_2 8$$

$$= 1\tfrac{3}{4}.$$

A simple binary code using 01, 10, 11, 00 for the four messages requires 2 bits per message. A more efficient code is constructed as follows. The two messages of smallest probability are each coded by three bit words, with a difference only in the last bit. For example the codes for a_3 and a_4 may be 110 and 111. Message a_2 is coded by a two bit word which must differ from the first two bits of the words for a_3 and a_4: a suitable code for a_2 is thus 10. Finally, a_1 is coded into the single bit 0. We may now estimate the average number of bits per codeword as shown in Table 8.1 and find that it equals H.

An important feature of this code is that no codeword is a 'prefix' of any other code word. This enables any message sequence to be uniquely resolved into its constituent codewords. For example, the sequence

TABLE 8.1

Message	p_i	Code	No. of bits in code, n_i	$p_i n_i$
a_1	$\frac{1}{2}$	0	1	$\frac{1}{2}$
a_2	$\frac{1}{4}$	10	2	$\frac{1}{2}$
a_3	$\frac{1}{8}$	110	3	$\frac{3}{8}$
a_4	$\frac{1}{8}$	111	3	$\frac{3}{8}$

$$p_i n_i = 1\tfrac{3}{4}$$

10110010 can only correspond to a_2, a_3, a_1, a_2. If however the code for a_3 had been 101, it would not be possible to decide whether 1010 corresponds to a_3, a_1, or to a_2, a_2: note this ambiguity arises since 10 is a prefix of 101. Codes which can be decoded uniquely need not satisfy the prefix condition, but optimum codes satisfying the prefix condition may always be found.

A general procedure is available for establishing a binary Huffman code. Arrange the messages in order of decreasing probability. The last digits of the code words for the two least probable messages are taken as 0 and 1. These two messages are now regarded as a single message of probability equal to the sum of their probabilities and the process is repeated. The process can be tabulated as shown for the examples in Tables 8.2 and 8.3.

TABLE 8.2

Message	Probability				Code word
a_1	0·5			0	0
a_2	0·25		0	0·50	10
a_3	0·125	0	0·25	1	110
a_4	0·125	1	1		111
$H = 1\tfrac{3}{4}$					

Average number of bits per codeword = $1\tfrac{3}{4}$.
Efficiency of code = 100 per cent.

Table 8.2 refers to the example already considered in Table 8.1. Note that the digits in the codewords are obtained by reading the entries in the table from right to left.

The efficiency of a code is given by the ratio (expressed as a percentage) of the source negentropy, H to the average number of bits per codeword.

TABLE 8.3

Message	Probability		Code word
a_1	0·30		00
a_2	0·18	0·60 0	10
a_3	0·15	0·30 1	010
a_4	0·15	0·40 1	011
a_5	0·10		110
a_6	0·06	0·22	1110
a_7	0·06	0·12	1111

$H = -0.3 \log_2 0.3 - 0.18 \log_2 0.18 - 2 \times 0.15 \log_2 0.15 - 0.10 \log_2 0.10 - 2 \times 0.06 \log_2 0.06$

$\quad = 2.60$

Average number of bits per code word

$\quad = (0.3 + 0.18) \times 2 + (2 \times 0.15 + 0.10) \times 3 + 2 \times 0.06 \times 4$

$\quad = 2.64$

Code efficiency

$\quad = 2.60 \times 100 / 2.64 = 98.3$ per cent.

For the example considered in Tables 8.1 and 8.2, the efficiency of a simple binary code is $(7/8) \times 100 = 87.5$ per cent while the Huffman code is 100 per cent efficient.

8.3. Channel Capacity

In the previous section we have indicated that the messages originating from a message source with average message content H bits, may be coded in such a way that the average number of codewords per bit is H. It is important to remember that the coding may only be possible if very long groups of messages are considered. This result may now be related to the requirements to be met by a communication channel capable of handling the output from such a message source. The message source rate will be the product of H and the number of messages per second, and is expressed in bits per second. We have seen in Section 2.5 that a communication channel of bandwidth W can transmit up to $2W$ independent signal samples per second. Suppose that each sample may be selected as any one of m distinguishable signal levels. The number of bits per sample is then $\log_2 m$ and the channel can therefore handle $2W \log_2 m$ bits per second. This quantity is called the channel capacity and is denoted by C, i.e.

$$C = 2W \log_2 m \qquad (8.4)$$

The output from a message source of information rate R bits per second may be coded in such a way that it is transmitted without error over a channel with capacity C greater than or equal to R. A result essentially similar to this statement was first propounded by Hartley in relation to telegraphy channels. The feature of most interest is that an exchange can be made between the bandwidth and the number of distinguishable levels. For a binary channel, $m = 2$ and the channel capacity is $2W$ bits per second. A four-level channel ($m = 4$) will have the same capacity for half the bandwidth since $2 \times (W/2) \times \log_2 4 = 2W$. The number of signal levels which can be distinguished depends on the signal distortion occurring within the channel.

The most serious restriction in practical channels is the noise added within the channel and Shannon extended his analysis of channel capacity to cover this situation. His argument is an extension of the one mentioned in the previous section in relation to coding the outputs of an information source. The result which Shannon obtained is that the maximum capacity of a channel of bandwidth W and signal-to-noise ratio S/N is

$$C = W \log_2 \left(1 + \frac{S}{N} \right) \text{ bits/second.} \qquad (8.5)$$

The conditions under which this maximum can be attained are important. A more formal statement is that information can be transmitted over the channel at a rate of up to C with negligibly small probability of error, provided that a suitable coding technique is used. The coding technique requires that blocks of messages extending over very long periods should be represented by a set of noise-like waveforms. A final condition for the validity of the result is that the noise should be Gaussian and white.

The conditions on Shannon's result imply a very long delay and this makes it very improbable that practical systems will ever attain the limiting channel capacity. An additional practical drawback is that the use of noise-like waveforms for conveying the information may lead to the build-up of large signal voltages, greater than the overload limits of communication equipment. As already mentioned, however, Shannon's result is particularly important in providing a rigorous theoretical basis for a study of the exchange between signal-to-noise ratio and bandwidth.

The argument summarized in this section refers to baseband channels but may easily be extended to bandpass channels. We note that the use of SSB techniques translates a baseband channel extending from 0 to W Hz to a bandpass channel extending from f to $(f + W)$ Hz. The information capacity is not affected by this change and we conclude that equation (8.5) may be applied to any channel of width W Hz irrespective of whether or not it is a baseband channel.

SSB methods are often difficult to use and it is relevant to ask what is the capacity of a DSB channel. At first sight, the capacity will be halved since a bandwidth $2W$ is required to accommodate both the upper and lower sidebands. If, however, coherent detection (Section 6.5.1) is used, it is possible to transmit simultaneously two AM signals over the same channel. The carrier waves for these two signals must be in phase quadrature. It is a simple matter to show by extending the argument in Section 6.5.1 that each signal may be detected independently of the other if the local oscillator phase is correctly adjusted. Similar analysis of the effect of noise leads to the conclusion that the full channel capacity predicted by equation (8.5) may, in principle, be realized.

8.4. Exchange of Signal-to-Noise Ratio for Bandwidth

We have encountered the possibility of such an exchange on two earlier occasions. Firstly, the analysis of frequency modulation (and in particular the discussion of the noise behaviour in Section 6.6) showed that FM provided a signal-to-noise ratio at the discriminator output which improved as the frequency deviation increased, with all other relevant parameters unchanged. Secondly, we have obtained an expression (equation (7.25)) which provides the channel capacity of a PCM system:

PCM channel capacity

$$= W \log_2 \left(1 + \frac{12S}{k_e^2 N} \right) \tag{8.6}$$

The similarity between equation (8.5) and (8.6) is very striking and indeed the equations are identical if $k_e = \sqrt{12}$. Such a low value for k_e leads to an error probability of about 0.4 in the PCM samples and this is much too high to be acceptable. When k_e is 8.8 (10^{-5} error probability) as suggested in Section 7.2.3, the signal-to-noise ratio required for PCM is about 8 dB above Shannon's limit. However, PCM does offer the same exchange possibilities between signal-to-noise ratio and bandwidth as does Shannon's formula. The possibilities are illustrated by Fig. 7.8.

We now explore this exchange further by examining equation (8.5). For large signal-to-noise ratios, it is quickly established that the product of bandwidth in Hz and signal-to-noise ratio in dB must remain constant to provide a specified channel capacity. For example the capacity of a system with bandwidth W and signal-to-noise ratio 40 dB is the same as that of one with bandwidth $2W$ and signal-to-noise ratio 20 dB. How does this affect the signal power required? For the cases we have studied, we have assumed white Gaussian noise and Shannon's result is only valid for noise of this type. A realistic practical assumption is that the noise power per unit bandwidth is constant, and so N will double if W is doubled. For the example above, the signal power required in the second case is therefore 17 dB (20 dB—3 dB to allow for increased noise) which is less than that in the first case.

If we pursue this exchange, we will eventually be working with small values of S/N at very large bandwidths. If S/N is less than 0·1, we may approximate as follows:

$$\log_2 (1 + S/N) = \log_2 e \times \log_e (1 + S/N)$$

$$\doteq 1{\cdot}45 \; S/N$$

Giving

$$C \doteq 1{\cdot}45 \; SW/N$$

$$\doteq 1{\cdot}45 \; S/N_1 \tag{8.7}$$

where N_1 is the noise power per unit bandwidth. Equation (8.7) therefore shows that the channel capacity is independent of the bandwidth when the signal-to-noise ratio falls below -10 dB. At first sight this seems of little interest, but practical systems working over very wide bandwidths with very low signal-to-noise ratios have been developed to provide very reliable communication. The binary digits 1 and 0 are each represented by long sequences of digits and the receiver uses a correlation technique to provide detection. The correlation of the incoming signal with each of the two possible message digit streams is measured and a decision made as to which message is the more likely. It is obviously necessary that these two digit streams should be uncorrelated and it is reasonable to expect that neither stream will be correlated with the random noise added during transmission. More than two message streams may be used with a corresponding number of correlation detectors at the receiver. The design of such detectors will be examined in a later section.

We now return to the FM case and examine the extent to which the bandwidth used influences the required channel signal-to-noise ratio. We restrict our attention to wideband FM for which the channel bandwidth, W, is approximately twice the frequency deviation. The result we require is obtained by a simple modification to equation (6.77). We consider the fully modulated case, $i.e.$ $m = 1$, and take $f_d = \frac{1}{2}W$ as noted above. The quantity S_{AM} in equation (6.77) is equal to $S/2bN_1$ where S is the mean power in the channel, N_1 is the noise power per unit bandwidth and b is the width in Hz of the baseband. Equation (6.77) therefore becomes:

$$S_{FM} = \frac{3W^2 S}{8b^3 N_1} \tag{8.8}$$

If the values of S_{FM} are kept constant, $i.e.$ b, the signal-to-noise ratio of the output signal obtained from the FM receiver discriminator and

N_1 are kept constant, then the trade-off between channel bandwidth, W, and channel signal power, S, is simply

$$W^2 S = \text{constant} \tag{8.9}$$

When the channel bandwidth is doubled, the signal power required drops by 4 or -6 dB. This compares unfavourably with the -17 dB change obtained earlier for PCM. We conclude that although FM does allow a trade-off between bandwidth and signal power, the extent of this trade-off, as indicated by equation (8.9), is much less favourable than for PCM.

8.5. Source Rate and Channel Selection

The general ideas described in this Chapter may be used as a guide to the selection of an appropriate channel and modulation method to be used in a specified communication requirement. The input to the communication link acts as a message source and its information rate may be calculated according to the arguments in Section 8.1. If the input signal is a continuous one, a choice of the number of quantizing levels needed may be made on the basis of an acceptable quantizing signal-to-noise ratio. The result for the source message rate is then readily obtained by using equation (8.4). A particular example will make this clear.

Consider a facsimile set which sends a 10 inch × 10 inch picture in about 15 minutes with a definition of 100 lines per inch. We assume that a photograph to be transmitted is satisfactory if it can be represented by 64 different levels of shade between black and white and that there are 100 dots per inch transmitted whilst scanning a line. Then we have:

Total elements in picture $= 10 \times 10 \times (100)^2$

No. of elements per second $= \dfrac{10^6}{15 \times 60}$

$= 1100$

Minimum bandwidth(ideal) $= \frac{1}{2} \times 1100 = 550$ Hz

Each element has 64 or 2^6 possible levels.

$\therefore$ Source rate $\quad = 2 \times 550 \times \log_2 64$

$= 1100 \times 6$

$= 6600$ binary digits/sec

If such a picture is sent over a speech channel of 3000 Hz bandwidth and signal-to-noise ratio 30 dB, we get, for the channel capacity based on Shannon's formula modified for good signal-to-noise ratios:

$C = 2W \cdot \log_2 (E_s/E_N)$

$= 2 \times 3000 \times 30/6$

$= 30,000$ binary digits per second where

E_s and E_N are the signal and noise voltages respectively. A signal-to-noise voltage ratio of 2 represents one binary digit and is normally referred to as a 6 dB ratio. Hence it is only necessary to divide the actual signal-to-noise ratio in decibels by 6 in order to get the corresponding number of binary digits. Alternatively, we may note that a 30 dB ratio is equivalent to about a signal-to-noise voltage ratio of 32 or 2^5 i.e. 5 digits.

If we accept that Shannon's formula produces signal-to-noise ratios that are about 9 dB less than is required in practice, we might say that our 30 dB channel is equivalent to an ideal channel of 21 dB signal-to-noise ratio. This is a channel which could be coded into 21/6 or rather more than 3 binary digits (8 levels). In this case the capacity would be

$$C = 2 \times 3000 \times 21/6$$

$$= \text{approximately 20,000 bits per second}$$

Comparing this with the source rate, the channel should be adequate to pass the information content of message, but direct connection of the source without some sort of signal transformation or matching of the source to channel would produce poor results. This follows because we wish to get 64 different levels in our picture, whereas the channel will only transmit about 8 levels with low error rates. This matching is, of course, not to be confused with impedance matching. What is required is some means of raising the bandwidth of the source so that it can tolerate a lower signal-to-noise ratio in the channel.

As a first step we might turn each 64-level digit from the source into two 8-level digits. The source would now emit, per second, 2200 samples each of 8 possible levels, i.e. the same rate of 6600 bits per second. Such a signal could be transmitted happily by the channel postulated, although in practice such factors as low frequency response and stability would have to be considered. As a second step, each 64-level digit could be turned into three digits of 4 levels, i.e. the source is coded into 3300 samples per second, each of 4 levels. Full coding to binary PCM would not be feasible in this instance, because it would require 6600 bits per second, or 3300 Hz bandwidth, which is higher than postulated. Furthermore, this brings in the question of channel frequency response, which would not be ideal. However, with a postulated channel of 30 dB signal to-noise ratio full PCM would not be necessary. In practice, this problem would probably be solved by employing frequency modulation of a carrier of about 1500 Hz but the above reasoning has been given to illustrate the subject of matching source and channel.

The subject is not capable of full quantitative treatment but it is possible to formulate general rules. If bandwidth is reasonably easy to obtain, but noise is high or power available is small, then PCM systems are worth considering. Such conditions often apply with microwave radio.

If bandwidth is at a premium, but noise is low or power is not restricted then AM or in the limit S.S.B. should be used. Such conditions apply with cables, particularly submarine cables.

The accurate transmission of a slow variable is a particular problem which needs careful consideration for the choice of form of signal. This problem occurs, for example, in transmitting the co-ordinates of the position of an aircraft. Direct analogue transmission would not be feasible because the signal would be a slowly varying direct current and time stability of amplitude to high accuracy, say 1 part per 1000, would be difficult to achieve. To represent each co-ordinate by a frequency would be an improvement, but accuracy of the transducers would present a problem. The use of PCM would enable the accuracy requirement to be met simply by using enough digits. Transducers converting shaft rotation to digital-code can be constructed without difficulty up to at least 12 digits. However, the source information would need to be sampled and the question arises as to how to determine the time between samples and how to obtain present position at the receiving end at a particular instant after the receipt of the last sample. The problem is one of predicting or extrapolation. The time between samples is based on a study of the *a priori* information relating to the source. In the case of aircraft motion, the acceleration limit on the aircraft and the geometry of the radar situation provides a graph of, say, X co-ordinate against time which would have only very low frequency components, much less than 1 Hz. Similar considerations to those in the sampling theorem determine the sampling rate. For prediction or extrapolation the use of the reconstituting filter of sampling theory is not feasible owing to the low frequencies. Both analogue and digital methods of extrapolation may be employed. Briefly, with linear prediction the last two points are extended linearly, *i.e.* the velocity is assumed constant, and with quadratic prediction the last three points are fitted by a parabola which gives the position until the next point is received, *i.e.* the acceleration is assumed constant. The extrapolation cannot, of course, be accurate; each sample received will generally show a small error against the extrapolation. However, by suitable design a slow variable may always be transmitted by PCM to a given accuracy.

8.6. Decision Theory applied to Communications

We have now referred in each of the last three chapters to a number of situations in which the function of a communication receiver is to make some kind of decision. Examples include the detection of radar pulses (Section 6.4), the regeneration of PCM (Section 7.1) and the use of correlation detectors (Section 8.4). The design of decision circuits may be optimized using the branch of statistics known as decision theory. We shall briefly discuss this theory and mention one or two applications in communications.

If a receiver must decide from the signal observed whether or not a pulse has been transmitted, he is choosing between *hypotheses*: the first hypothesis is that a pulse was transmitted, the second is that a pulse was not transmitted. From our knowledge of the communication system, we can specify probabilities to each of these hypotheses even before a signal is observed. We use the description *a priori* (the Latin for before) for such probabilities and in the present case suppose that they are p_1 and p_2 for the first and second hypotheses respectively.

After a signal, s, has been received, the probabilities will change because we have additional information from the signal. These *a posteriori* probabilities will be written $p(1 \mid s)$ and $p(2 \mid s)$ for hypotheses 1 and 2 respectively. The notation $p(1 \mid s)$ denotes the probability of hypothesis 1 given that the signal s has been received. Bayes Theorem in statistics provides values for these *a posteriori* probabilities in terms of the probability distributions for s. There are two such distributions depending on whether a pulse has been transmitted or not. For example, $p(s \mid 1)\mathrm{d}s$ is the probability that s lies between s and $s+\mathrm{d}s$ given that a pulse was transmitted. The joint probability distribution, $p(1, s)\mathrm{d}s$, is the probability that a pulse was transmitted and that s lies in the range s to $s+\mathrm{d}s$. We now have two relations

$$p(1, s)\mathrm{d}s = p_1 p(s \mid 1)\mathrm{d}s \tag{8.10}$$

(*i.e.* joint probability of hypothesis 1 and s = probability of 1 times probability of s given 1) and

$$p(1, s)\mathrm{d}s = p(s)\mathrm{d}s\, p(1 \mid s) \tag{8.11}$$

where $p(s)\,\mathrm{d}s$ is the probability of s lying in the range s to $s+\mathrm{d}s$. From equations (8.10) and (8.11),

$$p(1 \mid s) = \frac{p_1 p(s \mid 1)}{p(s)}$$

Similarly,

$$p(2 \mid s) = \frac{p_2 p(s \mid 2)}{p(s)}$$

The result we shall use is

$$\frac{p(1 \mid s)}{p(2 \mid s)} = \frac{p_1 p(s \mid 1)}{p_2 p(s \mid 2)} \tag{8.12}$$

which provides the ratio of the *a posteriori* probabilities of hypotheses 1

and 2. A basis for deciding which hypothesis to select is the Bayes
decision rule:
select hypothesis 1 if

$$\frac{p(1 \mid s)}{p(2 \mid s)} \geq 1$$

and select hypothesis 2 if

$$\frac{p(1 \mid s)}{p(2 \mid s)} < 1$$

The ratio $p(1 \mid s)/p(2 \mid s)$ is called the likelihood ratio.

We shall now use this rule in the detection of a pulse in the presence
of noise, under the same conditions as were used in Section 6.4. The
pulse amplitude is E and the noise x is Gaussian with r.m.s. value σ.
We then find:

$$p(s \mid 1)ds = p(s-E < x < s-E+ds)$$

since $s = x+E$ if the pulse is transmitted,

i.e. $p(s \mid 1)ds = \dfrac{1}{\sigma\sqrt{2\pi}} \exp\left[-\dfrac{(s-E)^2}{2\sigma^2} \right] ds$

Similarly,

$$p(s \mid 2)ds = \frac{1}{\sigma\sqrt{2\pi}} \exp\left[\frac{-s^2}{2\sigma^2} \right] ds$$

since $s = x$ if no pulse is transmitted.

$$\frac{p(1 \mid s)}{p(2 \mid s)} = \frac{p_1}{p_2} \frac{\exp\left(-(s-E)^2/2\sigma^2\right)}{\exp\left(-s^2/2\sigma^2\right)}$$

The decision that a pulse has been transmitted is made if

$$p_1 \exp\left[-(s-E)^2/2\sigma^2\right] \geq p_2 \exp\left(-s^2/2\sigma^2\right)$$

This may be rewritten after some algebraic manipulation as:

$$2sE \geq E^2 + 2\sigma^2 \log_e\left(p_2/p_1\right) \tag{8.13}$$

The decision criterion given by equation (8.13) reduces to $s \geq E/2$ if
$p_1 = p_2$, the same result as derived in equation (6.33).

The advantage of using the general approach based on the Bayes rule
rather than the direct method of Section 6.4 lies in the possibilities of
generalization. Firstly, the method is equally applicable for cases with
more than two hypotheses (*e.g.* more than two possible signals). The
hypothesis with the largest *a posteriori* probability is selected. Secondly,
the decision rule given by equation (8.13) incorporates the *a priori*

information given by p_1 and p_2. A further possible extension is to base the decision on more than one sample of the observed waveform at the receiver.

The decision rule based on equation (8.12) defines a threshold level, s_0, such that hypothesis 1 (pulse present) is accepted if s exceeds s_0. The probability that a wrong decision is made is then obtained as follows:

Probability that a pulse is wrongly indicated

= probability that $s \geqq s_0$ when no pulse is present

$$= \int_{s_0}^{\infty} p(s \mid 2)ds$$

Similarly, the probability that a pulse is missed

$$= \int_{-\infty}^{s_0} p(s \mid 1)ds$$

The calculations carried out in Section 6.4 show that these error probabilities depend on the ratio E/σ, i.e. on the signal-to-noise ratio. We now examine the possibility of processing the signal waveform to maximize the signal-to-noise ratio at the instant when the decision is made.

Suppose the signal waveform is $s(t)$ and that the noise is white Gaussian with density N_1 per unit bandwidth. We now examine the effect of a filter of transfer function $A(\omega)$ on the signal-to-noise ratio. The output from this filter corresponding to the signal is given by equation (5.3):

$$v(t) = \frac{1}{2\pi} \int_{-\infty}^{\infty} A(\omega)g(\omega) \exp(j\omega t)d\omega$$

where $g(\omega)$ is the spectrum corresponding to $s(t)$. The output voltage $v(t)$ will be a maximum at some instant, t_1, and so

$$v_m = \frac{1}{2\pi} \int_{-\infty}^{\infty} A(\omega)g(\omega) \exp(j\omega t_1)d\omega \tag{8.14}$$

The total noise power at the filter output is

$$N = \frac{N_1}{2} \cdot \frac{1}{2\pi} \int_{-\infty}^{\infty} |A(\omega)|^2 \, d\omega \tag{8.15}$$

the factor $\frac{1}{2}$ arising since N_1, the noise power per unit bandwidth, is split equally between positive and negative frequencies. The output signal-to-noise ratio at the decision instant, t_1, is

$$\left(\frac{S}{N}\right)_0 = \frac{\left| \int_{-\infty}^{\infty} A(\omega)g(\omega) \exp(j\omega t_1)d\omega \right|^2}{N_1\pi \int_{-\infty}^{\infty} |A(\omega)|^2 \, d\omega} \tag{8.16}$$

The function $g(\omega)$ corresponding to the waveform, $s(t)$, and N_1, the noise power density, are known. We therefore wish to choose $A(\omega)$ in such a way that $(S/N)_0$ is as large as possible. The choice is made using Schwarz's inequality which leads in the present case to the result:

$$\left| \int_{-\infty}^{\infty} A(\omega)g(\omega) \exp(j\omega t_1)d\omega \right|^2 \leqq \int_{-\infty}^{\infty} |A(\omega)|^2 \, d\omega \int_{-\infty}^{\infty} |g(\omega)|^2 \, d\omega$$

$$(8.17)$$

The largest possible value of $(S/N)_0$, given by using equation (8.17) with the equality sign, is therefore

$$\left(\frac{S}{N}\right)_0 = \frac{\int_{-\infty}^{\infty} |g(\omega)|^2 \, d\omega}{N_1 \pi}$$

But

$$\frac{1}{2\pi} \int_{-\infty}^{\infty} |g(\omega)|^2 \, d\omega = E_s$$

where E_s is the total energy of the signal. The maximum value of $(S/N)_0$ thus reduces to $2E_s/N_1$.

The only condition which leads to the equality sign in equation (8.17) is

$$A(\omega) = g^x(\omega) \exp(-j\omega_1 t)$$

$$(8.18)$$

and so the filter characteristic for maximum signal-to-noise ratio is defined. Such a filter is called a 'matched filter' since it is matched to the signal waveform in use. An alternative to specifying the transfer function, $A(\omega)$, is provided by equation (5.5) which gives the filter impulse response:

$$h(t) = \frac{1}{2\pi} \int_{-\infty}^{\infty} A(\omega) \exp(j\omega t)d\omega$$

$$= \frac{1}{2\pi} \int_{-\infty}^{-\infty} g^x(\omega) \exp(j\omega t - j\omega t_1)d\omega$$

Since $h(t)$ must be a real function,

$$h(t) = \frac{1}{2\pi} \int_{-\infty}^{\infty} g(\omega) \exp(j\omega t_1 - j\omega t)d\omega$$

$$= s(t_1 - t)$$

$$(8.19)$$

in view of the definition of $g(\omega)$. The filter impulse response is thus the signal waveform with time reversed. Since the impulse function must vanish for $t < 0$, it is only possible to design matched filters for waveforms such that $s(t)$ vanishes if $t \geqq t_1$. This is equivalent to the condition that all the signal energy must be received before the instant t_1 at which the decision is made.

The improvement in signal-to-noise ratio gained by using a matched filter in place of those discussed in earlier chapters is only about 1 dB. There are however more important properties of matched filters which will be discussed below.

An alternative expression for $v(t)$, given by combining equations (5.4) and (8.19), is:

$$v(t) = \int s(\tau)s(t_1 - t + \tau)d\tau \qquad (8.20)$$

this integral being the auto correlation of $s(t)$. The range of integration is fixed by the time intervals during which $s(\tau)$ and $s(t_1 - t + \tau)$ exist. The received waveform is contaminated by noise so that the matched filter effectively correlates this waveform with the expected uncontaminated waveform. The matched filter may thus be regarded as a correlation detector in the sense referred to in Section 8.4.

8.7. Estimation Theory in Radar

The results in the previous section are relevant to radar, but further problems arise and the statistical theory is further extended to provide estimates of parameters such as the time of arrival of the received pulse. Such estimates are obtained by examining the probability distribution of the quantity required in the light of the information gained from the received waveform. The theory is too complex to allow discussion here but reference will be made to some of the results.

It appears from the form of the output signal-to-noise ratio for the matched filter that the signal energy is concentrated at the instant when the decision is made. This implies that the output pulse will be sharper, *i.e.* shorter in duration, than the input pulse. A detailed examination shows that the output pulse provides the best estimate of the arrival time so that the range accuracy of radar may be improved. Even more important is the possibility of satisfying two conflicting requirements, those of transmitter pulse length and range resolution. Section 10.2 deals with radar operation in some detail. Anticipating the results, we find that for good range resolution a short transmitter pulse is required but such a pulse may imply a transmitter power requirement which cannot be met. However, if a longer pulse is used with an appropriate matched filter, the filter output can be sufficiently short to provide adequate resolution. A technique which is widely used is to apply linear FM to the transmitter pulse, since this facilitates the design of the matched filter. The acoustic pulses emitted by bats are believed to have a similar spectrum and the term 'chirp' pulse is often used to describe such pulses. The length of the transmitted pulse may be reduced by a factor of up to 1000, called the pulse compression ratio.

The basic relations for the chirp pulse and its matched filter are as follows. The transmitter pulse is taken to be of duration T, carrier frequency ω_0 and with a linear frequency modulation:

$$s(t) = \cos(\omega_0 t + \alpha t^2) \text{ if } -\tfrac{1}{2}T < t < \tfrac{1}{2}T \tag{8.21}$$

The angular frequency increases linearly from $\omega_0 - \alpha T$ to $\omega_0 + \alpha T$, the frequency change in Hz, Δf, thus being:

$$\Delta f = 2\alpha T / 2\pi \tag{8.22}$$

The matched filter output, given by equation (8.20), is

$$v(t') = \int_{\tau_1}^{\tau_2} \cos(\omega_0 \tau + \alpha \tau^2) \cos[\omega_0(\tau - t') + \alpha(\tau - t')^2] d\tau \tag{8.23}$$

where $t' = t - t_1$. The limits on the integration are obtained by noting that both τ and $(\tau - t')$ must lie within the range $-\tfrac{1}{2}T$ to $\tfrac{1}{2}T$ as required by the pulse definition in equation (8.21). We then find

$$\tau_1 = -\tfrac{1}{2}T \quad \text{if } \tau_2 = \tfrac{1}{2}T + t' \text{ and } t' < 0$$

$$\tau_1 = -\tfrac{1}{2}T + t' \text{ if } \tau_2 = \tfrac{1}{2}T \quad \text{and } t' > 0$$

and note that $v(t')$ will be zero if t' lies outside the range $-T$ to T. The integration in equation (8.23) may now be carried out by expressing the integrand as a sum of two cosine terms one of which has an effective frequency of $2\omega_0$ and may be ignored assuming that it will be filtered out. The other leads to the result:

$$v(t') = \frac{\sin \alpha t'(T - |t'|)}{2t'} \cos(\omega_0 t') \text{ if } -T < t' < T \tag{8.24}$$

The output from the matched filter is thus a pulse of carrier frequency ω_0, with an envelope given by $\dfrac{\sin \alpha t'(T - |t'|)}{2t'}$. The envelope is very similar to the $\sin(x)/x$ waveform, the only difference being the factor $(T - |t'|)$ in the argument of the sine. This factor is the constant T for small values of $|t'|$ and the behaviour of the envelope may be deduced from $\sin(x)/x$ for such values of t'. In particular, the first zeros of the envelope occur when $\alpha t' T = \pm \pi$ and the interval between these zeros is $2\pi/\alpha T$.

A reasonable measure of the pulse width is half this value, i.e. $\pi/\alpha T$, and so:

$$\text{Pulse compression factor} = \frac{T}{(\pi/\alpha T)}$$

$$= T\Delta f \qquad (8.25)$$

on using equation (8.22).

An example will indicate the numerical values likely to occur in practice. Suppose $T = 10\ \mu s$, and a compression factor of 10^3 is desired. Then $\Delta f = 10^3/10^{-5} = 10^8$. A frequency change of this magnitude, 100 MHz, is quite realistic in relation to a carrier frequency of 10 GHz, typical of a centimetric radar.

We now examine the requirements to be met by the matched filter. The spectrum of $s(t)$ is:

$$g(\omega) = \int_{-\frac{1}{2}T}^{\frac{1}{2}T} \cos(\omega_0 t + \alpha t^2) \exp(-j\omega t)dt \qquad (8.26)$$

An argument on the lines of that used in Section 5.4.1 shows that for positive values of ω:

$$g(\omega) = \tfrac{1}{2} \int_{-\frac{1}{2}T}^{\frac{1}{2}T} \exp j(\omega_0 t + \alpha t^2 - \omega t)dt$$

with

$$g(-\omega) = g^{\times}(\omega) \quad \text{(equation 5.17)}$$

Hence

$$g(\omega) = \tfrac{1}{2} \int_{-\frac{1}{2}T}^{\frac{1}{2}T} \exp j[\alpha(t - t_0)^2 - \alpha t_0^2]dt$$

where

$$t_0 = (\omega - \omega_0)/2\alpha$$

i.e.

$$g(\omega) = \tfrac{1}{2} \exp(-j\alpha t_0^2) \int_{-\frac{1}{2}T - t_0}^{\frac{1}{2}T - t_0} \exp(j\alpha t^2)dt$$

The integral on the right-hand side may be expressed in terms of Fresnel integrals for which extensive tables are available. A crude approximation which is adequate for our present purpose is:

$$\int_{-\frac{1}{2}T - t_0}^{\frac{1}{2}T - t_0} \exp(j\alpha t^2)dt = k \quad \text{if} \quad -\tfrac{1}{2}T < t_0 < \tfrac{1}{2}T$$

$$= 0 \quad \text{if} \quad |t_0| > \tfrac{1}{2}T$$

where k is a complex constant.

We now express $g(\omega)$ in terms of ω by substituting for t_0 and find:

$$g(\omega) \doteqdot \tfrac{1}{2}k \exp[-j(\omega - \omega_0)^2/4\alpha] \quad \text{if} \quad |\omega - \omega_0| < \alpha T$$

$$= 0 \qquad\qquad\qquad \text{if} \quad |\omega - \omega_0| > \alpha T \qquad (8.27)$$

Since $\alpha T = \pi \Delta f$ (equation 8.22), we see that the spectrum of $s(t)$ lies between $f_0 - \frac{1}{2}\Delta f$ and $f_0 + \frac{1}{2}\Delta f$, f_0 being the carrier frequency.

The transfer function of the matched filter, equation (8.18), is

$$A(\omega) = \tfrac{1}{2}k^x \exp\left[j(\omega - \omega_0)^2/4\alpha - j\omega t_1\right] \text{ if } |\omega - \omega_0| < \alpha T \quad (8.28)$$

and therefore has a uniform amplitude response from $f_0 - \frac{1}{2}\Delta f$ to $f_0 + \frac{1}{2}\Delta f$. The significance of the phase response is most easily seen by considering the group delay (Section 4.4.2), which equals:

$$t_d = t_1 - (\omega - \omega_0)/2\alpha \quad (8.29)$$

The group delay thus decreases linearly with frequency. The constant t_1 must be sufficiently large to ensure that the delay, t_d, remains positive for all frequencies which are accommodated.

The group delay relation provides a simple physical explanation for the operation of the chirp system. The frequency of the transmitter pulse at time t is $\omega_0 + 2\alpha t$, and this frequency is subject to the delay $t_1 - t$. The arrival time at the output of the matched filter is therefore $t + t_1 - t = t_1$, irrespective of the value of t. All frequency components are thus delayed to arrive simultaneously.

The above discussion suffices to show the importance of matched filters in relation to radar operation. The basic ideas provided by estimation theory can be used further to investigate the problem of extracting velocity information from the Doppler shift experienced by radar waveforms. Such an investigation leads to the conclusion that a bank of matched filters tuned to different frequencies provides the most efficient method of measuring simultaneously the position and velocity of a radar target.

Chapter 9

Communication Systems

9.1. Introduction

The term system engineering is applied to the overall planning and design of a communications system to meet a specified performance. It is an important field of activity because the engineer concerned must know the detailed performance of individual sub-systems or units, any interaction which may occur when these are connected together to form the whole system, and he must be responsible for the final performance as it affects the user. There is now available such a wide variety of systems that the choice of the most suitable one for a particular problem is almost bewildering. The logical approach would be to determine from the customer the requirements for baseband width and signal-to-noise ratio and then to carry out a study of costs for the various methods of transmission. The theory of modulation, noise, and the properties of channels are an essential preliminary to an understanding of the problems, but a great deal of practical knowledge is necessary before the correct choice of system can be made. All that can be done here is to give a somewhat superficial summary of the factors involved and then to consider a few examples taken from the various branches of the subject.

It is necessary to have some appreciation of the viewpoint of the user of a communication system, and we start therefore with a brief discussion of the advantages and disadvantages of the various methods of conveying intelligence.

9.2. Comparison of the various Methods of Conveying Intelligence

9.2.1. TELEGRAPHY

The relative merits of telegraphy and telephony are well established and widely appreciated, at least so far as public systems are concerned. For private communication systems the situation is a little different, because the organization for preparing telegraph messages can then be concentrated near the telegraph instrument. Nevertheless, the choice between telegraphy and telephony depends very greatly upon the nature of the message traffic. The transmission of large volumes of alpha-numeric data (*i.e.* letters and figures) for which no answers are required such as with press traffic, is a clear case for telegraphy, as the cost figures will show. If there are large volumes of traffic requiring answers, but not immediately, then this will also be a case for telegraphy.

The typical machine telegraph is the teleprinter and in the U.K. this normally operates at a signalling speed of 50 bauds (bits per second), resulting in a speed of 66 words per minute, if allowance is made for the start-stop synchronizing signals. It is possible to transmit the signals from a number of such telegraph instruments over one telephone line using FDM. Each teleprinter has its own voice frequency signal which is either keyed on and off or frequency modulated. It is therefore known as V.F. multi-channel telegraphy and over a high-grade telephone of 300 to 3400 Hz bandwidth it is possible to accommodate as many as 24 channels. This represents a total capacity of 1584 words per minute, each word averaging five letters plus one space between words. This capacity is at least 10 times greater than a fast talker can speak; indeed, it is doubtful whether many telephone conversations exceed 100 words per minute. The channel cost of telegraphy is therefore less than the channel cost for telephony, but, of course, the cost of messenger distribution at the terminals of a public system of telegraphy is relatively high. The telegraph system has the advantage of providing a paper copy of the message. Automatic error checking and correcting is not usual on systems working mainly over cable as these conform to a high standard (*i.e.* a low error rate) and some checking can be provided by the form of the message. For example, figures can be repeated at the end of the message, and important numerics can be spelt out. Where long-distance radio telegraphy is employed it is usual to employ some form of error detecting code and some systems employ automatic request for a repeat of a character received in error.

The telex service which covers the UK and is widely international, employs teleprinters and connections may be set up by dialling. Bothway transmission is possible on such connections, i.e. the service is essentially a 'conversational' teleprinter service.

9.2.2. TELEPHONY

The advantages of the telephone are well known. Two talkers having a telephone conversation can often come to immediate agreements and decisions, whereas the inevitable delays in preparing telegrams, passing them to machine operators and delivering them at the distant end would consume much more time. Where the traffic is urgent, involving immediate agreements and decisions, then telephony is essential.

It is now generally accepted that telephony and telegraphy are complementary forms of communication both of which are necessary for public and private systems. The facility of being able to provide telegraph channels over telephone circuits enables flexibility to be introduced into the planning and the telephone circuit is therefore often regarded as the basic unit of a communication system.

9.2.3. FACSIMILE

The use of the facsimile system for transmitting information is still strictly limited. It is largely employed for transmitting press photographs, but is has also proved valuable for transmitting weather maps and allied information. It is not an economical method of transmitting written information. The reason for this is that in the scanning method of sending printed matter much time is wasted in scanning empty paper. Furthermore, a large number of dot elements are required to transmit a single printed letter whereas in normal telegraphy one letter can be sent with a code having five dot elements (binary digits) for any letter of the alphabet. Most facsimile systems operate over a telephone-type channel by using a sub-carrier voice frequency tone, which is either amplitude or frequency modulated by the picture video signal. Use of a telephone channel also contributes to economy and flexibility in the design of a complete communication system.

9.2.4. DATA TRANSMISSION

With the growth of computer techniques and the need to convey accurate information to and from large digital computers or data handling systems, there is an increasing development of high-speed binary signal transmission. These mark-space signals are generally applied to a channel by using one frequency for a mark and another frequency for a space (frequency shift keying or frequency modulation). Such signals may be transmitted over a telephone channel, and signalling speeds of 200, 600 and 1200 bauds have been adopted as international standards. A simple application of the Shannon formula (Chapter 8) suggests that much higher speeds are possible, but the necessary complications of delay equalization and single sideband are generally not justified. Although uncorrected error rates of 1 in 10^5 bits are to be expected on a high-grade telephone channel, there is likely to be a need for error detection and correction with this form of communication. This facility might take the form of a parity bit at the end of each block of digits, with a facility for repeating a block when lack of parity occurs. By comparison a 24-channel VF telegraph system has a capacity of 1200 bits per second, but allowing for teleprinter start and stop signals this reduces to about 800 bits per second. The actual information capacities of the two systems are therefore not very different from one another. The high-speed binary system can be used to request information from a data-handling system, without the need for message copy and manual intervention. The high speeds of data-handling systems enable information to be returned immediately.

The method of digit synchronizing in such systems would generally be of the flywheel type described in Chapter 2 but indexing of blocks of digits to ensure correct decoding necessitates special arrangements.

Indexing may be carried out by introducing a third signal level, indicating the start of a block of digits, but this arrangement detracts from the efficiency of the binary system and an alternative is to use binary digits with a special code to indicate the start of a block. Steps then have to be taken to prevent the occurrence of this code in the information digits. This might be done by breaking up the information digits with fixed digits which would prevent this occurrence.

9.2.5 HIGH-QUALITY AUDIO CIRCUITS

So far we have discussed communication facilities based on a telephone circuit or its sub-divisions. The next facility of increased bandwidth is generally the music circuit, and a response of 30 Hz to 12 kHz or higher may be required with a peak signal-to-noise ratio of at least 50 dB. Such a channel is required between a broadcast studio and a remote broadcast transmitter. It calls for no special comment except that in a flexible communication system it should be possible to derive such circuits when they are required. This leads to a general preference for frequency division multi-channel systems, a topic which will be discussed presently.

9.2.6. TELEVISION

Although there are several special methods of communication of greater bandwidth, the next important case is the television channel. The approximate channel requirements of various picture definitions are given in Table 9.1.

TABLE 9.1

Number of lines, definition	405	625	819
Nominal upper cut-off MHz	3	5	10
Signal/noise ratio dB	50	50	50

For a given method of transmission, *e.g.* cable, or microwave relay, it is true to say that cost increases with bandwidth, but there is no simple relation between the two factors. When providing communications over territory for the first time there are heavy costs, for providing lines or cables, or aerial sites, and the cost of providing increased bandwidth is small compared with these first costs. Nevertheless, the cost of providing television channels is likely to remain high, and although it is an admirable method of communicating, it is not likely to challenge telephony. A valuable measure of flexibility is achieved if a wideband channel can be made to accept television or a large block of telephone channels. If the peak of telephone traffic occurs during the daytime and the need for television occurs in the evening, then good economics can result from this flexibility.

9.3. Choice of the Transmission System

The choice of the transmission system requires consideration of a number of factors, technical, economic and many others. Between fixed locations there is a choice between open wire line, cable, and many forms of radio. Between mobile locations and between fixed and mobile locations the choice is limited to radio, although there are many forms of radio system.

For fixed communications we have to consider distance, the territory between the terminal stations, the bandwidth, and the signal-to-noise ratio of the baseband and of the transmission medium. If the communication channel or channels form part of an integrated system, *i.e.* the channels have to be connected to other channels, then there are important questions of stability. The question of reliability in relation to cost must be considered.

The scope for open wire line is generally rather limited owing to the availability of improved systems using cable and radio. It is limited to subscribers' telephone lines and public telephone lines in sparsely populated territory. The application of 12-channel FDM is successful over well-constructed open wire telephone routes, although there may be crosstalk problems if there are many adjacent pairs carrying FDM, as the frequencies involved extend up to 60 KHz. The spacing between repeater stations depends upon the conductors used but a typical distance is 100 miles.

The open wire telephone line over sparse territory has a serious competitor in the form of a radio relay using VHF or UHF frequencies. Each hop of the relay path is limited in length by earth curvature. Over flat country towers for the aerials may be used but economics would limit the height to 100 or 150 ft, giving a hop length of about 30 miles. If suitable high ground can be used this figure can be raised to 50 miles or more. Although the relay stations are closer than the repeater stations with open wire, maintenance of the line is avoided. The radio system needs transmitter powers of only 10 watts or so, and capacities of 12, 24 or 48 telephone channels are easily obtained.

A method of radio propagation, giving a greater path length of, perhaps, 100 to 200 miles, is known as tropospheric scatter. It requires high transmitter power (about 10 KW) and is usually carried out in the UHF band. Large aerials (30 or 60 ft diameter) and low noise receivers are required. The channel capacity depends upon the reliability required. The medium has ample bandwidth capability. The method of propagation is such that there may be periods in which high-grade communication fails, and the statistics of propagation over the path must be studied to determine the best arrangement of the communications. The method is economical over sea gaps and in the frozen North.

For overland systems with requirements extending to several hundreds of telephone channels, there are multi-pair cable systems, each pair having 12, 24 or 48 telephone channels, coaxial cable systems and microwave relay systems. If there is a requirement for television this narrows the choice to co-axial cable and microwave relay. The choice is not easy to make and detailed local factors must be considered.

Over very large ocean distances the submarine cable has now been developed to convey 36 telephone channels by using submerged amplifiers or repeaters fed by power over the cable itself. Although the distances across oceans of several thousand miles can easily be bridged by HF radio using ionospheric reflection, the channel capacity in this band is strictly limited and the channels are liable to occasional failure due to ionospheric disturbance. Using one radio frequency, the number of telephone channels is usually limited to about 4. Single-channel HF radio is eminently suitable for ship-to-shore working. Communications in this band are liable to fading and interference and it is necessary for each part of the system to be optimised within the limit of economics. The use of single sideband modulation is highly desirable because of its better signal-to-noise ratio for a given transmitter peak power, and its lower audio distortion when multipath propagation exists.

Turning to communications between fixed and mobile points, these are usually limited to one audio channel, although one telegraph channel may be preferred in some cases. Typical systems are aircraft ground-air communications, ship-to-shore communications, and police and military systems for communicating to and from vehicles, etc. For line of sight systems or distances 10 or 20 per cent beyond, there is a good choice of radio, using VHF or UHF frequencies, and AM or FM modulation may be used. There is a tendency to prefer AM where the radio path is subject to many reflections. When the distance is much beyond line of sight HF radio must be used, for the shorter distances by ground wave propagation, and for the long distances by ionospheric reflection. If one station is fixed it is possible to employ high power, but the aerial on the vehicle is likely to limit the system performance. The return path from vehicle to fixed station is at a disadvantage because of the limit to transmitter power.

An important factor in the choice of radio systems is that of frequency allocation. The transmission medium, the aether, is common to all users and unless careful control and planning is carried out at all levels from international down to local, then serious crosstalk and interference can occur. The factors determining whether two separate radio systems will interfere with one another are transmitter powers, geographical distances between transmitters and receivers, aerial polar diagrams, propagation (*i.e.* whether ionospheric or line of sight) and frequencies used. Each system must confine its frequencies as far as possible to the allotted band. The line of sight propagation systems are much easier to plan because the coverage of a particular transmitter can be predicted with fair accuracy.

The ionospheric systems are more difficult to plan because of the variable nature of the ionosphere, and the difficulty of achieving good polar diagrams at the lower frequencies due to the size and cost of the aerials. When ionospheric conditions are good, extremely long ranges are possible via multiple reflections, and this can prove an embarrassment because of the interference that can be brought in from long distances. Round the world echoes frequently occur when ionospheric conditions are most favourable.

9.4. Two-way circuits

For telephony it is necessary to provide two-way circuits. A simple two-wire telephone line provided this facility, the microphone and earpiece at each terminal being effectively connected in series. This means that the transmitted signal from the microphone is heard in the local receiver as sidetone. This is objectionable and telephone circuits usually employ a bridge circuit arrangement to minimize the level of side-tone. Fig. 9.1 illustrates the principle, the network N being an impedance which approximates the impedance of the line over the telephony band of frequencies. The signal from the microphone is shared between the line and the network, the earpiece being connected across points of equal potential. In fact, a close balance of impedance is impossible to achieve, but the power in the earpiece is reduced by an appreciable amount. For signals incoming from the line the power is shared between the earpiece and the microphone. If the transformer ratio and impedances are correctly chosen the network is connected across points of equal potential and negligible power appears in the network. This provides a satis-

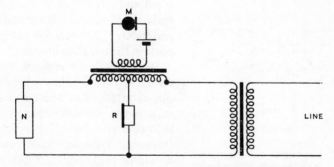

Fig. 9.1. Principle of 2-way, 2-wire telephone with anti-sidetone transformer.

factory telephone circuit, provided the attenuation of the two-wire line does not exceed about 20 dB. However, this is not satisfactory for large switched telephone networks, where it is necessary to have long-distance circuits of zero or small loss, so that they can be connected in tandem.

The provision of two-way amplification on two-wire lines presents very great difficulties which will be explained later. In a similar manner there are extreme difficulties in obtaining a send channel and a receive channel in a two-way radio system by using only one aerial and a common carrier frequency. Because of these difficulties it is usual in telephone practice to use 4-wire circuits, *i.e.* the two oppositely directed channels are formed by using two separate pairs of wires. In the radio case two separated frequencies for the two directions of transmission are necessary and generally separate aerials for transmission and reception, although this is not essential if adequate separating filters can be constructed. It is usual to call telephone circuits derived in this way by the same term, namely, 4-wire circuits.

Such a method of providing two-way telephony enables each channel to be lined up to zero loss, and tandem connection of several such links provides an overall zero loss two-way telephone system. This implies 4-wire switching at the telephone exchanges. However, such a method of switching is not yet common practice, and individual telephone subscribers' lines use two wires only. It is necessary, therefore, to have a means of reducing a 4-wire circuit to a 2-wire circuit for switching purposes. This is achieved by a hybrid transformer which is similar to the anti-side tone circuit described above. The full theory will not be given here but only the broad principles.

Consider a device with 4 branches, as in Fig. 9.2. When these branches are correctly matched, say all with the same impedance, the device has the property that power applied at any branch such as A, is equally divided and appears half in branch B and half in branch D. If the matching and balance are ideal no power appears in the opposite branch C.

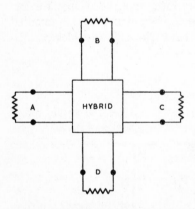

Fig. 9.2. The principal of the Hybrid. Power in at any branch divides equally between adjacent branches, with no power into opposite branch.

Such a device is employed to reduce a 4-wire circuit to a 2-wire circuit as in Fig. 9.3. The networks *N* must balance the impedance of the two-wire line for ideal operation. Power in at the left hand two-wire line is divided, one half being amplified in channel A, the other half being dissipated in the output of channel B. The signal power is delivered to the distant end where half is delivered to the 2-wire line and half is dissipated in the network N. If the balance condition is ideal there is no power across the hybrid into the input of the return channel B. This prevents feedback and any tendency for the circuit to oscillate or sing. In practice, the balance condition is never perfect and some power is fed into the return channel.

There is a 3 dB loss in the signal path through the hybrid at each end of the circuit. To provide a zero loss circuit the channels A and B must each have a net gain of 6 dB. It should be noted that if the right-hand two-wire line is disconnected the signal power that would have been delivered to it is fully reflected and after being divided again, one half is delivered to the input of channel B. Under this condition the hybrid is unbalanced and power appears in the opposite branch at a level of 6 dB below the input. If this open circuit condition exists at each end the total gain in the loop is 12 dB and the total loss 12 dB, causing the circuit to be unstable, *i.e.* liable to sing. However, with precautions to ensure that reasonable impedance terminations exist, this method of operating telephone circuits is common practice. The two wire ends may be connected to the two wire ends of other 4-wire circuits or to 2-wire telephone lines.

The same technique is applied to 4-wire radio systems. The stability problem outlined above shows that the gain stability of each radio channel must be very good. That is, the gain of each channel must be capable of being set to $+6$ dB, and it must remain within a small tolerance of that figure. If the radio path itself is subject to large changes of attenuation then some means of gain control, either manual or, preferably, automatic, must be used. Alternatively, where frequency modulation may be employed the gain stability requirement is transferred to the stability of the modulating and de-modulating circuits. In this case changes of radio path attenuation cause a change in signal-to-noise ratio, but the design must be such that the signal-to-noise ratio does not deteriorate below the necessary standards for the communication system.

The above description of the hybrid enables an insight to be obtained into the problem of 2-wire amplification. Consider the arrangement shown in Fig. 9.4. This shows a hybrid at the middle of a 2-wire telephone line. An amplifier having a gain comparable to the overall line loss is connected across opposite branches. Then a signal from A is divided, half into the amplifier input and half dissipated in its output internal impedance. The amplified signal is then divided, half being sent on to the end B and half being returned to A. Any unbalance between the two opposing line impedances will cause some of the amplifier output power to be applied to the amplifier input with a consequent risk of singing. This balance of impedance must be obtained over the band of frequencies to be amplified and over the frequency regions where the amplifier gain is being reduced to a low value. In practice this balance condition is difficult to attain and maintain. This type of amplified two-way telephone line was employed at one time, but these difficulties have caused the 4-wire circuit to be almost universally employed.

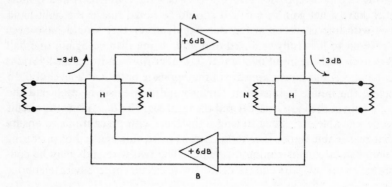

Fig. 9.3. A 4-wire circuit with 2-wire terminations.

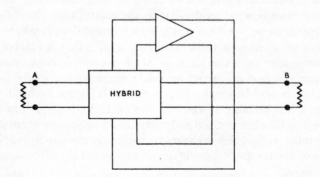

Fig. 9.4. A method of 2-wire amplification.

If these principles are applied to the problem of connecting a radio transmitter and receiver to one aerial, using a common frequency, the loss across opposing branches of the hybrid must be fantastically large. This follows because of the high power-level difference between a radio transmitter and a receiver, figures in excess of 100 dB being quite usual. Balancing impedances over a band of frequencies, such that the loss across opposing branches of a hybrid is of this order, is quite impracticable.

9.5. Integrated Communication Systems

Public communication systems, such as the network of the Post Office, afford a good example of the application of a wide variety of communication techniques to provide a flexible system offering a variety of services to the public. For example, the network provides telephony and telegraphy, whilst music and television channels can also be provided to broadcasting authorities. A number of other classes of services can be provided, calling for flexibility in the technical planning of the facilities. The need for inter-connecting such networks on an international basis requires a measure of standardization of techniques and performance, and this is provided by a series of recommendations agreed by committees of the International Telecommunication Union (I.T.U.). The two main committees concerned are the International Telegraph and Telephone Consultative Committee (C.C.I.T.T.) and the International Radio Consultative Committee (C.C.I.R.). The two bodies work closely together, but C.C.I.R. has the additional task of radio frequency planning, *i.e.* organizing the allocation of radio frequencies to world-wide users such that interference is at a minimum.

The international telephone circuit is the most important unit on which standardization is necessary, but this must then proceed upwards to standardization of the interconnection of large blocks of channels, and downwards to the sub-division of a telephone channel into telegraphy and other services. The international telephone circuit must be of such a quality that it inserts the minimum of noise and distortion into long-distance telephone calls, and enables any telephone subscriber in one country to communicate with any subscriber in another country. In order that this can be achieved economically each country must have a system of telephone exchanges in areas and regions, the main exchange in each region having high-grade telephone circuits to the main international exchange. It is common for the latter circuits to be also planned to be of the same high quality as international circuits.

A typical international connection might follow the pattern in Fig. 9.5, although in large countries there may be further sub-division of areas into smaller areas, and there would be more than two trunk circuits in tandem to reach the main international exchange. The actual telephone line from the subscriber to his local exchange would be two-wire, *i.e.* his telephone would have the microphone and receiver earpiece matched as efficiently as possible via an anti-sidetone device to a both-way circuit. The majority of the trunks between exchanges, and certainly the international circuits, would be four-wire, although at the exchange switching points it is still common practice to unite the two directions into a two-wire circuit for ease of switching.

To ensure satisfactory communication between the two telephones shown in Fig. 9.5 the transmission loss should generally not exceed about 20 dB, with 30 dB as an absolute maximum, the frequency response of the several circuits in tandem should be preferably 300 Hz to 3000 Hz, and the signal-to-noise ratio as perceived at the earpiece should be 20 to 30 dB or higher. The total delay over the circuit should also be within limits such that the talkers are not unduly hindered. With these overall figures in mind, the performance of an international circuit should be such that it provides a high-grade interconnection, leaving the main losses in performance to occur in the extensive local networks.

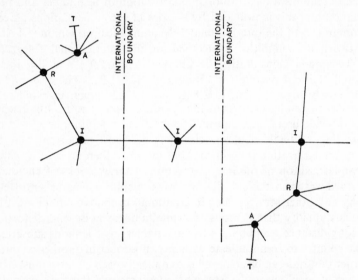

Fig. 9.5. International connection in integrated telephone system.

I = International exchange.
R = Regional exchange.
A = Area exchange.
T = Telephone subscriber.
I to I = International circuit.
R to I = National circuit to region.
A to R = Trunk circuit.

The standard international circuit has been defined by C.C.I.T.T. in their extensive proceedings. All that can be realized here is a rather broad interpretation of some of the salient performance requirements of such a circuit.

(1) Audio bandwidth

The audio bandwidth should be 300 Hz to 3400 Hz. The nature of the cut-off at each limit is not closely specified.

(2) *Group delay*

The present provisional limit for group delay is 0·1 sec. For radio propagation this represents some 25,000 km or 15,000 miles of circuit length allowing for typical delay in terminal equipment. This limit is to enable a number of circuits to be connected in tandem without the delay building up to a figure which would embarass the talkers in a two-way conversation.

(3) *Gain stability*

In principle, if circuits have a loss of 0 dB, any number can be connected in tandem and the overall loss is still zero. However, there must be a tolerance on this loss. If this tolerance is too wide, possibilities for instability or oscillation arise. This tolerance has not yet been specified for an individual circuit, but it should be of the order of $\pm\frac{1}{2}$ dB r.m.s.

(4) *Circuit noise and crosstalk*

If we consider a point at which the mean speech power is 1 mW in 600 ohms the total mean interference power should not exceed 10^{-8} watts, *i.e.* a signal-to-noise ratio not worse than 50 dB. This includes circuit noise and interference due to crosstalk. The latter might arise in the system from small non-linearities in wideband amplifiers carrying signals from many channels, or from magnetic induction or electrostatic unbalance between adjacent circuits in a cable. In a detailed treatment of circuit noise it is permissible to weight the noise spectrum according to its annoyance value, as perceived by the average ear.

(5) *Overload point*

The switching point of the two-wire input point to a 4-wire circuit is usually taken as the reference point for power levels. The impedance at this point, looking into the circuit, is made as near as feasible to a standard of 600 ohms, and, as in the case of circuit noise above, the reference power level is 1 mW. A normal telephone connected at this point would deliver with a normal talker about this average power in continuous speech. The circuit would, however, be tested with a tone of 1000 Hz at a power output level of 1 mW. To ensure that a loud talker did not unduly overload the channel at the peaks of his speech output it is usual to set the overload level some 6 to 10 dB above the 1 mW power level.

Having considered the individual channel we can now turn to the larger units of communication. The use of FDM for blocks of telephone demands is practically universal and standardization is based on this technique. It provides considerable flexibility for branching or dropping off or inserting blocks of channels at points en route. Typical arrangements favoured by C.C.I.T.T. are shown in Fig. 9.6, 12 channels form a Basic Group A and occupy 12 to 60 kHz. A Basic Group B is 12 channels, occupying 60 to 108 kHz. 5 Basic groups (60 channels) form a Super Group 1 between 60 and 300 kHz whilst 5 basic groups between 312 and 552 kHz form a Super Group 2. A block of 5 super groups (300 channels) is the next unit and is called a master group. Above this standardization is not complete but Fig. 9.6(c) and (d) show some typical arrangements.

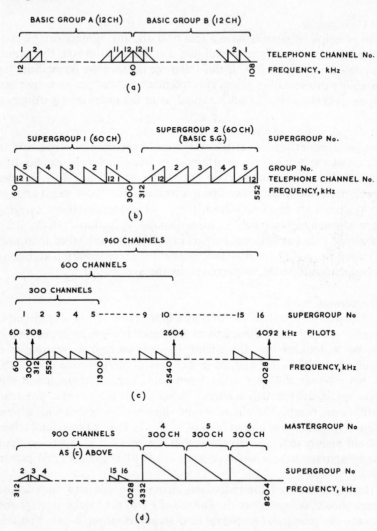

Fig. 9.6. Typical arrangements of f.d.m. telephony channels (a) 12- and 24-channel systems. (b) 60- and 120-channel systems. (c) 300-, 600- and 960-channel systems. (d) 1800-channel system.

The basic telephone channels are assembled as single sideband channels at 4 kHz spacing, with the minimum of guard space and necessitating high-grade sending and receiving filters. The Basic Groups B are assembled at 48 kHz spacing between 60 and 400 kHz with no additional guard spaces. Above this, small guard spaces are introduced to facilitate extracting and introducing supergroups, but remarkably little frequency spectrum is wasted. This enables maximum usage of cables within a given attenuation limit and maximum usage of radio paths.

9.6. Integrated Telegraph Systems

Telegraph channels may be arranged to provide a switched network communication system in a similar manner to the switched telephone network. The channels are connected together by manual or automatic switchboards, the instructions for switching being transmitted over the circuits themselves. The telegraph instruments are of the keyboard type capable of both transmitting and receiving, and channels are normally two-way so that acknowledgment signals can be sent.

An alternative method of operating a telegraph system is the tape relay system. Telegraph messages may be stored on paper tape, by punching holes according to the code across the tape. In the 5-unit code there would be 5 positions across the tape for each character. Such a tape may be fed into a tapereader which senses the holes and transmits the corresponding code over the channel. At the receiving end a re-perforator punches a new message tape. For ease of reading, the message may be overtyped on the tape, so that the address may be easily seen. The operator seeing the address chooses the next channel and feeds the tape into the corresponding tapereader on that channel. In this way a message is routed through the system which is thus known as the tape relay system.

Telegraph channels are usually provided by FDM applied to high-grade telephone channels, but they are also provided over long-distance HF radio. In the latter case both FDM and TDM are employed. In the case of subdivided telephone channels, the international standards usually specify channel frequency spacings of 120 Hz with 420 Hz as the lowest frequency. Thus, for 18 channels the top frequency would be 2040 Hz, and for 24 channels it would be 2760 Hz, both well within the recommended upper limit of a high-grade telephone channel. The method of modulating the tone carrier frequencies with mark and space digits may be AM (*i.e.* on-off) or FM (*i.e.* frequency shift). The signalling code is generally similar to that in Fig. 2.6 and the digit rate usually 50 bauds. The synchronizing system is usually of the start-stop variety described in Chapter 2. In this case the normal idle condition is 'mark' so that on-off modulation systems may derive automatically the correct half-amplitude bias level which acts as the threshold for determining mark or space. The start signal is then a single-space digit followed by the 5-digit code, and then by a $1\frac{1}{2}$-digit mark as the stop signal which allows the receiving mechanism to disengage so as to be ready for the next start signal. There are therefore $7\frac{1}{2}$ digits per character, and at 50 bauds signalling speed the system could send 400 characters per minute or, nominally, 66 words per minute.

Where FDM is used over long-distance radio the same system is employed, but where TDM is employed it is more usual to adopt fully synchronous working. In the latter case, start and stop digits are dispensed with, giving a higher rate of word transmission for a given digit signalling speed. In this system the digit signalling speed would be derived by the flywheel method described in Chapter 2, and indexing (*i.e.* ensuring that the 5 bits of each character are decoded correctly) would be carried out by an initial setting procedure. Up to 4 channels are usual with this form of TDM, although greater numbers have been used, the limit being set by multi-path propagation limits. In some TDM systems the 5-unit code is converted to a 7-unit error-detecting code of the type described in Chapter 7, and error correction by automatic request can also be applied.

9.7. One-way Reversible Circuits

The difficulties of providing two-way radio circuits, using a common aerial and a common frequency, were outlined above. Where the communication problem is of a simple point-to-point form and where the talkers at each end are suitably trained, the one-way reversible circuit may be used. Where there is only one radio frequency allocation and two-way working is required it is the only solution.

The arrangement is shown in Fig. 9.7. Normally the receivers are switched on awaiting an incoming signal, the aerial relay being in the receive position as shown. If station A wishes to make a call he presses a switch on his transmitter and changes the aerial relay to send. He may call the attention of the distant end by a number of means. The simplest means is a loudspeaker, but this has the objection that it may emit noise in the idle condition. If the radio path is stable enough the receiver may have a carrier-operated relay, which responds to the carrier from the distant transmitter and causes a suitable bell or other alarm to be operated. Station B acknowledges by switching on his transmitter and talking, but before he can do this he must observe that Station A has switched off his transmitter. To be able to converse comfortably with such a system the talker must indicate when he has completed each statement, usually by the term 'over', then switch to receive. The switch is usually incorporated in the telephone handset, and with a small amount of training satisfactory point-to-point communication may be achieved. The system does not easily lend itself to tandem connection (*i.e.* the connection of several links in series) owing to the complications involved in the directional switching. However, where the radio path is such that a carrier relay can be operated reliably it is feasible to operate two such links in tandem, by causing the carrier relay to switch on the transmitter of the next link. The two links must of course have different radio frequencies.

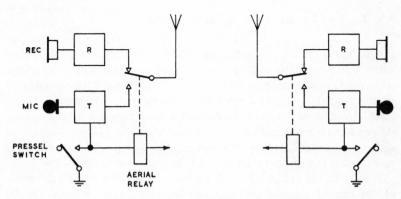

Fig. 9.7. One way reversible radio circuit.

In the case of single links it is possible to employ a voice-operated relay which switches on the transmitter and changes over the aerial relay. Voice signals from the microphone are amplified and rectified to cause the relay to operate. However, false operation of the relay can be caused by extraneous room noises unless extreme precautions are taken.

The system is also applicable to telegraphy. In fact, early single-wire earth return telegraph circuits were operated in a similar manner. If the telegraph instrument is of the keyboard type the send-receive switch is automatically switched to send each time a letter key is depressed.

This method of operating communication circuits is applicable to military and police problems and to communications in sparsely-populated areas.

9.8. The Net or Group System

This is a radio system in which a number of points have a means of one-way reversible communication, each with any other. In this system anyone point can broadcast to all the others at one time. Normally all stations are at receive. If anyone wishes to pass a message he switches to transmit and passes his message. This message will be received by all other stations, but if it is only intended for certain stations this information is included in the message.

This system is particularly suited to telegraphy or telephony between a number of mobile units. It has the advantage that only one radio frequency allocation is required. It does, of course, involve a strict discipline between the users, because only one user must talk or signal at a time. It is usual to nominate the senior member of a net as the control station, and one of his duties is to maintain this discipline.

9.9. The Two-frequency Net or Group System

In this radio system a central transmitter broadcasts to a number of surrounding stations, each having a receiver and transmitter. These stations are usually mobile. The receivers are, of course, tuned to receive signals from the central transmitter at a particular frequency f_1. The mobile transmitters are set to radiate a frequency f_2. At the central station there is a receiver set to receive the frequency f_2. Thus the central station can broadcast to all the surrounding mobile stations. Any one mobile station can pass a message to the central station.

If the receiver output at the central station is connected to the input of the central transmitter, then any message passed from a mobile station is re-broadcast to all other mobile stations. The result is rather similar to the simple net or group system, except that all messages are boosted by the central transmitter. This system is often used for police radio.

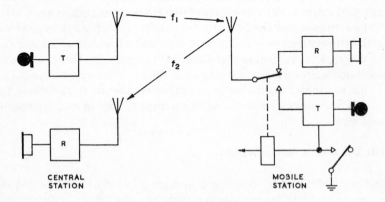

Fig. 9.8. The two frequency net system.

The general arrangement of the system is shown in Fig. 9.8. The central station generally employs aerials of appreciable height and a more powerful transmitter than the mobile stations. Frequencies in the VHF band are used. This type of system is employed for sending control messages to aircraft, and for receiving their position reports.

9.10. Broadcasting Systems

A radio broadcasting system is one in which a central transmitter radiates a signal for reception at a large number of remote points. The information transmitted is generally speech, music or television for entertainment purposes. The transmitter needs to radiate considerable power, and uses an aerial, which radiates uniformly in all horizontal directions. The radiation at high angles of elevation is generally reduced, unless transmission to aircraft is required, and the radiation at angles below the horizontal can also be reduced. The polar diagram of the aerial in the vertical plane can thus be made beam-shaped thus providing aerial gain and increasing signal strength at the distant receivers.

The methods of modulation used in practice are usually AM or FM. For sound transmitters in the LF, MF and HF bands the modulation system is invariably AM with double sideband and carrier in order that receivers may employ simple square law detectors. The congestion in these bands necessitates frequency allocation with the minimum of wasted spectrum, and this precludes the use of modulation systems such as FM which exchange bandwidth for signal-to-noise ratio. Frequency spacing between channels is often as small as 10 kHz for speech and 20 kHz for music. This necessitates receivers of extremely high selectivity, *i.e.* their channel response is a bandpass filter type with steep cut-off to avoid crosstalk from adjacent frequency stations.

High-quality sound broadcasting is usually in the VHF band and FM is often used with a baseband extending up to 15 kHz and a deviation peak of ± 75 kHz. Television broadcasting is invariably in the VHF and UHF bands and AM and FM are used. For stations in the lower VHF band, frequency band economy usually dictates AM and sometimes one sideband is partially removed. This economises in spectrum, but still permits simple detection, as described in Chapter 3.

The baseband channel in a broadcast system may be considered to have its input at the baseband terminals of the studio or transmitter, and its output at the loudspeaker terminals of a typical receiver. In the television case the video terminals at transmitter and receiver are points between which there is a baseband channel. The channel characteristics and channel response theory developed in Chapters 4 and 5 apply equally well to the baseband channel of a broadcast system.

9.11. Telecontrol Systems

The scope of telecontrol was given briefly in Chapter 1 and will now be illustrated by a few examples, using the concepts of transfer functions developed in Chapters 4 and 5. The latter concept was explained mainly in terms of 4-terminal electrical networks and channels, but it can usefully be extended to cover servo-mechanisms and many other processes in which there is a clear relation between output and input. Many such processes are non-linear, but a simple linear treatment can often be applied to give a useful representation of the performance. For example, if the control column of an aeroplane is pulled back the elevators are raised and aerodynamic forces are brought into play which act upon the airframe, causing it to pitch upwards slightly, and to increase the incidence on the main lifting surface so as to induce a climb or upwards movement of the whole machine. Care must be exercised in defining input and output quantities if the concept of transfer function is to be of value. Generally it is necessary to set up differential equations, taking into account the most important factors, which are usually terms involving inertia, elasticity and damping, and from these it is possible to arrive at an approximate transfer characteristic which relates certain output and input quantities when the system is excited by a sinusoidal input.

The type of system considered here is usually one in which the excitation frequencies are very low by comparison with those for electrical communications networks. As a rule it is necessary for the system to respond to zero frequency and small fractions of a cycle per second. Many systems cut-off at nominal frequencies below 1 Hz and most cut-off below 10 Hz. If the system has no connection, either electrical or mechanical, from the output back to the input, it is known as an 'open loop' system. It is characteristic of open loop systems that it is difficult to design the transfer function such that it is invariant with time. Thus, no matter how accurately the input is set, the output will develop an error which will accumulate with time. In the aircraft example, accurate locking of the input control for, say, horizontal flight, produces, over a period, inevitable changes of height. Air disturbances act on the airframe and cause further errors. The latter can be looked upon as a noise-like variable injected at some internal point of the transfer function. To overcome these drift difficulties the output must be sensed in some way. Any error is determined and then fed back to the input control in such a manner as to reduce the output error. Such systems are known as closed loop systems or feedback systems. Regulators and servomechanisms are examples of such systems. In the aircraft example the aircraft attitude may be sensed by a suitable gyroscope and the error obtained as, say, a voltage from a potentiometer.

Then, by suitable amplification, sufficient power may be made available to operate the control so as to reduce the error in attitude. Over a longer period it may be necessary to sense height by an altimeter. These principles, suitably developed, form the basis of the auto-pilot.

To develop these ideas a little further consider a torque amplifying system, in which an output shaft has to position a heavy load torque, so that it agrees closely with an input shaft, which may be driven by a low power electric motor which, in turn, may be controlled remotely over a communication channel. Let θ_i and θ_0 be the positions of the input and output shafts respectively. These quantities have to be imagined as sinusoidally varying angles, having amplitudes and phases, and are to be regarded as functions of the sinusoidal frequency. Then the relation between the output and input shafts under steady-state conditions can be expressed in transfer function form as:

$$\frac{\theta_o}{\theta_i} = A(\omega) \exp[-j\phi(\omega)] \tag{9.1}$$

The transfer function has been written thus, to show that the peak amplitudes of output and input do not necessarily agree and the output sine wave of angular motion lags the input. If this open loop system is ideal the right-hand side of the equation is unity at all frequencies, but this is clearly impossible. The method of torque amplification may be electrical, mechanical or hydraulic, but for an open loop system it is not possible to obtain a one-to-one correspondence between output and input, even at zero frequency. Errors in position of a few per cent are about the best that can be attained, and this is hopelessly inaccurate for some problems, for example, the remote position control (RPC) of a gun-barrel. The rapidity with which the output will respond to a transient at the input, such as a unit step, depends entirely upon the characteristic of A considered as a low pass filter, and this also determines the form of the response, *i.e.* whether overdamped, critically damped or with overshoot. A large delay through the system (*i.e.* a large slope of the phase characteristic $\phi(\omega)$) obviously causes the output transient to be delayed.

If the magnitude of A in the above equation can be increased to a large value or suitably modified, then full feedback of the output position to the input allows the transfer characteristic to be modified and it can be made close to unity, at least so far as low frequencies are concerned. It is assumed that this feedback path is ideal: for example, it may be a rigid light shaft carrying negligible load. A device is used which determines the error between input and output shafts and this error becomes the input to the torque amplifier. This is shown diagrammatically in Fig. 9.9, the circle with signs being the usual symbol for the error or differencing device. Then, if ε is the error,

$$\theta_i - \theta_o = \varepsilon \tag{9.2}$$

$$\frac{\theta_o}{\varepsilon} = A \tag{9.3}$$

Then

$$\theta_i = \varepsilon + \theta_o$$

$$= \theta_o\left(1 + \frac{1}{A}\right)$$

or

$$\frac{\theta_o}{\theta_i} = \frac{A}{1 + A} \tag{9.4}$$

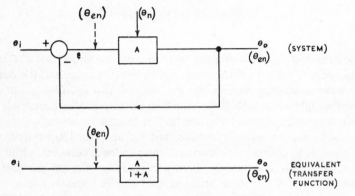

Fig. 9.9. Servomechanism with unity feedback.

As before, the quantities in this equation have to be considered as phasors at any angular frequency ω. Then, if A is large compared with unity, we obtain a close one-to-one correspondence between θ_0 and θ_i. This will generally apply at low frequencies, but at high frequencies A is small compared with unity, so that θ_o/θ_i tends to A. If the open loop function A is of general low pass form then under closed loop conditions the system behaves as a low pass filter, of near unity gain at low frequencies and cutting off in a similar way to A. In the analysis of servomechanisms it is usually necessary to determine transient response, particularly for a unit step input. The procedure is generally to express A and thence $A/(1 + A)$ as a ratio of two polynomials in $j\omega$. The Fourier transform may then be used to determine response to a unit step. The Laplace transform method described in Appendix A provides a useful alternative approach.

In this arrangement, the output is proportional to the error, provided the open loop transfer function A has a constant gain at low frequencies. It is therefore referred to as proportional control. A small error must exist to maintain the output at a value close to the input. Exact one-to-one correspondence between input and output under static conditions is not possible with such a simple open loop function, but can be achieved by placing some form of integrator after the error device. This follows because any small error is integrated to provide an input and, hence, an output, and the system cannot come to rest until the error is reduced to zero. This is known as integral control.

Thus

$$\theta_o = A \times \int \varepsilon \, \mathrm{d}t$$

and

$$\frac{\mathrm{d}\theta_o}{\mathrm{d}t} = A\varepsilon$$

The error is thus proportional to output velocity. With a constant velocity input the output follows at a constant velocity but there is a lag in position. This is then the weakness of integral control.

To overcome this defect second-order integration must be placed in cascade with A. This produces zero velocity lag and the process may be repeated to produce zero acceleration lag. However, it is not easy to achieve these integrations and there are serious problems of stability which have to be dealt with as integration is introduced. An alternative to integration in cascade is to provide suitable modification to the feedback and to provide inner loops.

The modification to equation (9.4) when the feedback is not unity can be deduced as follows:

$$\theta_i - B\theta_o = \varepsilon \tag{9.5}$$

$$\frac{\theta_o}{\varepsilon} = A \tag{9.6}$$

$$\theta_i = \varepsilon + B\theta_o$$

$$= \frac{\theta_o}{A} + B\theta_o$$

or

$$\frac{\theta_o}{\theta_i} = \frac{A}{1 + AB} \tag{9.7}$$

where B is the feedback transfer function as shown in Fig. 9.10. It should be noticed that the open loop transfer function is AB. If B is of the form $1+C$ then Fig. 9.10 can be replaced by Fig. 9.11(a) and this can be converted to Fig. 9.11(b) which is a unity feedback servo in which the forward transfer function can be modified by the factor C.

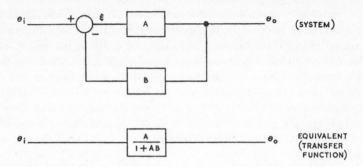

Fig. 9.10. Servomechanism with transfer function in feedback path.

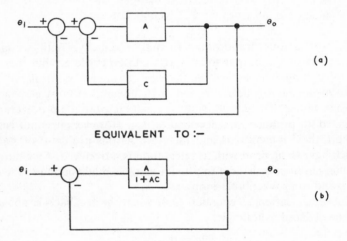

Fig. 9.11. System with inner loop.

Shaping of the open loop function by placing transfer function blocks in cascade with A, or by modifying the feedback, is a comprehensive study in itself and involves consideration of stability. If there is a frequency at which the open loop phase shift is 180° and the magnitude is unity, then the system will oscillate. For stability there must be a margin of phase below 180° when the magnitude is unity and a margin of magnitude below unity when the phase shift is 180°.

The advantages of closed loop systems may be summarized briefly as:

(a) Accurate correspondence between input and output under static and dynamic conditions.

(b) If the open loop has harmonic distortion this will be reduced under closed loop operation.

(c) If there are disturbances or noise acting on the load and affecting the output under open loop conditions, these will be reduced under closed loop conditions.

The last two follow because spurious movement at the output will be fed back in anti-phase, and after passing through the amplifying system will tend to cancel the disturbance. Referring to Fig. 9.9 we may consider a disturbing torque at the output which, with the loop open, produces a certain angular noise displacement dependent upon the load and the power stage of the torque amplifier. We can focus attention on a small segment of the noise spectrum at a certain angular frequency. If T_n is the noise torque and θ_n the angular noise it produces with the loop open then the ratio θ_n/T_n defines a type of source admittance. Closing the feedback loop reduces θ_n, and thus the effective source admittance. With the loop closed, the angular error noise θ_{en} will take up a value which satisfies the following equation:

$$\theta_{en} \times A + \theta_n = -\theta_{en}$$

or

$$\theta_{en} = \frac{-\theta_n}{1 + A}$$

Since the feedback ratio is assumed to be unity the output noise is equal to θ_{en} and it is, therefore, less than θ_n. The ratio by which the disturbance is reduced is $1/(1 + A)$ or since the disturbing torque remains unaltered the source admittance has been reduced by the same factor.

This necessarily brief introduction is not rigorous, but it has been included because of its close similarity to the problems of telecommunications, and because it illustrates transfer function theory. For a fuller treatment the reader is referred to the many excellent textbooks on servo-mechanism theory.

The problems of telecontrol are closely related to the study of servo-mechanisms. It is not economical to connect an accurate telecontrol channel to an inaccurate control amplifier system or vice versa. On the other hand, there may be a case, in some problems, for having a relatively inaccurate forward control system (i.e. telecontrol channel and control amplifier) and an overall feedback link such as a telemetry channel, to monitor the output, and enable the error to be deduced at the main control point. In such instances the transfer function of the feedback path, particularly its delay, must be allowed for, as it may influence the stability of the system or its tendency to oscillate or hunt. In many telecontrol problems there is a case for a telemetry or check-back link purely to prove the correct operation of the system. Generally, the actual process being controlled will have its own feedback loop and Fig. 9.12(a) is therefore typical of many problems. The loop may be converted to Fig. 9.12(b) in which A is the forward transfer function and B the feedback transfer function. It should be noted from equation (9.7) that when A is large and B is approximately equal to unity, that the overall accuracy is dependent upon $1/B$. Thus the feedback portion of the main loop may be required to be very accurate or invariant. The problems are thus transferred from the forward control channel to the feedback channel, *i.e.* the telemetry channel or the method of observing the output remotely.

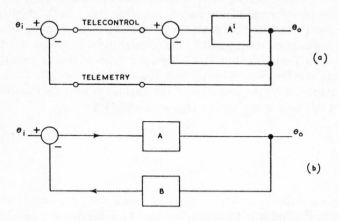

Fig. 9.12. Closed loop telecontrol.

Returning to the example of the height control of an aircraft (or guided missile), consider the control problem in a vertical plane which passes through the control point and assume the aircraft flies in this plane. Then there will usually be an elevator servocontrol system and an autopilot feedback system to maintain a steady attitude. The block diagram would be similar to Fig. 9.13, in which height is chosen as output quantity. There are transfer functions representing the elevator fin servo, the dynamics and aerodynamics of the aircraft. Suppose now that it is desired to remote-control the aircraft from the ground. A forward radio control link would be necessary. This link should preferably transmit a baseband signal of variable amplitude, symmetrical about zero, the output controlling the elevator control system. The baseband frequency response of the link should exceed the control responses but this would not be difficult over a radio path. So far this is open loop telecontrol and under completely blind conditions it would be impossible to achieve any particular height. Closing the loop visually would represent an improvement, but a height-finding radar would enable an absolute height measurement to be made and the loop could be closed, so that any height set into the input could be automatically maintained. In this case the accuracy of height maintained would depend mainly upon the radar which merely senses the output of the whole process. In this example the method of modulation of the radio carrier could be direct FM, or the baseband could be applied to deviate a tone frequency which then modulates the RF carrier by AM or FM. Alternatively, the control link could be combined with the radar which tracks the aircraft and measures position. In this event the baseband could be applied to deviate the radar pulse recurrence rate about a certain value which represents the zero point. Since the system is closed loop telecontrol the accuracy and stability demanded of the forward link is not high. A suitable transfer function unit would be incorporated on the

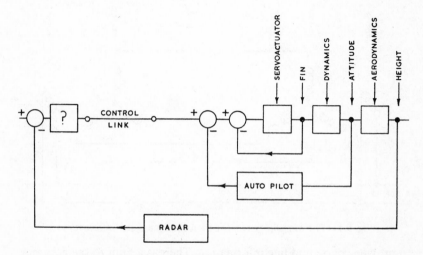

Fig. 9.13. Example of closed loop telecontrol for height control of aircraft.

ground to stabilize the system, as shown in the figure by the box with a question mark.

As a second example, consider the remote control of a large radio aerial for receiving signals from a satellite. The aerial could be under computer control from a large distance. High accuracy would be required. As a closed loop system the feedback telemetry would need to be of high precision. Apart from the advantage of a check-back facility, economy might dictate the use of a high-precision forward link, *i.e.* open loop control. The aerial itself would of course have its own closed loop servo systems. Such a telecontrol problem over distances of a few hundred yards would normally be solved by the magslip or selsyn type of system.

In this remote-control system the transmitter and receiver are similar, 3 winding stators with single winding two-pole rotors, connected as

shown in Fig. 9.14. Alternating current at, say, 50 Hz, is fed to the
stators. With a misalignment, currents will flow in the 3-winding system
and will create an AC field in the receiving stator, having its axis in
the same direction as the transmitter rotor. If the receiving rotor is
now released it will rotate so as to oppose this field, *i.e.* it will line up
with the transmitter. The receiver is therefore error actuated. The
method of transmitting the angular information is by variation of the
relative amplitudes (positive and negative) of the 50 Hz carrier current
which flows in the 3-winding system. These devices have been made to
have accuracies of a small part of one degree, and higher accuracies
have been achieved by having two systems arranged with gearing so as

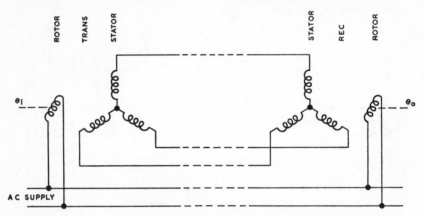

Fig. 9.14. The Magslip system for transmitting angular information.

to provide coarse and fine information. There is a limit to the accuracy
of such arrangements, and the best system for achieving the ultimate in
accuracy is the binary digital system, essentially a form of pulse code
modulation, as described in Chapter 7.

In this system a binary coded disc is attached to the controlled member,
in this example the azimuth (or elevation) shaft of the aerial. This
minimizes errors due to the main drive gearing. The coded disc may be
a precision device in which one rotation is divided into 2^n parts, where
n is the number of digits in the binary code. If $n = 14$ the least significant digit would represent 1 part in 16,384 of a revolution, or very
approximately 1 minute of arc. Such a disc would have 14 circular
tracks, the inner track having the most significant digit and the outer
track the least significant digit. The tracks could be read optically with
photocell pick-ups and the 14 digits representing the angular position
could be passed to a digital store. The computer at the sending end of the
communication link would send 14 digits representing the desired azimuth position of the aerial. At the receiving end these would be passed

to another store or statisizer. The digits representing the present aerial position would then be subtracted from the incoming digits to give the error. The error as a small binary number would then be converted to an analogue quantity (positive or negative) for amplification, and would act as error input to the main servo motor which positions the aerial. Fig. 9.15 illustrates the general principle. This is really a pulse-type servo because the error is only available at discrete sampling instants. The theory of such servos is rather specialized.

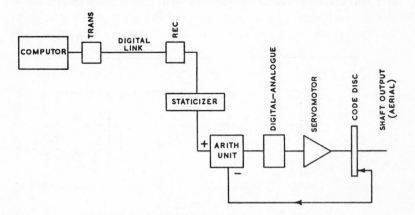

Fig. 9.15. Telecontrol by accurate digital method.

9.12. Telemetry Systems between Fixed Locations

The simplest type of telemetry system operates between fixed locations and the information is normally carried over lines, cables, or channels in fixed radio links. There is a wide variety of signals to be dealt with, including simple on-off indications, meter readings, transducer outputs, and, sometimes, transmission of precision quantities. Multi-channel operation is often required, particularly where many meter readings have to be sent. These signals generally involve very low baseband frequencies, relative to available channel bandwidth, and the main problems are those associated with accuracy. It is usually required that the accuracy is better than 1% of full scale and often 0·1% is necessary. Over short distances with physical wire lines it is possible to provide these facilities by d.c. signalling methods. Fig. 9.16 shows a typical method in which there are sending and receiving potentiometers of high accuracy arranged in a Wheatstone bridge type of circuit. The middle wire carries the error signal which may be picked up at the receiving end. The error-sensing device may be a polarized or galvanometer relay, or a d.c. amplifier. Alternatively it is possible to use a chopper amplifier, that is, a relay or solid-state device which converts the error signal to a.c. for easy amplification. In all cases the sense of the error must be obtained and used to drive a motor, which drives the receiving potentiometer so as to null the error. The motor may drive a pen recorder for continuous recording of the transmitted quantity which is, of course, set into the sending potentiometer. The device is clearly a form of servomechanism so far as the receiving end is concerned, and it forms an open telecontrol system. It shows in fact that there is little difference

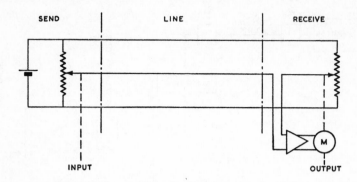

Fig. 9.16. D.C. telemeter system with recording meter.

between open loop telecontrol and telemetry except with regard to the amount of closed loop power amplification at the receiving end. Accuracy is mainly dependent upon the potentiometers. The line resistance must be small compared with the potentiometer resistance, but otherwise battery supply and line resistance variations do not prejudice accuracy to any serious degree. A variety of servo-operated recording instruments is available.

Where the distances are greater, the communications channels invariably have to conform to telephone practice and it is necessary to devise methods of accurate signalling, using tones. Methods using amplitude must generally be rejected and the most common methods are frequency modulation (FM) and pulse frequency modulation (PFM). In PFM the quantity to be transmitted is represented by the rate of transmitting pulses, each pulse being a fixed length burst of tone. To all intents and purposes it can be considered as FM, frequency in FM being equated to pulse rate in PFM. It is usually necessary to convey the steady value of meter readings and this necessitates a form of voltage-controlled oscillator at the sending end, and a counter type of frequency-measuring device or other frequency discriminator at the receiving end. Several types of electronic circuit for converting voltage to frequency are available, but it is not always simple to maintain an accurate relationship between these two quantities. If an accurate frequency discriminator is available it is possible to use a feedback circuit at the sending end which incorporates the same type of discriminator as at the receiving end. Fig. 9.17 shows the principle where VCO is a voltage-controlled oscillator. If the forward gain in the feedback loop is high, then the error signal is small and the two discriminators will be giving output voltages closely matching the input voltage. The same principle is sometimes employed for making an accurate analogue to digital encoder when only an accurate decoder is available. In this case the VCO would be replaced by the encoder which could be a reversible binary counter and the discriminator by the accurate decoder.

Reverting to the FM discriminator, this must be made insensitive to amplitude changes and in the case of PFM also insensitive to pulse length. If there are circuits for accurately maintaining these quantities, then conversion from frequency to voltage is a simple question of filtering or averaging. Providing the frequency to be measured is high compared with baseband quantity, a simple moving coil-meter will carry out the conversion since it reads average current. More elaborate discriminators involve counting techniques. If the quantity to be measured only varies slowly compared with a period of 1 second, then a simple count of pulses during every second gives the quantity required to a good order of accuracy if the frequency is of the order of 1000 Hz. The source of clock pulses must of course be accurate to a higher degree.

The transmission of simple on-off telemeter signals presents no problems and it may be considered as a case of very low speed telegraphy. Such signalling systems are often required to 'fail safe'. That is, a failure of the telemetry channel should give an indication. If the channel is of the double current or FM type then failure of signals represents a third state which enables the alarm to be given.

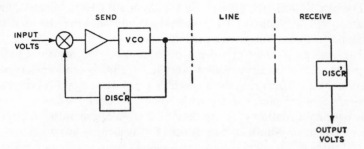

Fig. 9.17. FM or PFM telemeter system using feedback type of sending frequency control.

9.13. Telemetry from a Moving Location

The transmission of data from moving vehicles such as aircraft, rockets, road vehicles, ships, etc., is a valuable method of monitoring their performance on the ground under laboratory conditions. Such techniques of radio telemetry are very necessary, especially during the development phase of modern vehicles. Typical quantities which have to be telemetered are accelerations, structural strains, fuel rates, pressures, angular positions of control surfaces, voltages, currents, etc. The baseband frequencies are generally low, but many channels are required, and accuracies of 1 per cent or better are demanded. Ideally, the telemetry system should be flexible so that it can accept a wide variety of input devices, *i.e.* transducers or pick-ups.

The transducers used employ a wide variety of principles such as:

(*a*) Resistance variation
(*b*) Inductance variation
(*c*) Capacitance variation
(*d*) Voltage variation
(*e*) Frequency variation.

The commonest resistance variation pick-up is the strain gauge for measuring structural strains. It consists of a wire element cemented to the structure in such a way that strain stretches the element and changes its resistance. The change is small and bridge circuits are used to improve stability; a.c. input to the bridge is often used so that the weak out-of-balance voltage due to strain can be easily amplified.

The inductance variation transducer is suited to acceleration measurement, and consists of a spring-mounted mass controlling the air gap in the magnetic circuit of a solenoid. Acceleration then varies the air gap and thence the inductance, and this may be arranged to control the frequency of an oscillator. This frequency may be telemetered or it may then be converted to a voltage for telemetry in the ordinary way.

The capacitance pick-up is rather similar and is a form of capacitance microphone. It is suited for measuring pressures and may be used in a similar way to the inductance pick-up.

The voltage pick-up may be a potentiometer for measuring shaft position or other relative motion. Very low torque potentiometers have been made for measuring the angle of a gyroscope cage. For vibration measurement the piezo-electric pick-up may be used.

Frequency variation pick-ups may be small a.c. generators for speed measurement or vibrating wire devices for measuring gross strain.

The study of these devices to obtain good transducing characteristics is beyond the present scope, but they can be treated by the sinusoidal excitation and transfer characteristic method of Chapter 4.

The accuracy requirement generally necessitates that the information baseband (*e.g.* the variation of measured quantity with time) shall be transferred to FM at a tone or low-carrier frequency. This is also desirable because the change of range and therefore radio-carrier signal strength must not affect baseband accuracy. This then favours FDM methods of assembling channels. The multichannel signal may then be applied to the radio carrier, by AM or FM. The former are called FM/AM systems and the latter FM/FM. In either case the linearity of the final modulator must be high to avoid intermodulation and this generally leads to a preference for FM/FM. It also leads to a limit of about 10 in the number of channels which can be handled in this way. However, the channels, of very low rate of variation, can be assembled by TDM or connected to a suitable commutator, before being applied to the first FM stage, thus increasing the capacity of the system by a large factor. Such a system of FM/FM type is illustrated in Fig. 9.18

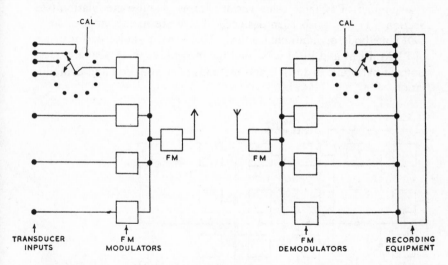

Fig. 9.18. General arrangement of mixed TDM/FDM system with FM/FM modulation.

and is largely self-explanatory after the treatment of Chapter 4. The received information may be recorded on magnetic tape at the subcarrier FM stage, or on various types of pen recorder or recording cathode-ray oscillograph. Alternatively it may be converted to PCM and recorded on magnetic tape for subsequent analysis.

Calibration voltages may be connected to separate channels, and in the TDM section some form of synchronizing and channel indexing is necessary. The latter may be a voltage outside the normal range of the input quantities.

In some systems the multi-chanelling may be entirely TDM, in which case Fig. 9.18 is read with the separate individual FM input modulators omitted, and replaced by a TDM commutator suitably geared to the sub-channel TDM commutator. In such a system the linearity requirement on the final radio modulator is removed and FM/AM or FM/FM may be used with no restriction on the number of channels, except those dictated by complexity of synchronizing and indexing.

A system specially suited for strain gauge work is shown in Fig. 9.19 and it employs mixed FDM and TDM. The strain gauges are fed with 5 different tone frequencies. The commutator samples 5 groups, each of 5 strain gauge outputs. The output should then preferably frequency modulate the radio transmitter to maintain amplitude stability through the system. This is then an AM/FM system with 25 channels. A sixth channel carrying a sixth tone provides a convenient method of synchronizing.

A number of systems using various forms of pulse modulation (see Section 3.1) have also been designed. These are mainly applicable to TDM methods of multi-channelling. The signal-to-noise performance of PDM, PFM and PPM is beyond the present scope, as also are questions of harmonic distortion and methods of modulation and demodulation.

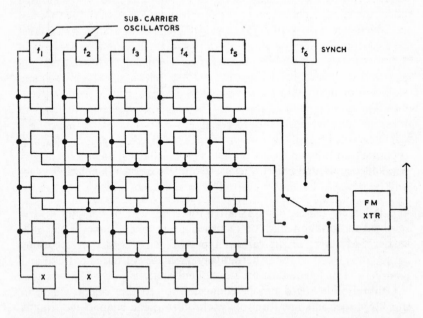

Fig. 9.19. 25-channel mixed FDM/TDM system with AM/FM modulation. X = strain gauge bridges fed by oscillators; unbalance output to TDM switch.

In most radio telemetry systems the transmitting end must be designed for small size and low weight and a satisfactory compromise must be achieved in the matter of performance. If accuracy is specified as, say, 1 per cent of full scale in any channel, then the inaccuracy must be shared between several parts of the system from transducer to recording device. The part of the inaccuracy which is allocated to transmission noise may be quite small, so that signal-to-noise ratio for each baseband may need to be high, and that for the main common channel will need to be exceedingly good. Questions of signal-to-noise ratio and crosstalk in multichannel systems were treated in Chapter 4.

9.14. Satellite Communication

A long distance method of transmission which is being widely used for providing high grade international multi-channel communications is radio relay via an earth satellite. Several system variants are possible but the simplest to consider is the synchronous satellite in which the satellite relay is placed in an equatorial orbit with a 24 hour period, so that if the direction in orbit is the same as that of the earth's rotation, it will appear to be stationary above a point on the equator. The radius of this orbit is 6.61 earth radii and the point on the equator should be at a longitude approximately halfway between the longitude of the stations requiring communication. The satellite must be spin-stabilised about an axis at right angles to the equatorial plane, and the aerial must be attached to a de-spun part of the satellite. The de-spinning rate must be such that the aerial is pointing to the earth. The frequencies usually employed are about 6 GHz for the earth to satellite direction of transmission and about 4 GHz for the satellite to earth direction. Given adequate stabilization the satellite may employ modest aerial gain at these frequencies. The ground stations consist of tracking dish aerials of 10 metres or more in diameter according to the number of channels required. The channels will generally be two way so that each ground station will have separate transmit and receive aerials. The satellite will however generally have one aerial used both for transmit and receive. The satellite repeater will be powered by energy derived from solar cells which collect radiation from the sun and convert it to electricity for operating the amplifiers for the two directions of transmission. The modulation method usually employed is FM and multi-channelling methods for TV, Telephony and Telegraphy are generally of the FDM type discussed in sections 9.5 and 9.6.

9.15. Wide-band Links

One of the advantages of PCM mentioned in Chapter 7 is the possibility of using the same communication link to handle all types of signals in digital form. The capacities of such links are most conveniently defined in terms of bits per second and from the capacity so defined it is a simple matter to derive the number of telephony, TV and data channels which can be handled simultaneously. The requirements for telephony and TV are respectively 56 k bit/s and 120 M bit/s while in the case of a data channel the bit rate is specified immediately by the volume of data to be transmitted. Videophones, currently under development, require 6 to 8 M bit/s, but removal of redundancy may reduce these figures.

The options currently available for wide-band links are coaxial cables, microwave line-of-sight systems and satellite systems. In each case, technical developments have led to a steady increase in the capacity of a single installation. Such an increase is desirable for two reasons. Firstly, the cost of providing communication at a given rate, *e.g.* the 56 k bit/s needed for a single telephone channel, decreases as the total system capacity increases. It is obvious that this should be the case, since many of the costs, both capital (such as digging trenches in which to accommodate cables) and recurrent (such as maintenance) are virtually independent of the system capacity. Secondly, a study of the statistics of the traffic handled by a system shows that the larger the system the more nearly it can be operated to full capacity without an intolerably high probability of becoming overloaded.

Coaxial cable systems with solid state regenerators at spacings of 2 km are capable of operating at rates up to 500 M bit/s. Microwave relays in the frequency bands around 2, 4 and 6 GHz carry up to 1800 telephone channels in present-day systems using FM. Developments in progress will lead to new systems in the 11 and 20 GHz bands, the higher frequencies being necessary to overcome the congestion which already exists in the radio spectrum but also offering digital rates of up to 500 M bit/s per microwave carrier. Repeater spacings are likely to be up to 30 km at 11 GHz and up to 10 km at 20 GHz. The possibility of using even higher frequencies is being intensively investigated, particular attention being given to methods, such as route diversity, of overcoming the high attenuation to which such frequencies are subject during rain.

Despite these advances in well-established systems, the growth of demand for communication facilities may require new techniques with even greater capacities. One such technique, low-loss waveguide, has already been demonstrated to be effective although it has not yet been used commercially. Low-loss waveguide exploits the behaviour of a particular mode of propagation, designated the TEO1 mode, in circular waveguide. Such a mode has the very unusual property that the attenuation decreases as the frequency increases. A guide of diameter 50 mm allows waves of frequencies within the band 30 to 100 GHz to propagate with attenuation less than 3 dB/km. Regenerators at intervals of up to 20 km maintain the signal level at an acceptable level. It is expected that such waveguides will be operated with carriers at about 1 GHz separation, each carrier being able to transmit information at 500 M bit/s. The potential capacity of the waveguide is thus over 35 G bit/s, equivalent to 300 TV channels or 600,000 telephone channels. Although a waveguide link of this kind is very expensive, largely because of the very tight tolerances which must be maintained on the straightness of the guide, the cost per channel is less than that of any existing system. Obviously, such guides will only be used on routes where the traffic density is sufficiently great to utilize a reasonable fraction of the available capacity.

Wideband systems inevitably require the use of large frequencies and it is natural to look towards the optical region of the electromagnetic spectrum where the available bandwidth is enormous in relation to that in the radio part of the spectrum. The most promising line of research at present is the investigation of glass fibres as 'waveguides' for optical frequencies. The guiding action is obtained by using a fibre with a central core in which the refractive index is a little larger than the remainder of the fibre—the outer cladding. The overall diameter, including the cladding, is less than 0.1 mm and cables containing 50 fibres are comparable in size to existing coaxial cables. Attenuation in a fibre of less than 20 dB/km has been measured under laboratory conditions. The capacity is limited by the dispersion of the light waves to about 50 M bit/s per fibre. Further development of this optical system will make it very strongly competitive with all existing systems and it is likely to play a major part in future communication networks.

Chapter 10

The Principles of Position Fixing by Radio

10.1. Introduction

Energy transmitted by the propagation of radio waves may be used to determine the position of moving stations, such as ships and aircraft, with respect to the position of other stations, generally fixed ground stations. The fundamental principles used are the measurement of the time delay of propagation from which range may be directly inferred, and the measurement of direction by locating planes of constant phase in a radio wave. All systems of position fixing use one or the other, or both these principles. The second principle was the first to be exploited for navigation purposes in the early forms of direction finding (DF) aerial. Such an aerial, in the form of a plane loop of wire, can be rotated until the signal in the receiver is nulled. The direction of arrival of the signal is then normal to the plane of the loop.

The measurement of range by time delay followed later, and the method first proposed used transmitters and receivers at each end of the range. A modulation applied to one transmitter was received at the distant receiver and then fed back into the distant transmitter and then to the local receiver. Suitable comparison of the transmitted and round trip received signal allowed the time delay and hence the range to be deduced. Allowance must of course be made for the known delays in the transmitting and receiving equipment. More recently it has become possible to transmit a radio signal and obtain reflections or echoes from objects such as aircraft and ships. The received echo can be compared with the transmitted signal and the time delay and range can be deduced. If this arrangement is combined with the DF principle, position fixing in two co-ordinates, range R and bearing θ can be established. This is known as radio direction and ranging or radar.

The earlier method involving a distant transmitter-receiver or 'transponder' as distance measuring equipment is sometimes called (DME) or, in some connections, it is known as secondary radar.

The principles of communications may be easily extended to cover systems of navigation and radar. All such systems involve transmitters and receivers, and the same relationships exist between the form of the signal and the channel transfer characteristics as those established in previous chapters. The problems of signal-to-noise ratio are very similar to those for the communication of intelligence. In this chapter the information is the positional co-ordinates which are to be determined by radio methods, and attention will be confined mainly to an elaboration of the above two basic principles.

10.2. Range Measurement

10.2.1. THE CLOCK METHOD OF RANGE MEASUREMENT

If we assume the existence of clocks of absolute accuracy, then the known velocity of propagation of radio waves enables range to be measured. Thus, in Fig. 10.1 there are two such clocks, A and B, and clock A can send a radio signal at a known instant of time. When received at B the time of receipt can be compared with the local clock and the time of difference t can be measured. Range R can then be deduced from

$$R = ct \tag{10.1}$$

where c is the known velocity of light or radio waves, or approximately 3×10^8 metres per second. This is roughly 186,000 statute miles per second or about 1000 ft/μsec.

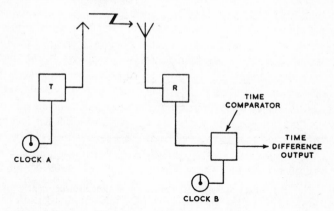

Fig. 10.1. Idealised method for measuring range using absolute clocks.

For measuring distances of the order of tens of miles, it is clear that such clocks must have exceptional absolute accuracy. Thus if one clock is in error by, say, 10 microseconds, the range measurement will be in error by about 2 miles. Even supposing that the need for absolute accuracy can be removed by employing some clock-setting procedure, then the need remains for the clocks to have exceptional stability or speed accuracy. Thus, if one clock gains say, 10 microseconds in 10 hours it will be producing a 2-mile range error in a significant navigation period. This is a stability of the order of 3 parts in 10^{10}. Such stabilities are feasible by using so-called atomic clocks, although they are not yet practicable for navigational use. Nevertheless, the method is of importance for describing the principles of range measurement.

10.2.2. RANGE ACCURACY

Quite apart from instrumental errors, the noise received with the signal will cause a certain inaccuracy of range measurement. So far the form of signal for range measurement has not been discussed, but it is clear that the pulse type of signal has much to commend it. The continuous wave signal in which several wavelengths exist in the space over which distance is to be measured, cannot be used unambiguously, because there is no way of identifying any particular wave.

We assume a rectangular pulse shape of envelope modulation on a fixed frequency carrier, although for individual applications better forms of signal may exist. It has been established in Chapter 8 that some form of receiving filter is necessary, and this involves a roughly sloping

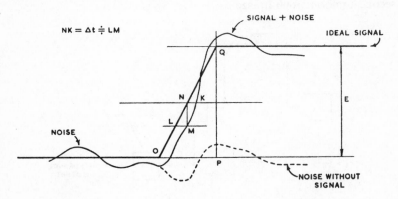

Fig. 10.2. Effect of noise upon accuracy of time measurement.

build-up of envelope as in Chapter 5. The criterion of measurement of time, and hence of range, may be taken as the time at which the signal build up crosses the half signal level as in Fig. 10.2. At this point a very short pulse may be generated and its time compared with the local clock.

Any noise with the signal will cause time displacement of this crossing point, as in Fig. 10.2. An approximate value for this time displacement may be deduced for the case of a baseband pulse edge.

Let E be the peak value of the pulse, $B/2$ be the cut-off frequency of the receiving filter, and v_n be the instantaneous value of the noise signal, at the instant when the half signal level is crossed. We will suppose that v_n is small compared with E.

In Fig. 10.2 the triangles OPQ and LMN are similar so that:

$$\frac{LM}{MN} = \frac{OP}{PQ}$$

Also, OP is the pulse build-up time, $1/B$

PQ is the peak value of the pulse, E,

LM $\doteqdot$ NK is the time error Δt caused by the noise signal

and MN is the noise voltage, v_n.

Hence,

$$\frac{\Delta t}{v_n} = \frac{1}{BE}$$

i.e.,

$$\Delta t = \frac{v_n}{BE} \qquad (10.2)$$

We cannot make any statement regarding the value of v_n on a particular occasion, but if a large number of occasions are considered, v_n will have a well-defined probability distribution. From the results of Chapter 6 we may assume that v_n has a Gaussian distribution with a mean value of zero and a mean square value given by:

$$\frac{\overline{v_n^2}}{R} = kTB/2$$

where T is the effective noise temperature of the system, and R is the resistance across which the voltages are measured.

Since Δt is directly proportional to v_n we see that it must also have a Gaussian distribution with zero mean and a mean square value:

$$\overline{(\Delta t)^2} = \frac{1}{(BE)^2} \cdot \overline{v_n^2}$$

$$= \frac{kT}{2B(E^2/R)} \qquad (10.3)$$

The standard deviation of the time error is therefore

$$\sigma_t = \sqrt{\frac{kT}{2BP_s}} \qquad (10.4)$$

where $P_s = E^2/R$ is the signal power in the load at the point of measurement.

The latter expression shows that increasing bandwidth improves accuracy. This process can only continue up to the point where false alarm and missed target probabilities as described in Chapter 6, remain acceptable.

10.2.3. RANGE MEASUREMENT BY RADAR

Range can be measured to a high order of accuracy by radar without the need for high stability clocks. If a pulse-type signal is used the transmitted pulse can start some type of time measuring circuit, which can be stopped by the receipt of the echo pulse. The circuits may be arranged to use the leading edges of both pulses. The time-measuring circuit may be a crystal oscillator whose cycles may be counted over the duration between transmitted and received pulses. If the accuracy of the oscillator is, say, 1 part in 10^6, then range error from this cause will be very small. If the time delay is t, then range is given by

$$R = ct/2 \qquad\qquad (10.5)$$

the factor of $\frac{1}{2}$ being necessary because the signal has covered the range twice.

The choice of recurrence rate of the transmitted pulses, *i.e.* the sampling rate of the range, is limited by the maximum range to be measured. If range ambiguity is to be avoided, the minimum time between successive transmitter pulses must exceed the time delay associated with the maximum range. This leads to the restriction that the pulse repetition frequency (p.r.f.) should be less than $c/2R_{max}$ pulses per second, where R_{max} is the maximum range to be measured.

10.2.4. RANGE RESOLUTION OF RADAR

Range resolution of discrimination is the ability of a radar system to detect as separate entities two objects separated in range. The range resolution may be defined as the minimum value of the separation between two objects, such that it is just possible to detect or recognize them as separate entities. Strictly speaking, the concept of probability of separate detection should be introduced into the definition, but the following simplified approach is usually employed.

Consider a radar with a rectangular pulse of fixed carrier frequency. If the pulse duration is τ sec the pulse length in space will be $c\tau$. It is assumed that each of the two objects to be resolved is small compared with $c\tau$. If the two objects are at the same range, it will not be possible to detect from the received signal the existence of two objects, excepting by signal strength and this will not in general be a reliable method. As the objects are separated in range, the received pulse will extend beyond the length $c\tau$, and there will be interference between the RF waves of the two separate echoes over the section where overlap occurs. When the two objects are separated in range by a distance just in excess of $c\tau/2$ the lagging pulse will have travelled twice the separation and a gap will just appear between the two pulses. It is usual to define the resolution of a pulse length τ on this argument, as a distance $c\tau/2$. In theory it should be possible to achieve a better resolution than this, because the known length of transmitted pulse can be compared with the received pulse. Any extension in length would indicate two or more targets or, perhaps, an extended target. However, as we have seen, the rectangular pulse requires infinite bandwidth, and any elaboration of the above theory is not justified.

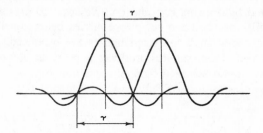

Fig. 10.3. Received pair of $(\sin x)/x$ pulses which can be easily resolved.

If the pulse is assumed to be of $(\sin x)/x$ shape the resolution may be defined as the spacing which would cause the two received pulses to be separated by a distance equal to the distance between the first two zeros. This is shown in Fig. 10.3. If the pulse duration between zeros is τ secs the resolution is again $c\tau/2$. Again it is possible in theory to achieve better resolution by comparing with the transmitted pulse.

If the transmitted pulse varies in frequency over its length, and the two received pulses are overlapping, a more complicated situation results. A more sophisticated approach to the provision of good resolution has been discussed in Section 8.7.

10.2.5. RANGE METHODS OF POSITION FIXING

The simplest of the position-fixing systems, using only range measurements, is the range–range method. For example, an aircraft carrying a DME transmitter-receiver can measure range to two known ground transponder stations. This clearly fixes the aircraft position, apart from the uncertainty as to which side of the baseline between ground stations the aircraft lies upon. This uncertainty can be resolved by measuring range to a third station.

A second system of position fixing, using the range principle, measures the difference in the distance of an object from two ground stations separated by a baseline. Consider two ground stations A and B emitting pulses on different frequencies at the same instant of time, so that a stationary observer midway between A and B receives pulses simultaneously from A and B. Suppose an object P, such as an aircraft or ship anywhere in the coverage receives these two pulses and measures the time difference between their instants of arrival. This difference can be converted to distance and the object can thus measure the distance $(AP - BP)$. The locus of a point of constant distance difference from A and B is a hyperbola with foci at A and B. The system of confocal hyperbolae for varying range differences divides the area covered by transmitters A and B into lanes as in Fig. 10.4. These hyperbolae may be constructed by drawing concentric circles with centres at A and B, the range difference between successive circles being constant. Where the circles intersect small foursided figures are produced. Joining the opposite corners of these figures produces hyperbolae and ellipses. The hyperbolae cut the baseline into equal segments.

The method just described only permits P to find out in which lane

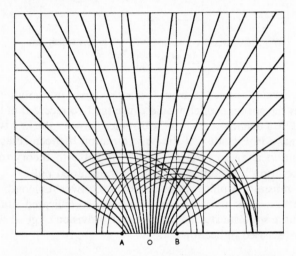

Fig. 10.4. Construction of system of confocal hyperbolae
with foci at A and B.

it is located. To obtain an unambiguous position-fix a third pulse-transmitting station C is required, preferably with its baseline BC approximately at right angles to AB. This enables the range difference $(BP - CP)$ to be measured, and sets up a second system of hyperbolic lanes, enabling P to fix itself in a two-co-ordinate hyperbolic system.

In practice, absolute synchronization of pulse emissions is difficult to achieve and it is usual to make the middle station, B, the Master, and A and C slave stations. The pulse transmitted by B is received at A and C, triggering the pulses from those stations. This merely adds constant known distances to the range differences being measured, and can be allowed for.

The range-difference system can be operated, using either pulse or continuous wave signals. If CW signals are used, the three transmitters must operate on different but related frequencies, to enable the receiver to distinguish between the three incoming signals. For a description of the principle we may consider the frequencies to be identical and all emitted in phase in the sense of absolute simultaneity. Then lines of constant phase difference for signals received from A and B will be hyperbolae. If there are several wavelengths in the distance AB then there will be many hyperbolae representing, say, zero phase difference. If the object P is measuring zero phase difference it cannot determine on which hyperbola it is lying. However, when P starts from a known position with known phase difference, it can follow an hyperbola by maintaining that phase difference. Any deviation from the hyperbola causes a deviation of the phase difference, and this can be measured by a suitable phase-meter so that information may be derived to indicate the passage from one hyperbola to another. This process may be carried out for the two systems of hyperbolae as with the pulse system. Thus the vehicle P can start from a known point and record continuously its deviation from that point on phase-meters. The phase-meter readings can be converted to distance by referring to maps having overprinted hyperbolae. Such a system is known as an integrating system because position is obtained by adding (*i.e.* integrating) displacements from the starting-point. The ambiguities associated with phase change of 2π radians are removed, since the motion of P is continuously recorded.

However, if the receiver on P fails to operate for a period, the inegration of displacements will not take place, and when the receiver becomes operative again, its reading will be ambiguous. This difficulty can be removed by applying some form of low frequency modulation to the signal. The low frequency modulation would enable coarse range differences to be established, and fine range differences, based on the phase of the carrier, can also be employed. The frequency of the LF modulation (which may be a pulse type modulation) should be low enough to avoid ambiguities occurring over the useful field of operation, but not so low as to prejudice the determination of the correct fine lane established by the carrier phase. Three stages of ambiguity resolution, coarse, medium and fine, may be used in practice.

10.3. Direction Measurement

10.3.1. MEASUREMENT OF DIRECTION OF ARRIVAL OF A RADIO WAVE

Study of the techniques used in radio direction-finding, and of the accuracy of DF measurements, necessitates an understanding of radio propagation and aerial theory. It is not possible to develop the full theory in this volume but the basic principles can be established.

The simplest aerial suitable for DF is one in which two parallel dipoles are used to determine the plane of a phase front in an incoming radio wave. The two dipoles (Fig. 10.5) are connected to the receiver so that their e.m.f.'s are in phase opposition. The spacing d between dipoles is small compared with the wavelength of the signal. The phase of the e.m.f. induced in dipole B will lag that in dipole A because it is further away from the transmitting source.

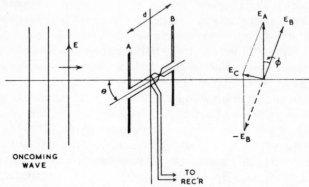

Fig. 10.5. Simple direction finding aerial with phasor diagram.

For an angle θ between the plane of the dipoles and the direction of arrival of the radio wave, the dipole B will be a distance $d \cos \theta$ farther away than dipole A. The phase lag corresponding to this distance is:

$$\phi = \frac{2\pi d \cos \theta}{\lambda} \qquad (10.6)$$

and will be small since d is much less than the wavelength λ.

In the phasor diagram, E_A is the e.m.f. in dipole A, and E_B the e.m.f. in dipole B lagging by angle ϕ. Because the dipoles are connected in opposition the resultant e.m.f. is given by the phasor E_C. Since the angle ϕ is small the resultant E_C is given approximately by:

$$E_C = E_A \frac{2\pi d}{\lambda} \cos \theta \qquad (10.7)$$

When the plane of the dipoles is normal to the direction of arrival, $\theta = 90°$ and $E_C = 0$, *i.e.* the plane of the dipoles is in the plane of the phase fronts. If E_C is plotted against θ we obtain a cosine wave. It is more usual to plot in polar co-ordinates as in Fig. 10.6(*a*). In this case we obtain a figure-of-eight diagram composed of two circles. Such diagrams, showing the dependence of aerial sensitivity on direction, are known as aerial polar diagrams.

Such a DF aerial is used by rotating the aerial about a vertical axis to vary θ in order to find the null direction where the signal is extinguished or lost in the noise. A point roughly halfway between points where the signal reappears gives the best estimate of the direction of arrival.

With this simple arrangement there is no method of determining sense, *i.e.* whether the wave in Fig. 10.5 is travelling from left to right

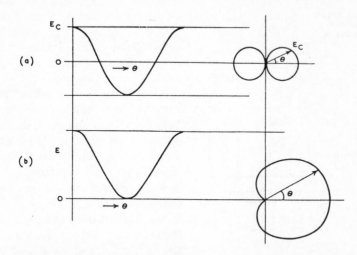

Fig. 10.6. (a) Polar diagram of simple D.F. aerial. (b) With sense aerial added.

or vice versa. A sense-finding aerial can be added which enables this ambiguity to be resolved by a second operation.

Consider a single dipole placed on the axis of rotation of the DF aerials. This will receive a signal roughly in phase with E_A and E_B. A circuit can be added to this sense aerial to shift the phase of the signal by 90° so that it is now either in phase or in opposition to E_C, since E_C is approximately in phase quadrature with E_A and E_B. The magnitude of signal from the sense aerial is made roughly equal to the maximum value of E_C, and is added to E_C by a suitable circuit before being applied to the receiver. This process is equivalent to adding a fixed voltage independent of θ to the original polar diagram and the new polar co-ordinate plot has a cardioid shape as shown in Fig. 10.6(*b*). Thus, by switching in the sense aerial and rotating the DF aerial a direction for maximum signal can be found and this direction can be arranged to point approximately to the transmitter. An accurate bearing can be obtained by locating the null position with the sense aerial switched out.

10.3.2. MEASUREMENT OF DIRECTION BY LARGE APERTURE BEAM AERIALS

The simple DF aerial described in the previous section is only capable of determining bearings to a limited order of accuracy, perhaps 5° or worse. If aerials can be constructed to accept a much larger area of the phase fronts, then high accuracy bearings can be obtained. At the higher frequencies, *i.e.* UHF and SHF, it is possible to construct aerials having apertures with dimensions of several wavelenghts. Such aerials radiate narrow beams if used for transmission. For reception they exhibit polar diagrams with beams of width of the order of a few degrees or even less.

A typical aerial capable of this performance is the microwave paraboloidal reflector shown in Fig. 10.7. It consists of a metal reflector with a surface shape formed by rotating a parabola about its principal axis. A small radiator, usually an open-ended waveguide, is placed at the focus. The path length from the focus to the reflector and then to the aperture plane is constant for

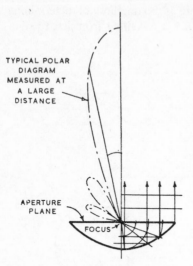

TYPICAL POLAR DIAGRAM MEASURED AT A LARGE DISTANCE

APERTURE PLANE

FOCUS

Fig. 10.7. The paraboloidal reflector aerial.

all rays, so that the radiated wave has a plane phase front. The phase and amplitude of the field in the aperture plane determines the polar diagram of the aerial, and the relationship may be deduced by considering the aerial as a transmitter. The reciprocity theorem for electromagnetic waves shows that the polar diagram as a receiving aerial is identical to the transmitting polar diagram.

The method of deriving this relationship is based on the Kirchoff-Huygens Principle, in which every small element in the aperture plane can be considered as a source emitting secondary spherical waves. If we consider a narrow strip along a horizontal diameter of the aperture plane of the paraboloid, we wish to calculate the distant field due to this strip in the plane containing the normal to the aperture and the diameter. The geometry is shown in Fig. 10.8 and although it is not essential to the argument we may consider the polarization of the field in the aperture to be at right angles to the plane of the figure. The field distribution across the diameter will determine the polar diagram in the plane, *i.e.* the variation of the distant field with θ. This distribution may be represented by $f(x)$. If the aerial has no phase error across the aperture, *i.e.* each element radiates in phase, then $f(x)$ will be real. If there is a phase error then $f(x)$ will be complex. Consider a point in a direction at an angle θ to the normal and sufficiently distant that rays r and s may be considered parallel. The field from an element dx will have to travel a distance x . $\sin \theta$ greater than the field from the origin, *i.e.* the

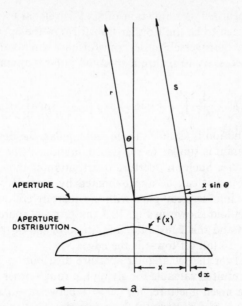

Fig. 10.8. Geometry for calculating polar
diagram from aperture distribution.

field contribution at the distant point will lag in phase relative to that
from the origin by

$$\frac{2\pi x}{\lambda} \cdot \sin \theta \quad \text{(radians)}$$

The field contribution at the distant point due to $\mathrm{d}x$ is proportional
to $f(x)\mathrm{d}x$, and is therefore given by

$$K \cdot f(x)\mathrm{d}x \cdot \exp\left[-j\frac{2\pi x}{\lambda} \cdot \sin \theta\right]$$

when allowance is made for the phase. K is a constant of proportionality
and depends on the distance of the observation point from the aerial.
The total field at the distant point can be obtained by integrating over
the aperture, thus

$$E = K \int_{-a/2}^{a/2} f(x) \exp\left[-j\frac{2\pi x}{\lambda} \sin \theta\right]\mathrm{d}x \qquad (10.8)$$

Since the aperture is finite, there will be no contribution from outside
the limits, so that we may integrate between infinite limits and obtain
the general expression:

$$E(\sin \theta) = K \int_{-\infty}^{\infty} f(x) \exp\left[-j\frac{2\pi x}{\lambda} \sin \theta\right]\mathrm{d}x \qquad (10.9)$$

provided $f(x)$ is taken as zero outside the aperture.

The field regarded as a function of $\sin \theta$ is related to the polar diagram, and is seen to be the Fourier transform of the aperture distribution. If we take the inverse Fourier transform we can obtain the aperture distribution necessary to realize a specified polar diagram. The inverse transform is:

$$f(x) = \frac{1}{K\lambda} \int_{-\infty}^{\infty} E(\sin \theta) \exp \left[j\frac{2\pi x}{\lambda} \sin \theta \right] d(\sin \theta) \quad (10.10)$$

If the distribution is assumed to be rectangular, the determination of the polar diagram is similar to the determination of the spectrum of a rectangular pulse. Such an aperture distribution cannot be realized in practice, although it can be approximated by a plane wave normally incident on an infinite metal plate with the aperture cut out of the plate. For the paraboloid shown in Fig. 10.7 the primary source has its own polar diagram and this causes the amplitude of the distribution in the aperture plane to fall off towards the edges.

The integral for the rectangular aperture distribution is easily evaluated and the result is valuable for giving the rough order of beamwidth provided by a given aperture. If $f(x) = 1$ between $-a/2$ and $a/2$ and zero elsewhere it can be deduced from equation (10.9) that:

$$E = \frac{\sin \left(\frac{\pi a}{\lambda} \sin \theta \right)}{\frac{\pi}{\lambda} \sin \theta} \quad (10.11)$$

The constant K does not alter the shape of the polar diagram and is taken as unity.

As θ approaches zero, i.e. in the normal direction, the value of E becomes a, showing that the distant field for a given aperture field intensity is proportional to the one-dimensional aperture. The first zero of the expression occurs when

$$\left(\frac{\pi a}{\lambda} \right) \sin \theta = \pi$$

or

$$\sin \theta = \frac{\lambda}{a}$$

$$\theta = \sin^{-1} \left(\frac{\lambda}{a} \right) \quad (10.12)$$

Thus, if an aperture is 10 wavelengths wide the angle out to the first null is about one-tenth radian or 6°. If the aperture length is of this order or greater we can replace $\sin \theta$ by θ and we see that the shape of the polar diagram is $\sin (n\theta\pi)/\theta$ where n is the aperture dimension measured in wavelengths. This is similar in shape to the general function $\sin \alpha / \alpha$ and differs only in the angular scale.

In the case of narrow beam aerials of this general type the value of E at one half the angle to the first null is approx.

$$a \cdot \frac{\sin \pi/2}{\pi/2} = \frac{2a}{\pi} = 0\cdot64a \qquad (10.13)$$

or about 3 dB below the peak value on the normal. It is usual to specify the beamwidth of an aerial as the total angle between the half-power points. This is therefore roughly the same as the angle out to the first null or:

$$\theta_a = \sin^{-1}(\lambda/a)$$

$$\doteqdot \lambda/a \quad \text{(radians)} \qquad (10.14)$$

A similar result can be obtained in a plane through the axis at right angles to that just considered. If the aperture is rectangular of dimensions $a \times b$, the $\sin \alpha/\alpha$ shape polar diagrams will exist in each plane and there will be a solid polar diagram, with elliptical sections at right angles to the normal or principal axis. In practice, the beamwidth will be greater than that given by the above simple relation owing to the amplitude taper towards the edges of the aperture. The distant field in the axial direction will be proportional to ab or the distant power density to $(ab)^2$. However, the transmitted power will be proportional to ab for a given aperture intensity, so that for a given unit power transmitted the distant power density is proportional to ab or the aperture area. The factor by which the distant power density is increased due to a beamed aerial, compared with the density from a radiator which radiates uniformly in all directions (isotropic radiator) is known as the power gain of the aerial. Thus power gain may be shown to be given by:

$$G = 4\pi ab/\lambda^2$$

$$\doteqdot \frac{4\pi}{\theta_a \theta_b} \qquad (10.15)$$

It can be shown that, in the receiving condition, when the aerial is pointing directly at the desired source, the signal-to-noise ratio is improved by this factor G. The signal power received is increased by G whereas the noise power is independent of G in the case where noise is assumed to arrive with equal intensity from all directions.

Thus we may conclude that using a beam aerial provides a better prospect for measuring direction accurately, first because of a better signal-to-noise ratio and secondly because of the rapid change of received signal with change of aerial direction. It will be shown in Section 10.3.4 that ultimate bearing accuracy is a function of signal-to-noise ratio, and since noise is generally a function of bandwidth, bearing accuracy can always be improved by reducing bandwidth. In other words, bearing accuracy can be improved with any aerial by averaging bearing over a longer time, subject of course to the assumption that the radio wave direction of arrival is not fluctuating.

Another important feature of beam aerials is their angular resolution and this will now be considered.

10.3.3. ANGULAR RESOLUTION

Angular resolution or discrimination is the ability of a system to detect as separate entities two objects separated in angle. Consider two objects emitting the same type of radio signal. If the angular separation of the two objects is small compared with the beam width of the aerial being used for direction finding, they cannot be resolved and the aerial will detect maximum signal when it is pointing somewhere between the two objects.

If the polar diagram is of $(\sin \alpha)/\alpha$ shape and it is sweeping through two objects separated by a small angle, the criterion of Fig. 10.3 may be used as a definition of angular resolution. If the beam width θ_b is defined as the angular width between half-power points, it is clear that the angular resolution on this definition is $2\theta_b$. As with the case of range resolution, it is possible in theory to resolve objects at a smaller angle than this by using the known shape of polar diagram and comparing it with the received signal versus angle.

10.3.4. ANGULAR ACCURACY

The angular accuracy that can be achieved by direction finding, using a narrow beam aerial, is certainly better than the beam width, but the actual amount by which it is better is dependent upon many factors. Methods which involve sweeping past the source and noting the point of maximum signal amplitude enable an accuracy better than a quarter of a beam width to be realized, depending upon the signal to noise ratio, but much better accuracies can be obtained by using the equivalent of the null method, and, of course, by using longer observation periods.

Consider two aerials, each of beam width θ_b set at an angle of θ_b to one another, so that the resultant polar diagram is as shown in Fig. 10.9(a). The two aerials can be connected in phase opposition so that the difference in the received signals can be derived. The sum can also be derived by branching off some of the power from each aerial into a summing circuit.

Thus, if E_a and E_b are the r.m.s. voltages from the two aerials we have available $E_a - E_b$ and $E_a + E_b$. It may be assumed that the source on which the two aerials are to be trained for direction finding is a CW oscillator, although the result can easily be adapted for a pulse source

The polar diagram may be assumed to be of $\sin \alpha/\alpha$ shape, and the difference voltage in the region of the angle where the polar diagrams cut one another can be determined from Fig. 10.9(b). One polar diagram is drawn negative so that the sum represents the difference. The slope of $\sin \alpha / \alpha$ at $\alpha = \pi/2$ may be derived by differentiation and will be found

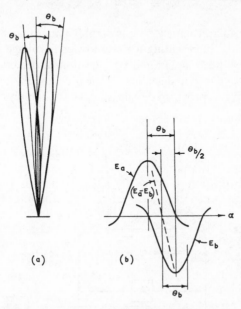

Fig. 10.9. Direction finding using null method with beam aerials. (a) Polar diagrams in polar co-ordinates. (b) (Sin α)/α polar diagram in cartesian co-ordinates.

to be $-4/\pi^2$. The slope of the difference will be twice this or $-8/\pi^2$ in volts per radian. If the maximum value of the $\sin \alpha / \alpha$ shape curve is E volts r.m.s., $i.e.$ when one aerial is pointing directly at the source, then the slope at the null will be

$$\frac{-E.8}{\pi^2} \text{ volts per radian}$$

The first zero of the actual polar diagram occurs at an angle θ_b whereas the first zero of $\sin \alpha / \alpha$ is at an angle $\alpha = \pi$. We must therefore multiply the slope by π/θ_b, giving finally

$$\frac{-E.8}{\theta_b \pi} \text{ volts per degree}$$

if θ_b is in degrees.

When the split between the two aerials is pointing accurately at the source, only the noise from the two aerials and the receiver will appear in the output because the two aerial signal voltages balance. The value of this noise voltage when converted to angle, using the above scale factor, represents the ultimate angular accuracy. If the r.m.s. value of the noise is σ_n volts then angular r.m.s. error is given by

$$\sigma_\theta = \frac{\sigma_n}{E} \cdot \frac{\pi}{8} \cdot \theta_b \quad \text{degrees} \tag{10.16}$$

Thus the angular accuracy is considerably better than θ_b, particularly if the first factor representing a noise-to-signal voltage ratio is small. In deriving this we have ignored the question of detection. If the sum signal is used for coherent detection, efficient demodulation will result, as shown in Chapter 6, provided the sum signal is reasonably free from noise.

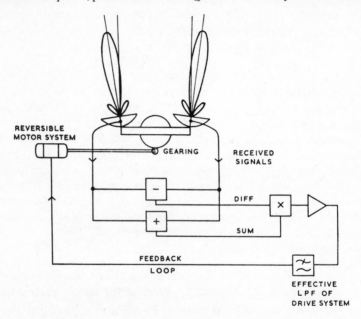

Fig. 10.10. Method of automatic tracking with narrow beam aerials.
(diagrammatic only).

The sign of the difference signal can be employed to control a servomechanism which rotates the aerial, so as to reduce the error signal, as shown diagrammatically in Fig. 10.10. In this way the aerial can be pointed automatically at the source. This is known as automatic tracking. The servosystem functions as a low pass filter of low cut-off frequency, so that the value of σ_n in the above equation can be quite small. The actual value of the cut-off frequency of the low pass filter is determined by the maximum expected value of the angular acceleration of the object being tracked.

Referring to Fig. 10.10, if the object being tracked can move an angle $\theta_b/2$ in a time short compared with the build-up time in the low pass filter,

THE PRINCIPLES OF POSITION FIXING BY RADIO

then there is a possibility of automatic tracking being lost. The tracking loop may become unlocked. If the maximum angular acceleration of the object, say, an aircraft, is known, the time to move $\theta_b/2$ can easily be estimated. The build-up or response time of the effective low pass filter constituted by the servosystem should be appreciably less than this period. Generally speaking, for aircraft, a filter cut-off of around 1 Hz can be achieved and is adequate to ensure lock being maintained. Such a low bandwidth reduces the noise in the system and enables accurate tracking to be realized.

Similar techniques enable automatic tracking in range to be achieved. Thus, if the pulse energy is split into early and late gates, error signals can be derived which drive the gates to follow the movement of the pulse in range. Similar considerations with regard to smoothing of the range information are applicable.

10.3.5. D.F. METHODS OF POSITION FIXING

Two fixed direction-finding stations can locate the position of a moving object carrying a suitable radio transmitter. Provided the DF stations can find true bearing with sense, there is no ambiguity. Generally it is necessary to communicate the position back to the moving object. In the early days of radio navigation this was a very satisfactory system, particularly for ships. It has been made semi-automatic for aircraft position fixing. When the aircraft transmitter is switched on the DF stations automatically record the bearings and transmit them to a central point where simple triangle-solving equipment locates the aircraft position, which is then sent by radio to the aircraft.

It is not, however, very satisfactory (except possibly in emergencies) for aircraft navigation, where rapid position fixing is essential owing to the high speeds involved. This has led to the development of ground-based flying aids which enable an aircraft to locate its own position. These devices employ the same fundamental principles which have been described here.

Conversely, a ship or aircraft can carry a DF aerial and by taking bearings on known fixed stations it can locate its position. Two bearings are necessary to provide a fix, and any bearings in excess of this can be used to improve accuracy. The size of aerial, which is limited, particularly with aircraft, and the proximity of metal structures, both tend to reduce the accuracy of bearings obtainable from moving vehicles.

10.3.6. THE RADIO DIRECTIONAL LOCALIZER

Any radio navigational aid which provides its information inside an aircraft is known as pilot-interpreted aid. Such systems employ some form of fixed radio transmitter and aerial on the ground (generally known as a beacon), and a receiver in the aircraft for establishing the direction to the beacon. To measure range a transmitter in the aircraft must be used to send a signal to the beacon which returns another signal to the aircraft, where measurement of delay gives range. The receiver-transmitter on the ground in this case is known as a transponder.

A simple method of establishing direction would be DF but, as explained in the previous section, this is often inaccurate when used on ships and aircraft. If an aircraft is homing on to a beacon using DF, and there is a fixed angular error, it is easy to show that a spiral path will be followed. A more usual method is to employ a beacon with a rotating aerial arrangement, or a fixed aerial arrangement in which the phase of a low frequency modulation is equal to the angle of any bearing line from the beacon. A separate modulation from a plain omni-directional aerial on the beacon is used to establish reference phase. For example, if an aerial has a polar diagram, such as Fig. 10.6(b) and it is rotated at 25 r.p.s. when transmitting a steady CW signal, a receiver will detect a 25 Hz sine wave whose phase is dependent on the bearing line between beacon and receiver. If a separate aerial radiates a carrier with 25 Hz reference phase modulation, a phase comparator at the receiver can establish the bearing line with respect to North. This can be displayed to the pilot of an aircraft on a meter. He can then fly on any radial path, or by measuring distance with a transponder he can fix his position.

When exceptionally high angular accuracy at a fixed azimuth is required such as for example, when making an instrument landing with an aircraft, the method of Fig. 10.9 can be adapted to the situation. Two aerials at the end of the runway would generate polar diagrams of the form shown. One aerial would transmit a carrier with modulation at one frequency and the second aerial would transmit the same carrier with a second modulation frequency. Careful control of carrier amplitudes and identical depths of modulation are necessary because the method depends upon amplitude comparison. An aircraft would then carry a receiver which compared the amplitudes of the two tones after demodulation and filtering. The larger of the two tones would deflect a meter so as to indicate to the pilot which way to turn to equalize the two-tone amplitudes. The course line where the two tones are equal would be the accurate direction, halfway between the aerial beams leading to the runway. The same technique may be used in elevation so as to delineate the glide path.

10.4. Radar Methods of Position Fixing

10.4.1. GENERAL PRINCIPLES

Radar employs the two basic principles of position fixing, enabling fixes to be obtained in polar co-ordinates from one station. In the general form of search radar a high gain narrow beam aerial is rotated about a vertical axis, so that the beam sweeps the horizon. The aerial polar diagram in elevation is generally shaped to cover from the horizon to an elevation of about 30°. The aerial functions as a transmitting and receiving aerial, a high-speed electronic switch being employed to switch it from transmitter to receiver. The transmitter emits pulses of RF energy at a recurrence rate determined by the maximum range required, as explained in Section 10.2.3. After the emission of each pulse, the receiver is switched to the aerial to receive echo pulses from aircraft or other objects.

If T = time for 1 rotation of aerial
θ_b = azimuth beam width in degrees
f_r = pulse recurrence rate

Then time to move one beamwidth

$$= \frac{\theta_b}{360} \times T$$

and number of pulses per beamwidth

$$= \frac{\theta_b}{360} \times T \times f_r \qquad (10.17)$$

The value of this figure varies widely, a typical figure being about 10 pulses per beamwidth, *i.e.* the aircraft would be hit by 10 pulses as the beam swept past its angular position.

The simplest method of displaying the position data from such a radar is to employ a cathode-ray tube with a radial time base rotating in synchronism with the aerial. When a pulse is transmitted the cathode-ray beam starts from the centre of the tube and is deflected radially outwards at an angle corresponding to the aerial position. The cathode-ray beam moves at a linear rate so that the bright spot it would produce represents range from the centre of the tube. Normally the beam is partially suppressed and it only makes a bright spot when an echo pulse is received. The several pulses corresponding to a beamwidth paint a small arc on the tube, at a range corresponding to the target giving the echoes, and the centre of the arc gives the best value of bearing. Such a cathode-ray tube is known as a plan position indicator (P.P.I.). Strictly speaking, it can only display slant range to the target, and without other features it cannot give true plan position. The broad principles just described are shown diagrammatically in Fig. 10.11.

This type of search radar can be applied to shipboard use as an aid to navigation in conditions of bad visibility. In the simple form described above it will give successive positions of other vessels relative to its own vessel. Coastlines will appear somewhat blurred if the P.P.I. has afterglow, because the coastline will appear to move relative to own vessel at the centre of the tube. If the course and speed of one's own vessel is known, it is possible to apply correction voltages to the P.P.I. so that one's own vessel moves at the true velocity and coastlines remain stationary. This is the true motion display which has advantages when navigating in coastal waters.

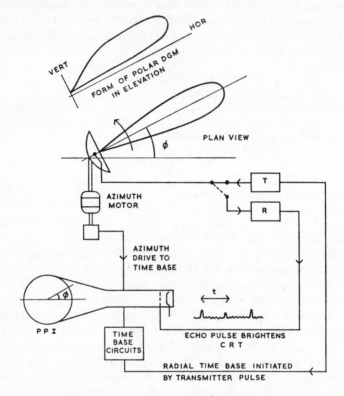

Fig. 10.11. General principle of search radar.

This general type of search radar may also be applied to airborne use. If we imagine a small search radar to be inverted and mounted underneath an aeroplane, we have a form of ground-mapping radar. Different kinds of terrain will return varying proportions of incident energy. Thus smooth water will return only a small proportion of energy, rough land a higher proportion and buildings a higher proportion still. Again, it is only possible to measure the slant range $h \csc \theta$ where h is the aircraft height and θ is the angle of depression to the object returning energy. If the radial velocity of the spot on the P.P.I. is linear it is evident that the ground map produced by the returned energy will have a distorted scale. If the radial velocity is varied according to a simple law, using the known aircraft height, a map to a nearly true scale can be obtained, and by comparing this with known maps, position fixing may be established.

10.4.2. THE RADAR EQUATION

The equation giving the echo power received from a radar target is important enough to merit a simplified presentation. If the transmitter power could be radiated uniformly in all directions the power density at the target would be

$$\frac{P_T}{4\pi R^2}$$

where R is the range of the target. The transmit aerial has a gain G_T relative to the isotropic radiator so that the power density at the target is

$$\frac{P_T G_T}{4\pi R^2}$$

Each target has a fictitious area σ which intercepts this density of radiation and re-radiates it uniformly in all directions. This area can be calculated for certain geometrical shapes but for aircraft it fluctuates widely, depending upon the direction of the incident wave. For a medium-sized aircraft it has a minimum value of about 20 m². The power re-radiated or scattered by the target is:

$$\frac{P_T G_T}{4\pi R^2}\,\sigma$$

and so the power density at the receiver is:

$$\frac{P_T G_T}{4\pi R^2} \cdot \frac{\sigma}{4\pi R^2}$$

If the receiving aerial has an aperture of area A_R the total received power is then:

$$P_R = \frac{P_T G_T}{4\pi R^2} \cdot \frac{\sigma}{4\pi R^2} \cdot A_R \qquad (10.18)$$

The important feature to note is that the received power is inversely proportional to the fourth power of range.

If the signal is of pulse type and P_T and P_R are peak powers, then for a receiving bandwidth B the signal-to-noise ratio for a single pulse is

$$\frac{S}{N}\ (\text{power}) = \frac{P_R}{kTB} \qquad (10.19)$$

The value of T depends upon the effective noise temperature of the receiving system. k is Boltzmann's constant.

If the pulse length is τ and the recurrence rate f_r, the duty cycle is $\tau \times f_r$ and the relation between peak and average power is

$$P_{av} = \tau f_r \times P_{peak} \tag{10.20}$$

Using the criterion of Chapter 5 for total bandwidth, namely $\tau = 1/B$

$$\frac{P_{av}}{f_r} = \frac{P_{peak}}{B} \tag{10.21}$$

showing that for the same noise density, the signal-to-noise ratio for a single pulse is the same as that deduced from the average power in a bandwidth equal to the recurrence frequency.

The probability of detection and false alarm rate for a single pulse are dealt with in Chapter 6. In systems where there are several pulses received from one target, an improvement in detection probability can be obtained by adding traces or pulses together as described in the next section.

10.4.3. INTEGRATION OF SUCCESSIVE TRACES IN RADAR

In the simple radar system just described, an aircraft is illuminated by the number of pulses that are emitted as the beam sweeps across it. The intensity of the pulses as measured at the aircraft will vary according to the transmitting aerial polar diagram. The reflected pulses will again be subjected to amplitude modification by the receiving polar diagram. Thus the effective polar diagram is the square of the one-way diagram. This assumes no change in the echoing area of the aircraft during one sweep.

To a first order the receiver on the ground will receive the number of pulses n in a nominal beamwidth as deduced in the previous sections, and these pulses will be within 6 dB (for the two-way polar diagram) of the maximum pulse value. The signal-to-noise ratio of these pulses may be improved by addition of the last n traces, and this may be shown to be equivalent to passing the video signal through a type of filter known as a comb filter.

The network equivalent of the process is the arrangement of parallel delay lines shown in Fig. 10.12. The top line with no delay delivers the latest pulse to the summing network. The second line of delay $\tau = 1/f_r$ stores the previous trace and delivers the previous pulse at the same time as the latest pulse. This process is continued over $(n - 1)$ delay lines, the last having a delay $(n - 1)\tau$. It is assumed that there is no appreciable range movement during the sweep of the aerial. Then the new sequence of pulses at the output of the summing network will have improved signal-to-noise ratio. The middle output pulse of a sequence (all delay lines delivering pulses) will have a voltage approximately n times that of one pulse at the input. The noise power at the output will be n times that for one trace because the noise voltages on successive traces are un-correlated. The peak pulse power has been increased n^2 times whilst the noise power has been increased n times, or an n-fold improvement in signal-to-noise power ratio. This is known as signal-to-noise ratio improvement by the integration of a repetitive signal, and as described here it is an efficient process.

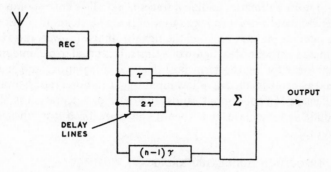

Fig. 10.12. Network equivalent of system for adding last n pulses of a radar receiver.

The transfer function of a line of delay $n\tau$ is $\exp[-j\omega n\tau]$ so that the output of the system is given by the geometric series

$$\Sigma = 1 + \exp[-j\omega\tau] + \exp[-j\omega 2\tau] + \ldots + \exp[-j\omega(n-1)\tau]$$

$$= \frac{[1 - \exp(-j\omega n\tau)]}{[1 - \exp(-j\omega\tau)]}$$

$$= \exp[-j\omega(n-1)\tau/2]\frac{\sin \omega n\tau/2}{\sin \omega\tau/2} \qquad (10.22)$$

The first term is the delay of the filter system and the second represents a comb filter as shown in Fig. 10.13. If we ignore the ripples between peaks, then the peaks constitute pass filters around each frequency

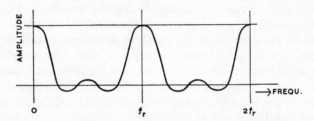

Fig. 10.13. Comb filter equivalent of radar pulse integrating system.

component of the video pulse train, the first being at the recurrence frequency $f_r = 1/\tau$ and thereafter at each harmonic.

If the radar receiver employs square law detection, as is usually the case, it should be noted that integration cannot regain the detection loss which occurs when the signal-to-noise ratio before detection is small compared with unity. Thus, if the input RF signal-to-noise (volts) ratio x is small, the ratio after the detector is x^2. On a power basis the signal-to-noise ratio after adding n traces is nx^4. This corresponds to an input RF signal-to-noise (power ratio) of $(\sqrt{n})x^2$ instead of x^2. In other words, integration has improved the signal-to-noise (power) ratio by $\sqrt{n}$, as compared with the ideal figure of n times. It is usual to call integration after the detector by the term post-detector integration, and it is recognized as being inefficient at low input signal-to-noise ratios, although quite efficient for good input ratios. Pre-detector integration is always more difficult to achieve, but when it can be realized it is efficient for all input ratios.

10.4.4. FREQUENCY MODULATED RADAR

The pulse type of radar just described is very convenient for accurate range measurement and good range resolution between adjacent targets. It lends itself to simple protection of the receiver from the high power of the transmitter by means of an electronic switch. Other forms of modulation of the RF can, however, be used for range measurement, besides pulse modulation. The theory of radar waveforms giving the relation between range and velocity resolution[1] cannot be given here, but a brief description of one important form of radar modulation, namely, FM, will be given. Several modulating waveforms such as sawtooth shapes may be used, as well as ordinary sinusoidal modulation in frequency of the RF carrier.

With such a radar the transmitter and its aerial must be separated from the receiver and its aerial such that the direct signal between the two is reduced to a suitably low value. This may be achieved by separation, screening or balancing out by means of a signal fed from the transmitter. The echo signal received by the receiver will now be a weaker signal of the same type of FM as that radiated, but delayed by a time corresponding to the two-way path to the reflecting object.

In Fig. 10.14(a) the triangular waveform at the top, shown in solid, is the graph of transmitted frequency against time. The waveform (in dotted) is the graph of the received frequency against time, the delay τ being the time for the radio signal to cover the go and return path to the object giving the echo. At any instant of time the difference between these two waveforms will be the beat note obtained by mixing the returned signal with the transmitter signal. This is shown in the graph below, negative frequency merely indicating that the phasor representing

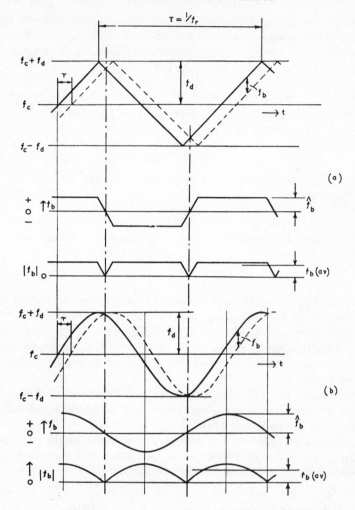

Fig. 10.14. Frequency modulated radar. (a) Triangular modulating
waveform. (b) Sinusoidal modulating waveform.

the beat frequency has reversed its direction of rotation. The next graph
shows beat frequency as a numerical against time, and the area under
this curve over a certain interval of time would give the number of beat
cycles in that time.

   Fig. 10.14(*b*) shows the same graphs for sinusoidal frequency modulation. The peak value of the beat frequency in either case can be used to
deduce the delay τ or we may count beat cycles, with an electronic
counter, over several modulation periods to determine the delay. The
area under the numerical beat frequency curve over a given period is

also equal to the area between the full and dotted curves. For small values of τ in comparison with the modulation cycle period T, the latter area is to the first order, independent of the waveform. If Δf is the beat note at time t, the area between the full and dotted curves is:

$$\Sigma \, \Delta f \cdot \tau$$

by dividing the area into trapezia of height Δf and base τ.

Over a full modulating cycle:

$$\Sigma \, \Delta f = 4 f_d$$

where f_d is the peak deviation of the carrier f_c, and the area between the curves is:

$$4 f_d \tau$$

The number of beat cycles counted or averaged over one second is therefore:

$$f_b(\text{av}) = 4 f_d \cdot \tau \cdot f_r \tag{10.23}$$

The number of beats can be counted over a period depending upon the desired data or sampling rate, by means of a suitable electronic counter. Then, with known values of deviation frequency f_d and recurrence (or modulation) frequency f_r, the delay τ and, hence, the range, may be determined. With large values of f_d small delays and ranges can be measured.

This type of radar is suitable for aircraft radar altimeters. The terrain below the aircraft presents a large single target, and the aerials may be arranged in the underside of the wings.

Considerable refinement is necessary if accurate height is required down to touch down. The above argument assumes that there is no doppler shift of the returned radio frequency.

10.5. Velocity Measurement by Doppler Frequency Shift

Any system which measures range from a fixed point can measure radial velocity relative to that point. There is, however, an important method, based upon the Doppler effect with radio waves, which enables velocity to be measured directly, and often with great accuracy. Consider a ground transmitter emitting a radio signal of frequency f and an aircraft approaching the transmitter with a radial velocity v. If the aircraft were stationary it would receive f cycles per second, but because it is approaching with velocity v it will in fact receive an additional v/λ cycles per second, where λ is the wavelength. Thus apparent frequency received in the aircraft:

$$f_a = f + v/\lambda$$

$$= f + \frac{v}{c} \cdot f$$

$$= f\left(1 + \frac{v}{c}\right) \tag{10.24}$$

Now consider the radar case in which the signal of frequency is f_a re-radiated back to the transmitting point. The frequency received back on the ground will by the same argument be

$$f_g = f_a + \frac{v}{\lambda_a}$$

$$= f_a\left(1 + \frac{v}{c}\right)$$

$$= f\left(1 + \frac{v}{c}\right)^2$$

$$\doteqdot f\left(1 + 2\frac{v}{c}\right) \tag{10.25}$$

when the velocity ratio v/c is small, as is usually the case for terrestrial objects. For receding objects the frequency is lowered and the equation becomes:

$$f_g = f\left(1 - \frac{2v}{c}\right) \tag{10.26}$$

The change in frequency or the Doppler shift is given by

$$\frac{2v}{c} \times f$$

An aircraft travelling at 670 m.p.h or 0·186 miles per sec gives a velocity ratio of 1 part in 10^6. The frequency shift is therefore 2 parts in 10^6, and for a radio frequency of 1000 MHz the actual shift is 2000 Hz. By beating the received signal with the transmitted signal, the Doppler shift can be detected and filtered. The bandwidth of the filter can be calculated from the maximum expected acceleration. The latter determines df_g/dt and if B is the filter bandwidth its build-up time will be of the order $1/B$. The signal must not pass across the frequency band of the filter in a time less than $1/B$, otherwise it will not produce any appreciable output, thus:

$$\frac{df_g}{dt} \ll \frac{B}{1/B}$$

$$\ll B^2$$

or

$$B \gg \left(\frac{df_g}{dt}\right)^{\frac{1}{2}}$$ (10.27)

Generally speaking, the accelerations involved with aircraft enable quite small bandwidths to be used so that weak signals can be detected.

An interesting application of this principle is the Doppler navigation system for aircraft. An aircraft emits a beam ahead of it, as in Fig. 10.15. It strikes the ground and is scattered, some of the energy being received back on the aircraft. The component of the aircraft velocity along the beam is $v \cos \theta$, and the Doppler shift will be

$$\frac{2v \cos \theta . f}{c}$$

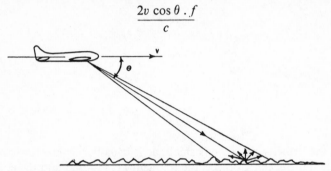

Fig. 10.15. Principle of Doppler navigation system.

If the shift is measured, the velocity v can be scaled from it. If the Doppler frequency is arranged to rotate a suitable a.c. synchronous motor the number of cycles can be integrated, and the distance travelled can be measured relative to the ground. In practice, it is necessary to elaborate the system to allow for attitude changes of the aircraft. The principal axes of an aircraft, the roll axis (fore–aft), the pitch axis (transverse to fore–aft), and the yaw axis (normal to other axes), do not always maintain an accurate fixed relationship to the path of the CG. For example, the roll axis may be at an angle to the CG path due to crosswind and varying angle of incidence. The pitch axis may not be horizontal, as in banking or rolling. To overcome these difficulties it is usual to employ 4 beams, 2 pointing forwards and downwards, and one each to either side. A suitable combination of the velocities determined from each beam enables the true ground speed to be derived with better accuracy. The positional accuracy is generally expressed as a percentage of distance run and something rather better than 1 per cent can be achieved, even over sea.

10.6. Position Measurement by Reference to a Satellite

A near earth satellite, *i.e.* one at a height of about 200 miles, enables position of a vehicle on or near the earth's surface to be determined, by monitoring the Doppler frequency change of a radio transmitter on the satellite. The satellite is assumed to move at constant height with a velocity V (about 5 miles per second) and for simplicity we assume the earth is flat. Let f be the frequency of the transmitter on the satellite, and d the distance of the point of nearest approach, $i.e.$ the point of transit, from the point on the earth's surface. Referring to Fig. 10.16 the radial velocity of the satellite relative to the point on the earth is:

$$\dot{r} = V \sin \phi$$

Then the Doppler shift is given by:

$$f_d = \frac{V \sin \phi}{c} \cdot f$$

The rate of change of Doppler is given by:

$$\dot{f}_d = \frac{V \cdot \cos \phi}{c} \cdot f \cdot \frac{\mathrm{d}\phi}{\mathrm{d}t}$$

$$= \frac{-V^2}{c} \cdot \frac{\cos^2\phi}{r} \cdot$$

$$\dot{f}_d = \frac{-V^2}{c} \cdot \frac{d^2}{(d^2 + x^2)^{\frac{3}{2}}} \qquad (10.28)$$

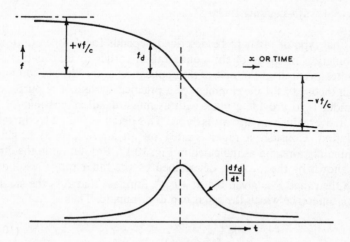

Fig. 10.16. Determination of distance from a satellite by measuring rate of change of Doppler frequency.

where x is the distance from the point of transit and is linearly proportional to time.

The ground vehicle will have a receiver which is designed to measure rate of change of Doppler. At the point of transit or the crossing point this will have its maximum magnitude which is from the above equation

$$f_a(\text{max}) = \frac{V^2}{c} \cdot \frac{f}{d}$$

or

$$d = \frac{V^2}{c} \cdot \frac{f}{f_a} \text{(max)} \tag{10.29}$$

This enables distance from the satellite at transit to be calculated. Prior knowledge of the height, velocity and position at various times of the satellite enables position of the ground vehicle to be determined. Thus if the time of transit is determined from a Doppler chart of the kind in Fig. 10.16 the satellite position at this time can be determined from tables together with its ground track. From the known height h we can find distance from the track from:

$$p = \sqrt{d^2 - h^2} \tag{10.30}$$

Approximate knowledge of position must be used to resolve the ambiguity as to which side of the track the vehicle is placed. Because of the high velocity of the satellite, it is necessary to work to very high orders of accuracy in time if position-fixing accuracy is to be useful.

10.7. Sideways-Looking Radar

The final type of radar to be described depends for its operation on a combination of many of the points raised earlier in this chapter and provides an excellent example of the application of the ideas of communication theory to the development of a practical system. The purpose of the radar is to provide a map, with as much detail as possible, of the terrain over which an aircraft is flying. This detail is limited by the radar resolution. Consider a radar located on an aircraft, A at height h, illuminating a region as indicated in Fig. 10.17. Resolution in the direction shown by the x-axis is determined by the radar range resolution, since the range R is given by $(x^2 + h^2)^{\frac{1}{2}}$. Suppose that R is the smallest range difference which the radar can discriminate. Then

$$\delta R = \frac{x\delta x}{(x^2 + h^2)^{\frac{1}{2}}} = \delta x \sin \alpha \tag{10.31}$$

from which we can deduce that δx is less than δR for α less than $\pi/2$.

The resolution in the x-direction is at least as good as the range resolution, which we have established in Section 10.2.4 to be $c\tau/2$, τ being the radar pulse length. The use of the pulse compression techniques, discussed in Section 8.7 makes it possible to use effective pulse lengths

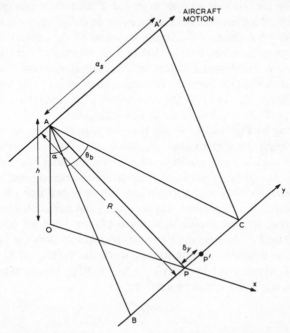

Fig. 10.17. Operating principles of sideways-looking radar. The plane OBC corresponds to the ground and AA′ is the aircraft flight path at constant height, h.

equal to the reciprocal of the bandwidth occupied by the radar pulse. For example, a radar bandwidth of 100 MHz gives an effective pulse length of 10^{-8}s, the corresponding range resolution being 1·5 m.

Resolution in the y-direction of Fig. 10.17 is determined by the angular resolution of the radar, which has been shown in Section 10.33 to be $2\theta_b$, where θ_b is the 3 dB beamwidth (radians) measured in the horizontal plane. We have also seen that this beamwidth is approximately (λ/a) radians (equation 10.14), so that the resolution in the y-direction is

$$\delta y \doteqdot 2R\theta_b \doteqdot 2R\lambda/a \qquad (10.32)$$

The value of δy is much larger than the 1·5 m found for δx in all practical situations. For example if $a = 100\lambda$ (giving a beamwidth of 10^{-2} radians), $\delta y = 200$ m for a range of 10 km and increases as the range increases. The question we now face is which technique, if any, can be used to reduce the y-resolution to a value comparable with the attainable x-resolution?

We note that the resolution is limited by the aerial size, a, and ask what possibilities exist for increasing this size. Although a is fixed by the size of the aircraft, we can make use of the aircraft motion to provide the equivalent of a much larger aerial. By storing the information received by the aircraft during a period T and then processing this information, the data provided is the same as could be extracted by an aerial of length equal to the distance travelled by the aircraft in time T. The processing required can be deduced by appropriate manipulation of the Fourier transform relation between aerial aperture distribution and radiation pattern. Such processing enables us to build up, or synthesize, a large aperture by the movement of a small one. The term 'aperture synthesis' is thus given to this technique.

Returning to Fig. 10.17, we see that we require an extension of the aerial aperture parallel to the y-direction and this is achieved if the aircraft motion is also parallel to this direction. Since the radar beam is radiated at right angles to the y-direction, the description 'sideways-looking radar' is obviously appropriate. We now examine the possible size of the synthetic aperture. Suppose θ_b is the horizontal beamwidth of the aerial, so that useful signals are obtained for all points lying between B and C. As the aircraft moves forward so does the beam, and a signal will continue to be obtained from the vicinity of C until the aircraft has reached A′ where $AA' = BC \doteqdot R\theta_b$. This distance equals the length, a_s, of the synthetic aperture, i.e.

$$a_s = R\theta_b \tag{10.33}$$

The 'beam-width' of the synthetic aperture is thus

$$\theta_s \doteqdot \lambda/a_s \doteqdot \lambda/R\theta_b \tag{10.34}$$

and the y-resolution becomes:

$$\delta y = 2R\theta_s \doteqdot 2\lambda/\theta_b \tag{10.35}$$

A remarkable feature of this result is that it is independent of R, the distance to the observation point. Further, since θ_b is λ/a, where a is the size of the actual aerial on the aircraft, then:

$$\delta y \doteqdot 2a \tag{10.36}$$

Again this is a remarkable result in that the resolution improves, i.e. δy becomes smaller, as a becomes smaller. This is the opposite behaviour to the earlier result in equation (10.32). The selection of a to provide a value for δy of the order of 1·5 m obviously presents no difficulty. The sideways-looking radar used in this way can thus achieve good resolution for both co-ordinates in the ground plane.

In view of the surprising nature of the above results, it is worth providing an alternative treatment which leads to a similar conclusion and also provides an indication of the nature of the processing of the radar signals which is required. The radar signals received from points on the line BC will be subject to Doppler shifts similar to those discussed in Section 10.5. The radial velocity of P as seen from the aircraft is zero when the aircraft position is at A. As the aircraft moves there will be a change in the radial velocity and hence in the Doppler shift. Suppose the aircraft passes through A at $t = 0$. The radar frequency is f_0 and we consider the signal to be reflected from the immediate vicinity of P. Then, the Doppler shift at time t will be

$$f_d = \frac{2v}{c} f_0 \sin \theta'$$

where θ' is the angular displacement of P, at time t, as seen from the aircraft, and v is the aircraft velocity. Under the conditions stated,

$$\theta' \doteqdot y/R$$

i.e.

$$f_d \doteqdot \frac{2vf_0}{c} \frac{y}{R} \qquad (10.37)$$

where

$$y = vt.$$

Now consider a point, P', very close to P and that $PP' = \delta y$. The Doppler shift associated with radar signals from P' will be

$$f_d' = \frac{2vf_0}{c} \cdot \frac{y + \delta y}{R}. \qquad (10.38)$$

The difference between f_d' and f_d is

$$f_d' - f_d = \frac{2vf_0 \delta y}{Rc} \qquad (10.39)$$

a constant value since δy, the separation between P and P', remains constant. This frequency difference may be observed during the time $T = a_s/v$ for which adequate signals can be received from P. The smallest frequency difference which can observed during a time T, i.e. the frequency resolution, is about $1/T$ Hz. Substituting this value in equation (10.39), we find

$$\delta y \doteqdot \frac{Rc}{2vf_0 T} \doteqdot \frac{Rcv}{2vf_0 a_s} \doteqdot \frac{Rc}{2f_0 R\theta_b}$$

$$\doteqdot \frac{ca}{2f_0 \lambda} \doteqdot \frac{a}{2}. \qquad (10.40)$$

The factor 4 by which this estimate differs from that in equation (10.36) arises from the rough approximations used in providing values for angular resolution, frequency resolution, etc.

The argument based on Doppler frequency shifts suggests that the signals received in the aircraft may be processed by two 'filtering' operations, one on the basis of radar range and one on the basis of Doppler shift, in which the samples thus obtained are stored for a period T and the distribution of radar targets in the ground plane are computed from the stored samples. The complexity of the equipment needed to do this can be seen by examining the volume of data to be stored. If $a = 1 \cdot 5$ m, we find that $a_s = R\lambda/a = 200$ m for $R = 10$ km and $\lambda = 3$ cm. The corresponding value of T is $0 \cdot 5$ s for an aircraft velocity of about 1500 km/h. During $0 \cdot 5$ s approximately 500 radar pulses will be received and each pulse must be 'filtered 'by range and Doppler frequency. If R extends from say $9 \cdot 5$ to $10 \cdot 5$ km, the number of range samples is $10^3/1 \cdot 5 = 670$. The largest Doppler frequency is given by substituting $y = 100$ m and $f_0 = c/\lambda$ in equation (10·37), and is 270 Hz. Frequencies in the range -270 to 270 Hz thus occur and may be resolved to 2 Hz, giving 270 samples. The total number of samples in $0 \cdot 5$ s is thus $500 \times 670 \times 270 = 9 \times 10^7$. The processing problem is thus a formidable one.

The first practical realisation of sideways-looking radar made use of an optical processing method, the radar signals being recorded on film and then analysed by using the transformation behaviour of optical lenses. Readers will by now not be surprised to learn that this behaviour is governed by the Fourier transform relation, but to explore all the details we should need a second volume.

10.8. Summary

Table 10.1 provides an indication of the operating principles of some of the radar and radio-navigation systems currently in use.

TABLE 10.1

Name	Where based	Where operated	General principles
Direction finding (DF)	Ground receivers	Ground	Bearing-bearing system from 2 stations, on ship or aircraft communications transmitter. Manual or automatic, sometimes with auto-triangulation. All frequencies
Radio compass	Ground transmitters	Ship or aircraft DF	Bearing-bearing on 2 ground stations. Manual or automatic. All frequencies
Consul (Sonne)	Ground transmitters	Ship or aircraft	Bearing-bearing on 2 ground stations. Transmitter produces multi-lobe polar diagram which can be rotated through 15°. Ambiguities resolved by coarse D.F. Frequency 250–500 kHz CW. Long range
Gee	Ground transmitters	Aircraft	Hyperbolic system using pulses in VHF band. Visual range
Decca	Ground transmitters	Ship or aircraft	Hyperbolic system using CW in LF band, with lane identification. Master transmitter with 3 slaves. Can give continuous display of position on ship or aircraft
Loran	Ground transmitters	Ship or aircraft	Hyperbolic long-range system using pulses in MF band (about 2 MHz). Later versions also use phase within pulse for greater accuracy
Distance-measuring Equipment (DME)	Ground responders	Ship or aircraft interrogator	Range measuring system using secondary radar principle, VHF + UHF frequencies
Oboe	Aircraft responder	Ground interrogator	Range-range system. Secondary radar principle
Tacan	Ground transmitter	Aircraft	Range and bearing. Range by air interrogator—ground responder. Bearing by rotating polar diagram and reference phase signal

TABLE 10.1 *continued*

VHF Omnirange (VOR)	Ground	Aircraft	Bearing only by rotating polar diagram and reference phase signal
VORTAC	Ground	Aircraft	Co-located VOR and TACAN
Doppler	Air	Air	Measures doppler frequency shift with radio beam from aircraft to ground, giving ground speed, drift angle, ground position and distance. Usually X-band frequency
Weather radar	Air	Air	Pulse radar with moveable pencil beam. Locates storms and coast lines. Usually X-band
Ground mapping radar	Air	Air	Pulse radar of search or scanning type
Ship radar	Ship	Ship	Pulse radar of search or panoramic type
Ground radar (GR)	Ground	Ground	Pulse radar of panoramic type, giving plan position. Also pencil-beam type for accurate tracking in azimuth elevation and range.
Ground controlled approach (GCA) Precision approach radar (PAR)	Ground	Ground	For aircraft landing. Pulse radar giving range. Azimuth and elevation by angular scanning
Instrument-landing system (ILS)	Ground	Air	Ground transmitter-aerial system provides overlapping pencil beams. Equi-signal course gives correct aircraft approach line for landing in azimuth and elevation
Standard beam approach (SBA)	Ground	Air	Earlier form of above for azimuth. One beam modulated with dots, other with dashes. Correct course when dots and dashes merge
Radio altimeter	Air	Air	Range-only radar on aircraft to measure height. Usually employs sinusoidal FM
Secondary surveillance radar (SSR)	Ground	Ground	Pulse-type secondary radar giving plan position. Aircraft responder also transmits coded information

The Use of Transforms in Circuit Analysis

A.1. Introduction

The analysis of any electrical circuit can always be reduced to finding the solution of a differential equation. A considerable simplification arises when periodic time waveforms are studied, for the operations of differentiation and integration do not materially alter the nature of the waveforms. The advantage so gained is most easily seen in practice by using the exponential waveform, $\exp(j\omega t)$. Differentiation and integration are then equivalent to the algebraic operations of multiplication and division by $j\omega$. The circuit differential equations thus reduce to algebraic equations and the solution is obtained much more easily.

The object of using transforms is to retain the possibility of solving circuit problems by algebraic methods even if the waveforms are not sinusoidal. Two types of transform are widely used—the Fourier and the Laplace. The Fourier transform is most useful when the waveform resembles a steady-state sinusoidal waveform and is the more widely used in communications work. The Laplace transform is particularly suited for solving transient problems.

Most of the material discussed in this volume is best handled by Fourier tranforms and the first part of this Appendix is devoted to their properties. The Laplace transform has many similarities to the Fourier transform and it is therefore advantageous to study both. The second part is devoted to Laplace transforms.

A.2. Fourier Series and Fourier Transforms

Any periodic function can be expressed as a sum of a fundamental and harmonic terms. We consider a function $f(t)$ which is periodic with period T and then:

$$f(t) = a_0 + \sum_{n=1}^{\infty} a_n \cos(n\omega t) + \sum_{n=1}^{\infty} b_n \sin(n\omega t) \qquad (A.1)$$

where ω is the fundamental angular frequency, given by:

$$\omega = 2\pi/T \qquad (A.2)$$

The amplitude constants in equation (A.1) are:

d.c. term:
$$a_0 = \frac{1}{T} \int_{-T/2}^{T/2} f(t) \, dt \qquad (A.3)$$

a.c. terms for harmonic n:
$$\begin{cases} a_n = \dfrac{2}{T} \int_{-T/2}^{T/2} f(t) \cos (n\omega t) \, dt & (A.4) \\[2mm] b_n = \dfrac{2}{T} \int_{-T/2}^{T/2} f(t) \sin (n\omega t) \, dt & (A.5) \end{cases}$$

In equation (A.1) the terms for any harmonic n can be combined in the form:

$$a_n \cos (n\omega t) + b_n \sin (n\omega t) = c_n \cos (n\omega t + \phi_n) \qquad (A.6)$$

where

$$c_n = (a_n{}^2 + b_n{}^2)^{\frac{1}{2}} \qquad (A.7)$$

$$\phi_n = \tan^{-1} (b_n/a_n) \qquad (A.8)$$

Yet another alternative expression can be obtained by writing the trigonometrical functions in terms of exponentials:

$$c_n \cos(n\omega t + \phi_n) =$$
$$= \tfrac{1}{2} c_n \exp j(n\omega t + \phi_n) + \tfrac{1}{2} c_n \exp -j(n\omega t + \phi_n) \quad (A.9)$$

This suggests that we write equation (A.1) in the form:

$$f(t) = \sum_{n=-\infty}^{\infty} d_n \exp(jn\omega t) \qquad (A.10)$$

in which the amplitudes d_n are in general complex.
    We have

$$\begin{aligned} d_n &= \tfrac{1}{2} c_n \exp(j\phi_n) \\ d_n &= \tfrac{1}{2} c_n \exp(-j\phi_n) \end{aligned} \qquad (A.11)$$

If $f(t)$ corresponds to a waveform which can be physically observed, then $f(t)$ must be a real function of t. It follows that the constants a_n, b_n, c_n, ϕ_n are all real quantities and so, from equation (A.11), we have:

$$d_{-n} = d_n{}^* \qquad (A.12)$$

where the asterisk denotes the complex conjugate.
    The constants d_n may be expressed in terms of $f(t)$ by using the results in equations (A.4)–(A.8) or, alternatively, by using the orthogonality properties of the functions $\exp(jn\omega t)$. By either method we obtain:

$$d_n = \frac{1}{T} \int_{-T/2}^{T/2} f(t) \exp (-jn\omega t) \, dt \qquad (A.13)$$

Equations (A.10) and (A.13) provide a convenient starting point for the discussion of the Fourier transform. The information given by equation (A.13) may be displayed graphically, by plotting the amplitudes and phases of the constants d_n against frequency as in Fig. A.1. This figure has been prepared for the function defined by:

$$f(t) = A: \quad \text{if} \quad nT < t < nT + \tau \atop = 0: \quad \text{if} \quad nT + \tau < t < (n+1)T \Big\} \; n = 0, \pm 1, \pm 2 \ldots \tau < T$$

The waveform is therefore the train of rectangular pulses shown in Fig. A.1(a). The constants d_n are calculated from equation (A.13) since the integrand is periodic with period T; the precise integration range is immaterial, provided it covers exactly one period and, for the present purpose, the range 0 to T is most convenient. Then:

$$d_n = \frac{1}{T} \int_0^\tau A \exp(-jn\omega t) dt = \frac{2A}{n\omega T} \exp\left(-\frac{jn\omega\tau}{2}\right) \sin\left(\frac{n\omega\tau}{2}\right)$$

$$= \frac{A}{n\pi} \exp\left(-\frac{jn\pi\tau}{T}\right) \sin\left(\frac{n\pi\tau}{T}\right)$$

since

$$\omega T = 2\pi$$

This expression is valid for both positive and negative values of n. When n is zero, the value can be found by a limiting process, or by returning to equation (A.13) and evaluating the integral for $n = 0$.

$$d_0 = \frac{A\tau}{T}$$

Fig. A.1 has been prepared for the case, $\tau = 0\cdot 1T$.

Both positive and negative frequencies appear in equations (A.10) and (A.13) and in Fig. A.1. This arises because of the use of the exponential function, $\exp(jn\omega t)$, as the basic time function: the trigonometrical factors $\cos(n\omega t)$ and $\sin(n\omega t)$ which arise in real waveforms can be expressed by suitable combination of $\exp(jn\omega t)$ and $\exp(-jn\omega t)$, i.e. as combinations of exponential functions with equal frequencies of opposite signs. The advantages of this apparently artificial device are clearly demonstrated in ordinary steady-state circuit analysis and also apply to Fourier analysis. In place of the three equations required if the form of equation (A.1) is used, we need only one, equation (A.13), from which to determine the constants. A further advantage will be apparent in a later example concerned with amplitude modulated waveforms. An interesting comparison can be made with the procedure used in steady-state analysis, in which a waveform such as $\cos(\omega t)$ is expressed as $\text{Re.}[\exp(j\omega t)]$. The problem is then solved for the complex waveform $\exp(j\omega t)$ and the required answer is the real part of the complex solution

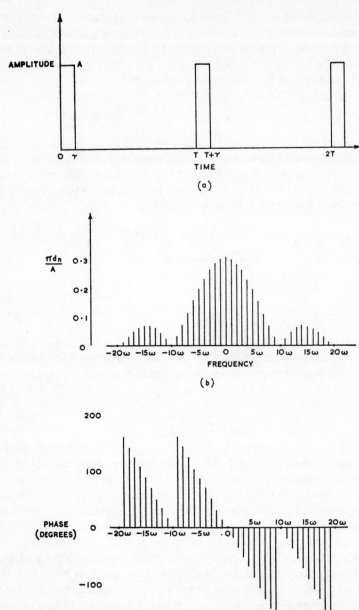

Fig. A.1. (a) Time waveform consisting of a train of rectangular pulses; (b) Amplitudes and (c) phases of the Fourier components of the time waveform in (a).

obtained. This is in effect a simple application of the Fourier series given in equation (A.10), for

$$\cos(\omega t) = \tfrac{1}{2} \exp(j\omega t) + \tfrac{1}{2} \exp(-j\omega t)$$

i.e. a special case of the series in which only two terms appear. In any steady-state problem, the required voltage or current must appear in the form:

$$f(t) = d_1 \exp(j\omega t) + d_{-1} \exp(-j\omega t)$$

This must be a real function and equation (A.12) gives:

$$d_{-1} = d_1{}^*$$

so that:

$$f(t) = d_1 \exp(j\omega t) + d_1{}^* \exp(-j\omega t)$$
$$= 2\mathrm{Re}.[d_1 \exp(j\omega t)]$$

This provides a justification of the use of $\exp(j\omega t)$ in steady-state problems.

It will be observed in Fig. A.1 that the amplitudes for the frequencies $n\omega$ and $-n\omega$ are equal, while the phases for these frequencies are numerically equal and of opposite sign. This is in agreement with the general result stated in equation (A.12). The phase cannot be specified at frequencies for which the amplitude is zero, *e.g.* the frequencies $2\pi/\tau$ and $4\pi/\tau$ in Fig. A.1. The values of the phases depend on the choice of the time origin and if the waveform is displaced along the time axis, the phases will be changed. For example, if $f(t)$ is defined by equation (A.10) then:

$$f(t - t_0) = \sum_{n=-\infty}^{\infty} d_n \exp(jn\omega t - jn\omega t_0)$$
$$= \sum_{n=-\infty}^{\infty} d_n{}' \exp(jn\omega t)$$

The magnitudes of d_n and $d_n{}'$ are equal, but the phase of $d_n{}'$ is $(n\omega t_0)$ less than that of d_n.

A figure, such as Fig. A.1, in which the dependence of the amplitude and phase upon frequency is shown graphically, is often referred to as the spectrum of the wave. This follows optical terminology where the spectral content of a light beam is related to frequency as well as to colour. In spectroscopic work, a spectrum in which the energy is restricted to discrete frequencies is called a line spectrum and this term could be applied to Fig. A.1: the term 'discrete spectrum' is also used. Continuous spectra, that is, spectra in which the energy is distributed over a continuous range of frequencies often arise in optics and this suggests that

continuous spectra may also occur for waveforms. A continuous spectrum can be considered as the limiting case of a line spectrum when
the separation between the lines becomes very small. In Fig. A.1, the
line separation is ω, the repetition frequency of the waveform: as ω becomes very small, the waveform period becomes very large and, in the
limiting case, is infinite. The waveform is then effectively non-periodic. The corresponding change in the Fourier series as this limiting
process takes place leads to the Fourier integral.

We start by supposing that T is very large and that the corresponding
fundamental angular frequency is a small quantity $\delta\omega$ given as in equation (A.2) by $2\pi/T$. Consider an angular frequency ω given by $n\delta\omega$
where n is an integer. Equation (A.13) can be written as:

$$d(\omega) = \frac{\delta\omega}{2\pi} \int_{-T/2}^{T/2} f(t) \exp(-j\omega t)\,dt \qquad (A.14)$$

in which the complex constant d is shown as a function of ω, the angular
frequency being considered. The interpretation of this constant requires
some care as we are changing over from discrete frequencies to a continuous range of frequency when we allow T to become infinite. The frequency interval, $\delta\omega$, appears on the right of equation (A.14) and this
suggests that when the spectrum becomes continuous, we interpret $d(\omega)$
as the amplitude and phase of the sinusoids in the small frequency interval ω to $\omega + \delta\omega$. We therefore define a frequency density by the equation:

$$d(\omega) = g(\omega)\,\delta\omega/2\pi \qquad (A.15)$$

the 2π being introduced to make $g(\omega)$ correspond to the amplitude for a
1 Hz band. Combining equations (A.14) and (A.15) and letting T tend
to infinity gives:

$$g(\omega) = \int_{-\infty}^{\infty} f(t) \exp(-j\omega t)\,dt \qquad (A.16)$$

where $g(\omega)$ is called the Fourier transform of the waveform $f(t)$. The
amplitude and phase of $g(\omega)$ can be plotted against ω to give the spectrum of the waveform. This has been done in Fig. A.2 for a waveform
consisting of the single pulse defined by:

$$f(t) = A \quad \text{if} \quad 0 < t < \tau$$
$$= 0 \quad \text{if} \quad t < 0 \quad \text{or} \quad t > \tau$$

Since $g(\omega)$ is a density function, the amplitude of the sinusoids between
frequencies ω and $\omega + \delta\omega$ is given by the area under this curve between
ω and $\omega + \delta\omega$: this corresponds to equation (A.14).

The calculation of the spectrum from a waveform would be of little

use in circuit problems unless the inverse operation of obtaining the waveform from the spectrum can be effected. To do this we return to equation (A.10) and replace d by its expression from equation (A.15)

$$f(t) = \sum_{-\infty}^{\infty} \frac{g(\omega)}{2\pi} \, \delta\omega \exp{(jn\delta\omega t)} \qquad (A.17)$$

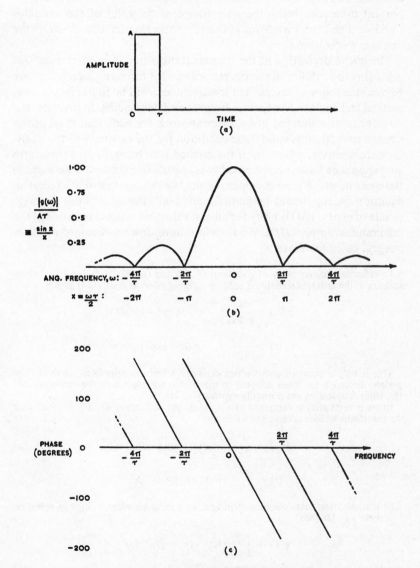

Fig. A.2. (a) Time waveform of a single rectangular pulse; (b) Amplitude and (c)
Phase of the Fourier transform of the time waveform in (a).

In the limit as T tends to infinity, $\delta\omega$ tends to zero, and ω equals $n\delta\omega$ and the summation can therefore be replaced by the integration*

$$f(t) = \frac{1}{2\pi} \int_{-\infty}^{\infty} g(\omega) \exp(j\omega t) \, d\omega \qquad (A.18)$$

an expression identical in all essentials to equation (A.16), the most important difference being the interchange of the roles of the variables t and ω. The time waveform is therefore the Fourier transform of the frequency spectrum.

The above derivation of the Fourier transform has ignored questions regarding the validity of the operations carried through: such points are beyond the scope of the present treatment and will be found in a mathematical text such as *The Fourier Integral* by Titchmarsh[2]. In practice, the waveforms encountered in engineering work are such that the Fourier transform method is valid. One condition for the existence of the transform does require attention: if the method is to be useful, the transform $g(\omega)$ should be finite, except possibly at specific frequencies. If we exclude the cases in which $g(\omega)$ becomes infinite, we require that the integral in equation (A.16) should be finite for any real value of ω. Since the magnitude of $\exp(-j\omega t)$ is unity for all real values of ω, the magnitude of the integrand is simply $|f(t)|$. We therefore have, from a standard result in integral calculus, that:

---

* To enhance the symmetry of equations (A.16) and (A.18) the factor $1/2\pi$, which appears in the latter, is sometimes split by defining $g_1(\omega)$ as $g(\omega)/\sqrt{(2\pi)}$ giving:

$$g_1(\omega) = \frac{1}{\sqrt{(2\pi)}} \int_{-\infty}^{\infty} f(t) \exp(-j\omega t) \, dt$$

$$f(t) = \frac{1}{\sqrt{(2\pi)}} \int_{-\infty}^{\infty} g_1(\omega) \cdot \exp(j\omega t) \, d\omega$$

This is only a point of detail which does not affect the principles involved. The present approach has been adopted to make $g(\omega)$ correspond to the amplitude per Hz, since frequencies are normally expressed in Hz.

In some texts $g(\omega)$ is expressed as amplitude per radian per second, in which case the transform and its inverse are written as:

$$g(\omega) = \frac{1}{2\pi} \int_{-\infty}^{\infty} f(t) \cdot \exp(-j\omega t) \, dt$$

$$f(t) = \int_{-\infty}^{\infty} g(\omega) \cdot \exp(j\omega t) \cdot d\omega$$

In mathematical texts the transform and its inverse are often written in terms of frequency f in Hz thus:

$$g(f) = \int_{-\infty}^{\infty} f(t) \cdot \exp(-j2\pi ft) \, dt$$

$$f(t) = \int_{-\infty}^{\infty} g(f) \cdot \exp(j2\pi ft) \cdot df$$

$$|g(\omega)| \le \int_{-\infty}^{\infty} |f(t)| \, dt \qquad (A.19)$$

and so $|g(\omega)|$ will be finite for any value of ω, provided that the integral on the right is finite. The integral $\int_{-\infty}^{\infty} f(t) \, dt$ is then said to be absolutely convergent. There are certain types of waveform of interest for which this condition does not hold, one example being the step function discussed in a later section. The difficulty is circumvented by a suitable limiting process.

A.3. Examples of Fourier Transforms

In this section, we will apply equation (A.16) to a number of waveforms of practical interest. In the solution of problems, it is usually possible to find the transform in a tabulated list or dictionary and the examples given will show how such a dictionary is prepared. Comprehensive dictionaries are available (see list of references) and greatly reduce the labour involved in using the Fourier transform. An abbreviated list, covering the results required in this volume, together with a few general rules, is given in Table A.2 at the end of this Appendix.

In practical problems, $f(t)$ must be a real function of time and so the transform $g(\omega)$ must satisfy a condition equivalent to equation (A.12). It follows at once from equation (A.15) that

$$g(-\omega) = g^*(\omega) \qquad (A.20)$$

i.e. the magnitudes of $g(\omega)$ and $g(-\omega)$ are equal and their phases are numerically equal but of opposite sign. In mathematical terms, the magnitude of $g(\omega)$ is an even function of ω, or $[|g(\omega)| = |g(-\omega)|]$ and the phase or argument of $g(\omega)$ is an odd function of ω, or $[\arg g(\omega) = -\arg g(-\omega)]$. This result can also be verified directly by taking the complex conjugate of equation (A.16):

$$g^*(\omega) = \int_{-\infty}^{\infty} f^*(t) \exp(j\omega t) \, dt$$

$$= \int_{-\infty}^{\infty} f(t) \exp(j\omega t) \, dt \qquad \text{(since } f(t) \text{ is real)}$$

$$= g(-\omega)$$

in agreement with equation (A.20).

This general result applies to all the examples given below.

A.3.1. RECTANGULAR PULSE

The single rectangular pulse has the waveform:

$$f(t) = A \quad \text{if} \quad 0 < t < \tau$$
$$= 0 \quad \text{if} \quad t < 0 \text{ or } t > \tau \tag{A.21}$$

and has already been mentioned in connection with Fig. A.2. From equation (A.16)

$$g(\omega) = \int_0^\tau A \exp(-j\omega t)\,dt$$
$$= A[\exp(-j\omega\tau) - 1]/(-j\omega)$$
$$= \frac{A}{j\omega} \exp(-j\omega\tau/2)\,[\exp(j\omega\tau\,2) - \exp(-j\omega\tau\,2)]$$
$$= \frac{2A}{\omega} \exp(-j\omega\tau/2) \sin(\omega\tau\,2) \tag{A.22}$$

The magnitude and phase of $g(\omega)$ are plotted in Fig. A.2.

This spectral function has many important properties and is considered below in some detail. Before doing this, we will make a slight modification to the result by using a standard relation.

A.3.2. SHIFT OF TIME ORIGIN

Suppose the waveform $f(t)$ has the spectrum $g(\omega)$. We wish to find the spectrum $g_1(\omega)$ corresponding to $f_1(t)$, identical to $f(t)$ but displaced in time, i.e.

$$f_1(t) = f(t - t_0) \tag{A.23}$$

where t_0 is a constant.

Then,

$$g_1(\omega) = \int_{-\infty}^{\infty} f(t - t_0) \exp(-j\omega t)\,dt$$
$$= \int_{-\infty}^{\infty} f(t') \exp(-j\omega \overline{t_0 + t'})\,dt' \quad \text{where} \quad t' = t - t_0$$
$$= \exp(-j\omega t_0) \int_{-\infty}^{\infty} f(t') \exp(-j\omega t')\,dt'$$
$$= g(\omega) \exp(-j\omega t_0) \tag{A.24}$$

A change in the time origin of a waveform therefore affects only the phase of the spectrum.

A.3.3. RECTANGULAR PULSE, SYMMETRICALLY POSITIONED WITH RESPECT TO TIME ZERO

The above result can be used to eliminate the phase term $\exp(-j\omega\tau/2)$ in equation (A.22). We choose t_0 to be $-\tau/2$ and obtain:

$$g_1(\omega) = \frac{2A}{\omega}\sin(\omega\tau/2) \qquad (A.25)$$

and this corresponds to:

$$
\begin{aligned}
f_1(t) &= A: \quad -\tau/2 < t < \tau/2 \\
&= 0: \quad t < -\tau/2 \text{ or } t > \tau/2
\end{aligned}
\qquad (A.26)
$$

i.e. the pulse is symmetrically placed with respect to $t = 0$.

The magnitude of $g_1(\omega)$ is identical to that of $g(\omega)$ and the phase of $g_1(\omega)$ is 0 (or any multiple of 2π) if $\sin(\omega\tau/2)$ is positive, and is π (or any odd multiple of π) if $\sin(\omega\tau/2)$ is negative. The properties of $g_1(\omega)$ can most easily be examined if $\omega\tau/2$ is replaced by x, giving:

$$g_1(\omega) = A\tau \cdot \frac{\sin x}{x} \qquad (A.27)$$

The magnitude of the function $\sin x/x$ is shown in Fig. A.2(b). There is a main lobe centred on the zero value of x and a train of side lobes, infinite in number, with maxima becoming progressively smaller as x increases. The position and heights of these maxima are important in that they determine the bandwidth required for the transmission of the pulse without intolerable distortion. The principal maximum, at $x = 0$, has amplitude unity since

$$\operatorname*{Lim.}_{x \to 0} \frac{\sin x}{x} = 1$$

The $\sin x$ term is zero when x is any multiple of π, so that the positions of the zeroes can be found exactly. It may be noted that the absence of a zero at $x = 0$ causes the main lobe to be twice as wide as the side lobes.

There is no corresponding simple expression for the positions of the maxima, but an adequate approximation can be found as follows.

The positions of the maxima are given by setting the derivative of the function equal to zero, *i.e.* by the roots of the equation:

$$\frac{\cos x}{x} - \frac{\sin x}{x^2} = 0$$

or

$$\tan x = x$$

The first root is clearly $x = 0$. The other roots may be obtained graphically and a rough sketch is sufficient to show that they are given approximately by:

$$x = \pm \frac{3\pi}{2}, \quad \pm \frac{5\pi}{2}, \quad \pm \frac{7\pi}{2}, \ldots$$

i.e.

$$x_n = \pm \left(n\pi + \frac{\pi}{2} \right) : n = 1, 2, 3, \ldots$$

There is no root near $\pi/2$ since $\tan x$ always exceeds x in the range $0 \leqslant x \leqslant \pi/2$. A better approximation is given by letting

$$x_n = n\pi + \frac{\pi}{2} + \delta_n$$

where δ_n is assumed to be small.

Then,

$$\tan x_n = \tan \left(n\pi + \frac{\pi}{2} + \delta_n \right)$$

$$= -\cot \delta_n \doteqdot -\frac{1}{\delta_n}$$

and since $\tan x_n = x_n$, we have

$$-\frac{1}{\delta_n} \doteqdot n\pi + \frac{\pi}{2} + \delta_n$$

Since δ_n is assumed small, it may be neglected on the right-hand side of the equation and we obtain:

$$x_n \doteqdot n\pi + \frac{\pi}{2} - \frac{1}{n\pi + \pi/2}$$

This can be used as the starting-point for a better approximation and repeating the process gives:

$$x_n \doteqdot n\pi + \frac{\pi}{2} - \frac{1}{n\pi + \pi/2} - \frac{2}{3} \times \frac{1}{(n\pi + \pi/2)^3}$$

Further terms can be obtained in the same way but the expression given is adequate for numerical work. The values of $(\sin x)/x$ at the maxima are found numerically and give the results shown in Table A.1.

TABLE A.1
Properties of $(\sin x)/x$

Position of maximum	Amplitude of maximum
0	1
4·436	0·2169
7·735	0·1282
10·904	0·0913
14·066	0·0709

A.3.4. IMPULSE FUNCTION

In many problems it is convenient to consider an idealization of the symmetrical pulse by supposing that the duration of the pulse becomes infinitesimally small while at the same time the amplitude becomes infinitely large. A measure of the strength of the pulse is then given by the product of the pulse length and the pulse amplitude. We therefore examine the results given in the previous section for the case defined by a function $\delta(t)$, such that

$$\delta(t) = \mathop{\mathrm{Lim.}}_{\substack{\tau \to 0 \\ A \to \infty}} f_1(t) \tag{A.28}$$

where $f_1(t)$ is defined by equation (A.26) and the limit is evaluated subject to the condition that the product $A\tau$ has the value unity. The function $\delta(t)$ is a unit impulse. It is sometimes referred to as the Dirac delta function and the following properties can be deduced from the definition:

(i) $\delta(t) = 0$ for all values of t except $t = 0$

(ii) $\int_{-\infty}^{\infty} \delta(t)\,dt = 1$

The function $\delta(t - t_1)$ is an impulse function occurring at time t_1 and an extension of the result in (ii) shows that

$$\int_{t_2}^{t_3} \delta(t - t_1)\,dt = 1$$

provided t_1 lies within the integration range t_2 to t_3. Further, since $\delta(t - t_1)$ is zero at all times except $t = t_1$, we may see that

$$\int_{t_2}^{t_3} f(t)\delta(t - t_1)\,dt = f(t_1) \tag{A.29}$$

again provided that t_1 lies within the integration range.

The Fourier transform of $\delta(t)$ can be obtained either by applying the limiting process to the results in the previous section or by using equation (A.29). Suppose the transform is $g_\delta(\omega)$ then, from equation (A.25) we have:

$$g_\delta(\omega) = \mathop{\mathrm{Lim.}}_{\substack{\tau \to 0 \\ A \to \infty \\ A\tau = 1}} \frac{2A}{\omega} \sin(\omega\tau/2) = \mathrm{Lim.} \frac{A\omega\tau}{\omega}$$

$$= \mathrm{Lim.}\ A\tau = 1$$

Alternatively, if we proceed from the definition of the Fourier transform, equation (A.16), we have:

$$g_\delta(\omega) = \int_{-\infty}^{\infty} \delta(t) \exp(-j\omega t)\,dt$$

and this becomes, on using equation (A.29),

$$g_\delta(\omega) = 1 \qquad (A.30)$$

as given by the limiting process.

The Fourier transform of the impulse function is therefore a constant, *i.e.* the frequency spectrum is flat. The inverse transform relation, equation (A.18), gives the result:

$$\frac{1}{2\pi} \int_{-\infty}^{\infty} \exp(j\omega t)\,d\omega = \delta(t) \qquad (A.31)$$

These results relating to the impulse function are useful not only in their immediate application to problems involving impulse excitation but also in deducing certain general theorems regarding Fourier transforms.

A further pair of transforms can be obtained by considering the impulse function $\delta(\omega - \omega_0)$ as a spectrum. The corresponding time waveform is:

$$f(t) = \frac{1}{2\pi} \int_{-\infty}^{\infty} \delta(\omega - \omega_0) \exp(j\omega t)\,d\omega$$
$$= \frac{1}{2\pi} \exp(j\omega_0 t)$$

Since the trigonometric functions $\cos(\omega_0 t)$, $\sin(\omega_0 t)$ can be expressed in terms of exponentials, this result can be used to find their transforms. For example, if:

$$f(t) = \cos(\omega_0 t) = \tfrac{1}{2} \exp(j\omega_0 t) + \tfrac{1}{2} \exp(-j\omega_0 t)$$

then

$$g(\omega) = \pi\delta(\omega - \omega_0) + \pi\delta(\omega + \omega_0)$$

which represents spectral lines at $\pm\omega_0$.

A.3.5. THE CONVOLUTION THEOREM

This theorem relates to the transform of the product of two functions and is used in showing the equivalence of the Fourier Integral and the Duhamel Integral methods discussed in Chapter 5. Suppose $f_1(t)$, $f_2(t)$ are two time functions with Fourier Transforms $g_1(\omega)$, $g_2(\omega)$ respectively. We will determine the nature of the time function whose transform is the product $g_1(\omega)g_2(\omega)$, *i.e.*, we wish to find $f(t)$ defined by:

$$f(t) = \frac{1}{2\pi} \int_{-\infty}^{\infty} g_1(\omega)g_2(\omega) \exp(j\omega t)\,d\omega \qquad (A.32)$$

where

$$g_1(\omega) = \int_{-\infty}^{\infty} f_1(t) \exp(-j\omega t)\,dt$$

$$g_2(\omega) = \int_{-\infty}^{\infty} f_2(t) \exp(-j\omega t)\,dt$$

Substituting the expressions for $g_1(\omega)$, $g_2(\omega)$ in equation (A.32), and changing the integration variables to avoid confusion gives:

$$f(t) = \frac{1}{2\pi} \int_{-\infty}^{\infty} \int_{-\infty}^{\infty} \int_{-\infty}^{\infty} f_1(x)f_2(y) \exp[j\omega(t - x - y)] \, dx \, dy \, d\omega$$

Provided that the functions satisfy the conditions for the integral to be convergent, the integrations can be carried out in any order and we first consider the integration with respect to ω. This can be evaluated immediately by using equation (A.31):

$$\frac{1}{2\pi} \int_{-\infty}^{\infty} \exp[j\omega(t - x - y)] \, d\omega = \delta(t - x - y)$$

so that

$$f(t) = \int_{-\infty}^{\infty} \int_{-\infty}^{\infty} f_1(x)f_2(y) \, \delta(t - x - y) \, dx \, dy$$

Either the x-integration or the y-integration can now be carried out, using equation (A.29), and we find the following two equivalent expressions for $f(t)$:

$$f(t) = \int_{-\infty}^{\infty} f_1(x)f_2(t - x) \, dx$$

$$= \int_{-\infty}^{\infty} f_1(t - y)f_2(y) \, dy \tag{A.33}$$

The integral on the right is called the convolution of the functions $f_1(t)$ and $f_2(t)$, and gives the required expression for the time waveform whose Fourier transform is the product $g_1(\omega)g_2(\omega)$.

A particular case of this general result is of some interest. Suppose that

$$g_2(\omega) = g_1^*(\omega)$$

and, correspondingly, $f_2(t) = f_1(-t)$ for real waveforms.

Then, two expressions for $f(t)$, given by equations (A.32) and (A.33) lead to the result:

$$\int_{-\infty}^{\infty} f_1(x)f_1(x - t) \, dx = \frac{1}{2\pi} \int_{-\infty}^{\infty} |g_1(\omega)|^2 \exp(j\omega t) \, d\omega$$

If t is placed equal to zero on both sides of the above equation we have:

$$\int_{-\infty}^{\infty} |f_1(x)|^2 \, dx = \frac{1}{2\pi} \int_{-\infty}^{\infty} |g_1(\omega)|^2 \, d\omega \tag{A.34}$$

If $f_1(x)$ represents a voltage or current the left-hand side represents an energy and the right-hand side the integral of energy along a frequency scale.

The reciprocal nature of the relations between time waveforms and frequency spectra implies that an equation similar in form to equation (A.33) can be used to find the spectrum of the product of two time functions. Suppose $f_1(t)$ and $f_2(t)$ are waveforms with transforms $g_1(\omega)$ and $g_2(\omega)$ respectively. Then the Fourier transform of the waveform.

is
$$\left. \begin{aligned} f(t) &= f_1(t)f_2(t) \\ g(\omega) &= \frac{1}{2\pi} \int_{-\infty}^{\infty} g_1(\omega - w)g_2(w)\,\mathrm{d}w \end{aligned} \right\} \tag{A.35}$$

A.3.6. THE GAUSSIAN PULSE

The pulse defined by the equation

$$f(t) = \exp(-kt^2) \tag{A.36}$$

is termed a Gaussian pulse, the name being derived from the similarity of the function to the Gaussian probability function of statistics. Direct substitution of $f(t)$ into the Fourier transform relation leads to an awkward integral and we will therefore use an indirect approach to obtain the Fourier transform.

Differentiation of equation (A.36) with respect to t gives:

$$f'(t) = -2kt \exp(-kt^2)$$

so that

$$f'(t) + 2kt\,f(t) = 0$$

If $g(\omega)$ is the Fourier transform of $f(t)$, we have:

$$f(t) = \frac{1}{2\pi} \int_{-\infty}^{\infty} g(\omega) \exp(j\omega t)\,\mathrm{d}\omega$$

$$f'(t) = \frac{1}{2\pi} \int_{-\infty}^{\infty} j\omega\, g(\omega) \exp(j\omega t)\,\mathrm{d}\omega$$

and so

$$f'(t) + 2kt f(t) = \frac{1}{2\pi} \int_{-\infty}^{\infty} [j\omega + 2kt]g(\omega) \exp(j\omega t)\,\mathrm{d}\omega = 0 \tag{A.37}$$

The t which appears within the brackets under the integral can be removed by integration by parts.

$$\int_{-\infty}^{\infty} t\, g(\omega) \exp(j\omega t)\,\mathrm{d}\omega = \frac{1}{j} \int_{-\infty}^{\infty} g(\omega)\,\mathrm{d}[\exp(j\omega t)]$$

$$= \frac{-1}{j} \int_{-\infty}^{\infty} g'(\omega) \exp(j\omega t)\,\mathrm{d}\omega \tag{A.38}$$

provided that $g(\omega)$ vanishes when ω tends to plus or minus infinity.

Substitution of the result of equation (A.38) into equation (A.37) gives:

$$\int_{-\infty}^{\infty} [j\omega\, g(\omega) + 2jkg'(\omega)]\, \exp(j\omega t)\, d\omega = 0$$

Since the term in brackets now depends only on ω, we can use the inverse transform to show that it must vanish, *i.e.*

$$\omega g(\omega) + 2kg'(\omega) = 0$$

This differential equation for $g(\omega)$ has the general solution:

$$g(\omega) = A \exp(-\omega^2/4k) \qquad (A.39)$$

where A is an arbitrary constant, the value of which will be determined below. We note that $g(\omega)$ does tend to zero when ω tends to either plus or minus infinity, so that the step involved in equation (A.38) is justified.

The constant A can be found by using equation (A.34), derived in the previous section. This gives:

$$\int_{-\infty}^{\infty} \exp(-2kx^2)\, dx = \frac{1}{2\pi} \int_{-\infty}^{\infty} |A|^2 \exp\left(-\frac{\omega^2}{2k}\right) d\omega$$

Let

$$y = \sqrt{2k}\, x \quad \text{and} \quad u = \omega/\sqrt{2k}$$

Then,

$$\frac{1}{\sqrt{2k}} \int_{-\infty}^{\infty} \exp(-y^2)\, dy = \frac{|A|^2 \sqrt{2k}}{2\pi} \int_{-\infty}^{\infty} \exp(-u^2)\, du$$

The integrals in this equation are identical and so

$$|A|^2 = \pi/k$$

From the transform relation,

$$g(\omega) = A \exp(-\omega^2/4k) = \int_{-\infty}^{\infty} \exp(-kt^2 + j\omega t)\, dt$$

and by letting ω be zero, we have:

$$A = \int_{-\infty}^{\infty} \exp(-kt^2)\, dt$$

The integrand is positive for all values of t, and so A is clearly a real positive number. It follows from the expression for $|A|^2$ that:

$$A = \sqrt{\pi/k} \qquad (A.40)$$

and so the transform of the Gaussian pulse $\exp(-kt^2)$ is $(\pi/k)^{\frac{1}{2}}\exp(-\omega^2/4k)$. The result quoted in Chapter 2 follows by letting $k \doteq 2\pi/\tau^2$.

The limiting case of a Gaussian pulse as k tends to infinity is an impulse function. To show this we consider

$$f_1(t) = C\exp(-kt^2)$$

where C is chosen to make

$$\int_{-\infty}^{\infty} f_1(t)\,\mathrm{d}t = 1$$

Since

$$\int_{-\infty}^{\infty} \exp(-kt^2)\,\mathrm{d}t = A = \sqrt{\pi/k}\,,$$

$$f_1(t) = \sqrt{k/\pi}\,\exp(-kt^2), \quad i.e. \quad C = \sqrt{k/\pi}$$

When k tends to infinity, $f_1(t)$ becomes infinite at $t = 0$ and zero elsewhere. Also $\int_{-\infty}^{\infty} f_1(t)\,\mathrm{d}t$ is unity and so $f_1(t)$ tends to the impulse function $\delta(t)$. The corresponding transform is

$$\lim_{k\to\infty} g_1(\omega) = \lim_{k\to\infty} CA\exp(-\omega^2/4k) = \lim_{k\to\infty} \exp\left(-\frac{\omega^2}{4k}\right)$$

$$= 1$$

in agreement with the result obtained in Section A.3.4.

A.3.7. THE COSINE-SQUARED PULSE

The cosine-squared pulse has the time waveform:

$$\begin{aligned} f(t) &= \cos^2(\pi t/\tau) : -\tau/2 \le t \le \tau/2\\ &= 0 \qquad\qquad : |t| \ge \tau/2 \end{aligned} \tag{A.41}$$

The transform can be evaluated by direct substitution into the integral expression for $g(\omega)$, but a more direct derivation can be found by using the convolution theorem. The time waveform is expressed as:

$$f(t) = f_1(t)f_2(t)$$

where

$$f_1(t) = \cos^2(\pi t/\tau) = \tfrac{1}{2}[1 + \cos(2\pi t/\tau)]$$

and

$$\begin{aligned} f_2(t) &= 1 : -\tau/2 \le t \le \tau/2\\ &= 0 : |t| \ge \tau/2 \end{aligned}$$

The transform of $f_1(t)$ follows from the result at the end of Section A.3.4 and is

$$g_1(\omega) = \pi\delta(\omega) + \frac{\pi}{2}\delta\left(\omega - \frac{2\pi}{\tau}\right) + \frac{\pi}{2}\delta\left(\omega + \frac{2\pi}{\tau}\right)$$

The waveform $f_2(t)$ is the rectangular pulse examined in Section A.3.3 and its transform is given by equation (A.25) as:

$$g_2(\omega) = \frac{2\sin(\omega\tau/2)}{\omega}$$

Substitution of the expressions for $g_1(\omega)$ and $g_2(\omega)$ in equation (A.35) gives the following result for the transform of $f(t)$:

$$
\begin{aligned}
g(\omega) &= \frac{1}{2\pi}\int_{-\infty}^{\infty} g_1(\omega - w)g_2(w)\,\mathrm{d}w \\
&= \int_{-\infty}^{\infty}\left[\tfrac{1}{2}\delta(\omega - w) + \tfrac{1}{4}\delta\left(\omega - \frac{2\pi}{\tau} - w\right)\right. \\
&\qquad \left. + \tfrac{1}{4}\delta\left(\omega + \frac{2\pi}{\tau} - w\right)\right]\frac{2\cdot\sin(w\tau/2)}{w}\,\mathrm{d}w \\
&:= \tfrac{1}{2}\left[\frac{2\sin(\omega\tau/2)}{\omega} + \frac{\sin(\omega\tau/2 - \pi)}{\omega - 2\pi/\tau} + \frac{\sin(\omega\tau/2 + \pi)}{\omega + 2\pi/\tau}\right] \\
&= \frac{\sin(\omega\tau/2)}{2}\left[\frac{2}{\omega} - \frac{\omega + 2\pi/\tau + \omega - 2\pi/\tau}{\omega^2 - 4\pi^2/\tau^2}\right] \\
&= \sin(\omega\tau/2)\cdot\frac{-4\pi^2/\tau^2}{\omega(\omega^2 - 4\pi^2/\tau^2)} \\
&= \frac{4\pi^2\sin(\omega\tau/2)}{\omega(4\pi^2 - \omega^2\tau^2)}
\end{aligned}
\tag{A.42}
$$

The significance of this result is best seen by replacing $\omega\tau/2$ by x as in Section A.3.3. Then:

$$g(\omega) = \frac{\pi^2\tau}{2}\cdot\frac{(\sin x)}{x}\cdot\frac{1}{(\pi^2 - x^2)} \tag{A.43}$$

The spectrum is thus the $(\sin x)/x$ function, corresponding to a rectangular pulse multiplied by the factor $1/(\pi^2 - x^2)$. When x equals $\pm\pi$, this second factor is infinite but $g(\omega)$ remains finite since $\sin x$ also vanishes at these values. The first zeros of $(\sin x)/x$ are thus suppressed by the second factor. For increasing values of $|x|$, greater than π, the second factor decreases rapidly, thus reducing the importance of the lobes of the spectrum. This, of course, is the reason for the use of a cosine-squared pulse in preference to a rectangular pulse.

A.4. Applications of Fourier Transforms

A.4.1. NON-PERIODIC WAVEFORMS

One of the most important uses of the Fourier transform is in determining the behaviour of a linear circuit when it is excited by a non-periodic waveform. This type of problem arises in all aspects of communications and it is because of this that so much emphasis is placed on the Fourier transform. The restriction to linear circuits is essential since the Fourier method relies on the addition of sinusoidal waves; the principle of superposition is therefore implicit in the method and this principle is only valid for linear circuits.

The class of problem with which we are concerned is illustrated by Fig. A.3, where the block represents a general four terminal network. If the input voltage is taken as $\exp(j\omega t)$, the output voltage will be proportional to $\exp(j\omega t)$ and can be written as:

$$v_o = A(\omega) \exp(j\omega t) \qquad (A.44)$$

where $A(\omega)$ is a complex constant, the amplitude and phase both depending in general on the angular frequency, ω. $A(\omega)$ may be measured by steady-state experiments or may be calculated from the known form of the network.

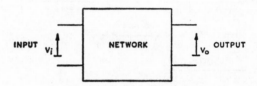

A.3. Basic four-terminal network.

Suppose now that the input voltage is

$$v_i = f(t) \qquad (A.45)$$

where $f(t)$ is an arbitrary function of time. We resolve $f(t)$ into its spectrum $g(\omega)$ by using equation (A.16). Each component sinusoid in the output voltage is given by the corresponding input component multiplied by the transfer function $A(\omega)$ and the output spectrum is therefore:

$$g_o(\omega) = A(\omega)g(\omega) \qquad (A.46)$$

The output voltage is therefore given by equation (A.18) as:

$$v_o(t) = \frac{1}{2\pi} \int_{-\infty}^{\infty} A(\omega)g(\omega) \exp(j\omega t) \, . \, d\omega \qquad (A.47)$$

We will now consider a few simple examples of this method.

Example 1

Consider the case of a single frequency input, for which we already know the answer.

Let

$$v_i = \cos(\omega_0 t)$$

Then

$$v_o = |A(\omega_0)| \cos(\omega_0 t - \phi)$$

where

$$A(\omega_0) = |A(\omega_0)| \exp(-j\phi)$$

Using the Fourier transform method, we have from Table A.2

$$g(\omega) = \pi[\delta(\omega - \omega_0) + \delta(\omega + \omega_0)]$$

Hence,

$$g_0(\omega) = \pi A(\omega)[\delta(\omega - \omega_0) + \delta(\omega + \omega_0)]$$

and

$$v_o = \tfrac{1}{2} \int_{-\infty}^{\infty} A(\omega)[\delta(\omega - \omega_0) + \delta(\omega + \omega_0)] \exp(j\omega t) d\omega$$

The integrals are easily evaluated, using the properties of the delta function, and so:

$$v_o = \tfrac{1}{2}[A(\omega_0) \exp(j\omega_0 t) + A(-\omega_0) \exp(-j\omega_0 t)]$$

Now, since v_o must be a real function of time, equation (A.20) shows that:

$$g_0^*(\omega) = g_0(-\omega)$$

i.e.

$$A^*(\omega)g^*(\omega) = A(-\omega)g(-\omega)$$

and

$$A^*(\omega) = A(-\omega)$$

since $g(\omega)$ is also derived from a real function of time. We now have:

$$v_o = \tfrac{1}{2}[A(\omega_0) \exp(j\omega_0 t) + A^*(\omega_0) \exp(-j\omega_0 t)]$$

$$= |A(\omega_0)| \cos(\omega_0 t - \phi)$$

agreeing with the answer given by the network properties at frequency ω_0.

Example 2

Suppose the network corresponds to a matched transmission line which has a phase velocity v, independent of frequency.
Then,

$$A(\omega) = \exp[-j\omega l/v]$$

where l is the length of the line.

From equation (A.47)

$$
\begin{aligned}
v_o(t) &= \frac{1}{2\pi} \int_{-\infty}^{\infty} g(\omega) \exp[-j\omega l/v + j\omega t] d\omega \\
&= \frac{1}{2\pi} \int_{-\infty}^{\infty} g(\omega) \exp[j\omega(t - l/v)] d\omega \\
&= v_i(t - l/v)
\end{aligned}
$$

The output voltage is therefore an exact replica of the input voltage but delayed by time l/v. The network behaves as an ideal delay line.

Example 3

Let

$$v_i(t) = \delta(t)$$

Then,

$$g(\omega) = 1$$

$$g_0(\omega) = A(\omega)$$

and

$$v_o(t) = \frac{1}{2\pi} \int_{-\infty}^{\infty} A(\omega) \exp(j\omega t) d\omega$$

This shows that if a very short pulse is applied to a network, the output voltage is the inverse Fourier transform of the frequency response of the network. It should be noted that $v_o(t)$ is the impulse response function, denoted $h(t)$ in Section 5.2.2.

Example 4

This example shows how the Fourier transform method may be applied to determine the transient behaviour of a network such as the RC combination in Fig. A.4. We wish to determine the output voltage if a step voltage is applied to the input. The input is therefore:

$$v_i(t) = 0 : t \leq 0$$

$$= 1 : t \geq 0$$

The Fourier transform of $v_i(t)$ is not defined for the integral is divergent but we may solve the problem for:

$$v_i(t) = 0 : t \leq 0$$
$$= \exp(-\alpha t) : t \geq 0$$

and then allow α to tend to zero.

We have,

$$g(\omega) = 1/(\alpha + j\omega)$$

$$A(\omega) = j\omega CR/(1 + j\omega CR)$$

$$g_0(\omega) = j\omega CR/[(\alpha + j\omega)(1 + j\omega CR)]$$

$$v_o(t) = \int_{-\infty}^{\infty} \frac{j\omega CR \exp(j\omega t)\,d\omega}{2\pi(1 + j\omega CR)(\alpha + j\omega)}$$

$$= \frac{1}{2\pi} \int_{-\infty}^{\infty} \left[\frac{\alpha}{\alpha + j\omega} - \frac{1}{1 + j\omega CR}\right] \frac{CR \exp(j\omega t)\,d\omega}{\alpha CR - 1}$$

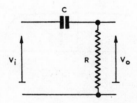

Fig. A.4. Simple RC network.

The use of relation 1 in Table A.2 gives:

$$v_o(t) = 0 : t \leq 0$$
$$= \frac{CR}{\alpha CR - 1}\left[\alpha \exp(-\alpha t) - \frac{1}{CR}\exp\left(-\frac{t}{CR}\right)\right] : t \geq 0.$$

We now let α tend to zero to correspond to the step function input, when:

$$v_o(t) = 0 : t \leq 0$$
$$= \exp(-t/CR) : t \geq 0$$

In practice, transient problems of this kind are solved more easily by Laplace transforms as discussed later.

A.4.2. THE SPECTRUM OF AN AMPLITUDE MODULATED CARRIER

The general expression for an amplitude modulated carrier is:

$$v(t) = A[1 + f_m(t)] \cos(\omega_c t + \phi) \tag{A.48}$$

where A, ω_c, ϕ are respectively the amplitude, the angular frequency and the phase of the unmodulated carrier, and $f_m(t)$ is the modulation waveform, which will be considered to be a real function of time.

We assume that the spectrum of $f_m(t)$ is $g_m(\omega)$ where:

$$g_m(\omega) = \int_{-\infty}^{\infty} f_m(t) \exp(-j\omega t)\, dt \tag{A.49}$$

and since $f_m(t)$ is a real waveform, $g_m(\omega)$ will have positive and negative frequency components satisfying the conditions discussed in Section A.3. The amplitude and phase of $g_m(\omega)$ therefore have the general form shown in Fig. A.5(a), it being assumed that the largest modulation frequency is ω_0.

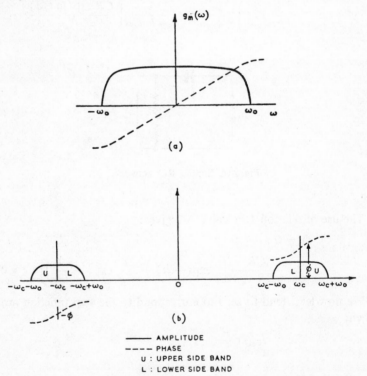

Fig. A.5. Spectrum of amplitude modulated carrier (a) Modulation spectrum: (b) modulated carrier spectrum.

The spectrum of the modulated carrier is:

$$g_c(\omega) = \int_{-\infty}^{\infty} v(t) \exp(-j\omega t)\,\mathrm{d}t$$

$$= \frac{A}{2} \int_{-\infty}^{\infty} [1 + f_m(t)][\exp(j\overline{\omega_c - \omega}\,t + j\phi) +$$

$$+ \exp(-j\overline{\omega_c + \omega}\,t - j\phi)]\,\mathrm{d}t$$

in which $\cos(\omega_c t + \phi)$ has been replaced by its equivalent in terms exponentials. The integration can be carried out by using equation (A.49) and result no. 3 in the dictionary, giving:

$$g_c(\omega) = \pi A \exp(+j\phi)\delta(\omega - \omega_c) + \pi A \exp(-j\phi)\delta(\omega + \omega_c)$$

$$+ \tfrac{1}{2}A \exp(j\phi)g_m(\omega - \omega_c) + \tfrac{1}{2}A \exp(-j\phi)g_m(\omega + \omega_c) \quad \text{(A.50)}$$

This spectrum consists of two lines, at ω_c and $-\omega_c$, corresponding to the unmodulated carrier and two continuous sections centred on ω_c and $-\omega_c$ as shown in Fig. A.5(b). These sections form mirror images of each other with respect to $\omega = 0$, and once again show the relation between positive and negative frequencies associated with real time functions. It suffices to consider one of them in detail. The spectrum $g_m(\omega - \omega_c)$ simply consists of the modulation spectrum $g_m(\omega)$, displaced by the carrier angular frequency, ω_c. The positive frequencies in $g_m(\omega)$ form the upper side-band of the modulated carrier, the negative frequencies in $g_m(\omega)$ form the lower sideband. The modulation process can therefore be considered as being essentially a frequency shifting process and demodulation consists of a frequency shift downwards by an amount ω_c. Any linear frequency shifting operation can be regarded as a translation of the frequency spectrum by an amount equal to the frequency difference.

The negative frequencies associated with any real time waveform can always be deduced from the positive frequencies by using the result $g(-\omega) = g^*(\omega)$, derived in Section A.3. In calculations, it is common practice to consider only the positive part of the spectrum in Fig. A.5. Examination of the above calculations shows that this is equivalent to taking the carrier wave as $\tfrac{1}{2}A \exp(j\omega_c t + j\phi)$. The factor $\tfrac{1}{2}$ is usually omitted for simplicity, the positive frequency spectrum then being doubled.

A.4.3. GROUP DELAY

The Fourier transform provides a convenient method for establishing the delay in the transmission of a pulse through a network. We consider a pulse which contains only a small range of frequencies, centred around the angular frequency, ω_0. A real waveform will of course have the complementary spectrum of negative frequencies centred on $-\omega_0$, but the argument in the previous section shows that we can investigate the behaviour in terms of positive frequencies only. We therefore assume that the spectrum of the input to the network, $g(\omega)$, differs from zero only for frequencies in the immediate vicinity of ω_0. The output voltage is given by equation (A.47) and only the small frequency range $(\omega_0 - \omega_1)$ to $(\omega_0 + \omega_1)$, where ω_1 is a measure of the bandwidth of the input signal need be considered, so that

$$v_o(t) = \frac{1}{2\pi} \int_{\omega_0 - \omega_1}^{\omega_0 + \omega_1} A(\omega) g(\omega) \exp(j\omega t) \, d\omega$$

The transfer function $A(\omega)$ can be approximated by its value in the vicinity of ω_0, and an examination of the form of the integral shows that a first approximation can be made by neglecting the variation of $|A(\omega)|$ and replacing $\arg A(\omega)$ by the first two terms of its Taylor expansion, *i.e.* we take:

$$A(\omega) \doteq |A(\omega_0)| \exp[-j\phi(\omega_0) - j(\omega - \omega_0)\phi'(\omega_0)]$$

where $\phi'(\omega_0)$ is the derivative of the argument, or phase, with respect to ω, evaluated at ω_0.

We now have for the output:

$$v_o(t) = \frac{1}{2\pi} \int_{\omega_0 - \omega_1}^{\omega_0 + \omega_1} |A(\omega_0)| g(\omega) \exp[-j\phi(\omega_0) - j(\omega - \omega_0)\phi'(\omega_0) \\ + j\omega t] \, d\omega$$

$$= \frac{1}{2\pi} |A(\omega_0)| \exp[-j\phi(\omega_0) + j\omega_0\phi'(\omega_0)]$$

$$\int_{\omega_0 - \omega_1}^{\omega_0 + \omega_1} g(\omega) \exp[j\omega\{-\phi'(\omega_0) + t\}] \, d\omega$$

The input signal is:

$$v_i(t) = \frac{1}{2\pi} \int_{\omega_0 - \omega_1}^{\omega_0 + \omega_1} g(\omega) \exp(j\omega t) \, d\omega$$

and comparison of the above two equations gives:

$$v_o(t) = |A(\omega_0)| \exp[-j\phi(\omega_0) + j\omega_0\phi'(\omega_0)]v_i[t - \phi'(\omega_0)] \quad \text{(A.51)}$$

Since the terms multiplying $v_i[t - \phi'(\omega_0)]$ are constant, the shape of the output waveform is the same as that of the input waveform, but is delayed by a time τ given by:

$$\tau = \phi'(\omega_o) = \left[\frac{d\phi}{d\omega}\right]_{\omega = \omega_o} \quad \text{(A.52)}$$

The quantity τ is called the group delay of the network, and is equal to the rate of increase with angular frequency of the phase shift through the network. The group delay is positive for all physical networks.

It is essential to recognize the restrictions which must apply before equations (A.51) and (A.52) are valid. They are: (*i*) The input pulse contains only a narrow range of frequencies and (*ii*) the amplitude response of the network does not vary too rapidly with frequency.

The second restriction shows that the result must be used with caution in situations such as the frequency band near a cut-off point in a filter.

It may be noted that the result obtained in Example 2 above follows at once from equation (A.52). The point about this example is that the above representation of $A(\omega)$ is true for all values of ω. This explains why there is no need in Example 2 to restrict the frequency spectrum of the signal involved.

A.5. Equivalence of the Fourier Integral and the Duhamel Integral Methods

In Chapter 5 two methods for determining the response of a linear channel are described. The first is the Fourier Integral method, as discussed in Section A.4.1. This leads to the results of equations (A.45) and (A.47), *i.e.* if

$$v_i = f(t) \tag{A.45}$$

$$v_o = \frac{1}{2\pi} \int_{-\infty}^{\infty} A(\omega)g(\omega) \exp(j\omega t)\,d\omega \tag{A.47}$$

where $g(\omega)$ is the transform of $f(t)$ and $A(\omega)$ is the complex response function of the channel.

The second method begins by regarding the input function as a succession of impulse functions. Mathematically this is equivalent to writing (see equation A.29):

$$f(t) = \int_{-\infty}^{\infty} f(t')\delta(t - t')\,dt' \tag{A.53}$$

Now, suppose the response of the network to the unit impulse $\delta(t)$ is $h(t)$. Then the response to the element $f(t')dt' \times \delta(t - t')$ of the time waveform is $f(t')dt' \times h(t - t')$ and the overall response obtained by summing all the contributions is the output voltage:

$$v_o = \int_{-\infty}^{\infty} f(t')h(t - t')\,dt' \tag{A.54}$$

It now remains to show that the results given by equations (A.47) and (A.54) are identical. The former shows that the Fourier transform of v_o is $A(\omega)g(\omega)$ and it will suffice to show that the Fourier transform of the integral on the right of equation (A.54) has the same value. This integral is identical in form to that (equation (A.33)) which occurs in the Convolution theorem and the use of this gives the Fourier transform as $g(\omega)$ $H(\omega)$, where $g(\omega)$, $H(\omega)$ are the transforms of $f(t)$ and $h(t)$ respectively. The time waveform $h(t)$ is the output of the network when excited by a unit impulse and it has already been shown in Example 3 of Section A.4.1 that the Fourier transform of this voltage is $A(\omega)$, i.e. $H(\omega) = A(\omega)$. The Fourier transform of the integral on the right of equation (A.54) is therefore $g(\omega) A(\omega)$, in agreement with the result in equation (A.46).

The result in equation (A.54) is mathematically correct, whatever the form of the response function $h(t)$. Clearly, no physical network can give an output prior to the input signal being applied. Since $h(t)$ is the response to the unit impulse $\delta(t)$, occurring at $t = 0$, $h(t)$ must be zero for t less than zero. Similarly, $h(t - t')$ is zero for t less than t', i.e. t' greater than t. Equation (A.54) may thus be written as:

$$v_o = \int_{-\infty}^{t} f(t')h(t - t')\,dt' \qquad (A.55)$$

The result given in equation (5.4) follows if $f(t')$ is assumed to be zero for t less than zero.

A.6. Other Uses of Fourier Transforms

There are many other uses of Fourier transforms in communication theory. These are beyond the scope of the present treatment. To exploit the full value of these transforms, it becomes necessary to permit the frequency variable to be complex. An account of the development which then proceeds will be found in references 6 and 8 listed at the end of this Appendix.

The idea of spectra can be applied with great advantage to the solution of electromagnetic field problems. Any electromagnetic field can be interpreted as a spectrum of plane waves, the variable corresponding to frequency being the direction of propagation of a constituent plane wave, and the variable corresponding to time being a spatial co-ordinate. The use of this concept has led to considerable advances in the solution of those electromagnetic field problems of most interest to communication engineers, namely, those concerned with radiation from aerials and radio wave propagation. The similarity of the mathematical techniques used for solving such problems to those used for handling circuit problems provides a most valuable conceptual link between the two subjects and enables advances in one subject to give corresponding improvements in the other. For example, the two topics of the design of bandpass filters and the production of aerial radiation patterns of desired shapes proceed on related lines. An elementary example is given in Section 9.3.2.

Another branch of communications in which Fourier transforms are widely used is the study of noise. A brief introduction to the methods used is given in Chapter 6. The principal difference in approach is that attention is switched from spectra derived from voltage or current waveforms to those showing the distribution of power throughout the frequency range. In this connection, the convolution theorem discussed in Section A.3.5 is a most valuable tool.

A.7. Laplace Transforms

A.7.1. DEFINITION

It has been seen that Fourier transforms can be used for the solution of transient problems, but the integration required to obtain the output time waveform from its frequency spectrum is often difficult to perform. A more convenient method in most cases is the use of the Laplace transform. An alternative, which in the algebraic details of the solution is very similar, is the operational method devised by Heaviside. The Laplace method is the easier to justify rigourously and the present tendency is for it to supersede the Heaviside method. In the present context, the emphasis on the idea of a transform makes the choice of the Laplace method preferable to the Heaviside one, but it should be noted that some of the results used are obtained as a consequence of Heaviside's work.

The first thing to be considered is the form of the transform to be used. In the discussion of Fourier transforms, the definition of the transform followed quite naturally from the idea of the Fourier series, which led to the idea of building up waveforms from sinusoidal functions. The selection of a suitable transform for dealing with transients is not so obvious,but the following considerations lead to a satisfactory definition:

(a) In the solution of transient problems we often encounter terms of the type $\exp(-\alpha t)$, where α is a real decay constant.

(b) We are interested in what happens after a switching operation, so that we need consider only the time interval, $0 \leqslant t \leqslant \infty$, by arranging for switching to occur at $t = 0$.

(c) The calculation of transients involves the solution of differential equations, so we seek a transform which converts the operation of differentiation to an algebraic operation.

The first two points suggest a transform defined by:

$$\mathscr{L}\{f(t)\} = \int_0^\infty f(t) \exp(-pt)\,\mathrm{d}t \qquad (A.56)$$

in which we are using the notation:

$$\mathscr{L}\{f(t)\} = \text{Laplace transform of } f(t)$$

The Laplace transform as defined by equation (A.56) is a function of the variable p, and an alternative notation is $\bar{f}(p)$. We will use both notations as convenient. We now examine the third point raised above, and consider the Laplace transform of the time derivative of $f(t)$.

$$\mathscr{L}\{f'(t)\} = \int_0^\infty f'(t)\exp(-pt)\,dt$$

where

$$f'(t) = \frac{df(t)}{dt}$$

Integrating the right-hand side by parts, we have

$$\mathscr{L}\{f'(t)\} = [f(t)\exp(-pt)]_0^\infty - \int_0^\infty f(t)\frac{d}{dt}[\exp(-pt)]\,dt$$

$$= -f(0) - \int_0^\infty f(t)(-p)\exp(-pt)\,dt$$

$$= p\mathscr{L}\{f(t)\} - f(0) \tag{A.57}$$

The term $f(t)\exp(-pt)$ vanishes at the upper limit of infinity, provided $f(t)$ is finite as t tends to infinity and p is a quantity with a positive real part. The right-hand side of equation (A.57) involves only the transform of $f(t)$ multiplied by p and a constant term $f(0)$ and is therefore obtained from $f(0)$ by a purely algebraic operation. We have therefore satisfied the third point, in that the operation of differentiating $f(t)$ corresponds to an algebraic operation on the transform.

Similar results apply for derivatives of higher orders. For example, considering the second derivative, we have:

$$\mathscr{L}\{f''(t)\} = p\mathscr{L}\{f'(t)\} - f'(0)$$

$$= p[p\mathscr{L}\{f(t)\} - f(0)] - f'(0)$$

$$= p^2\mathscr{L}\{f(t)\} - pf(0) - f'(0)$$

This process can be repeated indefinitely and the general result for the nth derivative is:

$$\mathscr{L}\{f^{(n)}(t)\} = p^n\mathscr{L}\{f(t)\} - p^{n-1}f(0) - p^{n-2}f'(0)\ldots$$

$$- p^{n-r-1}f^{(r)}(0)\ldots - f^{(n-1)}(0) \tag{A.58}$$

where $f^{(r)}(0)$ is the value of the rth derivative of $f(t)$, evaluated at $t = 0$.

A.7.2. RELATION OF LAPLACE TRANSFORMS TO FOURIER TRANSFORMS

Laplace transforms are only defined for the behaviour of the time waveform for positive values of t, whereas Fourier transforms depend on the behaviour for all values of t. We may evaluate both by considering a function $f_1(t)$ defined such that:

$$f_1(t) = 0 : t < 0$$
$$= f(t) : t \geq 0 \tag{A.59}$$

Then, from equation (A.56) we have:

$$\mathscr{L}\{f_1(t)\} = \bar{f}_1(p) = \int_0^\infty f(t) \exp(-pt)\,dt \tag{A.60}$$

and from equation A.16 we have:

$$g_1(\omega) = \int_0^\infty f(t) \exp(-j\omega t)\,dt \tag{A.61}$$

since $f_1(t)$ is zero for $t < 0$.

Examination of equations (A.60) and (A.61) shows that they differ only in that p is replaced by $j\omega$.

Hence we have:

$$g_1(\omega) = \bar{f}_1(j\omega) \tag{A.62}$$

$$\bar{f}_1(p) = g_1(-jp) \tag{A.63}$$

This shows that the Fourier and Laplace transforms of the function defined in equation (A.59) are very simply related.

Further, we have seen that if the Fourier transform is known, the time function can be evaluated by using equation (A.18), *i.e.*

$$f_1(t) = \frac{1}{2\pi} \int_{-\infty}^{\infty} g_1(\omega) \exp(j\omega t)\,d\omega \tag{A.64}$$

We can replace the Fourier transform by the Laplace transform and so obtain an expression for the inverse operation. We have:

$$f_1(t) = \frac{1}{2\pi} \int_{-\infty}^{\infty} \bar{f}_1(j\omega) \exp(j\omega t)\,d\omega \tag{A.65}$$

This is usually written in such a way that the integration variable is p, equal to $j\omega$.

If $p = j\omega$, then $dp = j\,d\omega$, and the integration for p must be performed from $-j\infty$ to $j\infty$, *i.e.*

$$f_1(t) = \frac{1}{2\pi} \int_{-j\infty}^{j\infty} \bar{f}_1(p) \exp(pt)\,\frac{dp}{j}$$
$$= \frac{1}{2\pi j} \int_{-j\infty}^{j\infty} \bar{f}_1(p) \exp(pt)\,dp \tag{A.66}$$

The imaginary values of the limits on this integral make it difficult to evaluate, and methods beyond the scope of the present treatment are used. We will rely on the use of a dictionary for the process of obtaining $f_1(t)$ from $\bar{f}_1(p)$, the required results being shown in Table A.3. Equation

(A.66) is, however, instructive in showing that there is a unique time waveform $f_1(t)$, which is zero for t negative, associated with any Laplace transform $\bar{f}_1(p)$.

The integral in equation (A.66) is most easily evaluated by using the method of contour integration, which requires a detailed knowledge of complex function theory. The reconciliation of Heaviside's operational method and the more formal Laplace Transform was first effected rigorously by Carson and Bromwhich by a study of this integral. It may be noted that the integral, which is often known as the Bromwich-Carson integral, provides the basis for a rigorous derivation of the Heaviside expansion theorem discussed in Section A.7.6.

A.7.3. THE APPLICATION OF LAPLACE TRANSFORMS TO THE SOLUTION OF
DIFFERENTIAL EQUATIONS

The type of differential equation which occurs in transient problems is:

$$L \frac{di(t)}{dt} + Ri(t) + \frac{1}{C} \int_0^t i(t) dt = e(t) - \frac{q_0}{C} \qquad (A.67)$$

where L, R, C are the circuit constants,

      $e(t)$ is the applied e.m.f., assumed known for $t \geqslant 0$,

      $i(t)$ is the unknown current,

and q_0 is the initial charge on the capacitor, C.

We may convert this equation to an ordinary second-order differential equation by replacing the current $i(t)$ in terms of a charge, $q(t)$, defined by:

$$q(t) = \int_0^t i(t) dt \qquad (A.68)$$

but since we are usually more interested in the nature of the current, we will solve for $i(t)$ directly. We therefore need the transform of $\int_0^t i(t)\,dt$.

$$\mathscr{L}\left\{\int_0^t i(t) dt\right\} = \int_0^\infty \left[\int_0^t i(t') dt'\right] \exp(-pt) dt \qquad (A.69)$$

Reverse the order of the integrations

$$\mathscr{L}\left\{\int_0^t i(t) dt\right\} = \int_0^\infty \int_{t'}^\infty \exp(-pt) dt\, i(t') dt'$$

$$= \int_0^\infty i(t')[\exp(-pt)/(-p)]_{t'}^\infty\, dt'$$

$$= \frac{1}{p} \int_0^\infty i(t') \exp(-pt') dt'$$

$$= \frac{1}{p} \mathscr{L}\{i(t)\} \qquad (A.70)$$

The operation of integrating the time function therefore corresponds to the algebraic operation of multiplying the transform by $1/p$.

The complete solution of any differential equation requires a knowledge of the initial conditions. In equation (A.67), one of these, the initial charge on the capacitor appears explicitly. A second is needed since the equation is of the second order and this will appear when transforms are taken because of the result in equation (A.57). One of the great advantages of the transform method is that the initial conditions are included at an early stage in the calculation.

We now take transforms of all the quantities in equation (A.67), using the results of equations (A.57) and (A.70). The compact notation, $\bar{i}(p), \bar{e}(p)$ for the transforms of $i(t)$, $e(t)$ is obviously now convenient and we obtain:

$$L\{p\bar{i}(p) - i(0)\} + R\bar{i}(p) + \frac{1}{pC}\,\bar{i}(p) = \bar{e}(p) - \mathscr{L}\{q_0/C\} \quad (A.71)$$

The transform of the constant, q_0/C can be obtained by direct integration of the defining equation (A.56), giving:

$$\mathscr{L}\{q_0/C\} = q_0/pC \qquad (A.72)$$

Equation (A.71) is now solved for:

$$\bar{i}(p) = \frac{\bar{e}(p) - \dfrac{q_0}{pC} + Li(0)}{pL + R + \dfrac{1}{pC}} \qquad (A.73)$$

To obtain $i(t)$, we must invert $\bar{i}(p)$.

A.7.4. THE INVERSION OF LAPLACE TRANSFORMS

The process of inverting a Fourier transform is basically identical to that of calculating the transform. For Laplace transforms this is not the case, and we have found in Section A.7.2 that the general method of inversion is rather complicated. In most practical problems, however, the need to use this general method can be avoided by following one or more of three lines of attack:

(a) The use of a dictionary compiled by calculating the transforms of those functions most likely to appear.

(b) A result taken from the Heaviside method, which reduces most of the transforms encountered to standard forms.

(c) Certain general theorems relating to Laplace transforms.

The results required are collected in Table A.3, and most of them can be proved by straightforward manipulation of the basic definition, equation (A.56). The more important results will be derived in later sections as they are required. The methods of attack are best illustrated by considering particular examples, and as a preliminary to this we consider the question of initial conditions in switching problems.

A.7.5. INITIAL CONDITIONS IN A CIRCUIT AFTER SWITCHING

Suppose we have a circuit in which currents are flowing prior to a switching operation assumed to occur at time, $t = 0$. To determine the currents after $t = 0$, we must know the differential equations for the circuit for times $t \geqslant 0$, and the 'initial conditions'. These initial conditions are mathematically the additional constants required to give a unique solution: electrically they must, of course, be related to the state of the circuit before switching and to the instantaneous changes caused by the switching The essential principle in determining the relevant constants is the fact that stored energy cannot change instantaneously: the energy stored in the circuit reactances is not therefore changed by the switching operation. This principle will always provide all the information needed. We can readily convert it to a more convenient form for application to transient problems by examining inductances and capacitances separately. We do not need to think about resistances since there is no stored energy associated with them.

(a) Inductances: The energy stored in an inductance L, through which a current $i(t)$ is flowing, is $\frac{1}{2}Li^2(t)$. The current in an inductance immediately after switching must therefore be the same as that immediately before switching.

(b) Capacitances: The energy stored in a capacitance C, which carries a charge $q(t)$, is $\frac{1}{2}q^2(t)/C$. The charge in a capacitance immediately after switching must therefore be the same as that immediately before switching. Since the voltage across a capacitance is directly proportional to the charge, it too must be unaltered by the switching.

These two statements cover all the situations which arise in normal transient problems. An apparent exception occurs if an uncharged capacitor is switched in parallel with a charged one, for then the voltage across each must become instantaneously equal. The difficulty is immediately resolved if a small resistance is assumed to exist in the connecting leads. Other similar anomalous cases can always be covered by the insertion of one or more additional resistors in the circuit. These resistances can be equated to zero in the final answer.

Example 5 (See Fig. A.6)

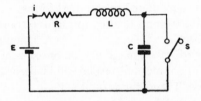

Fig. A.6. Circuit for Example 5.

Switching with a d.c. generator in an oscillatory circuit

The circuit is operating under steady-state conditions with S closed. At $t = 0$, S is opened. Find $i(t)$

(a) Initial conditions: Up to $t = 0$, $i = E/R$ and the voltage across C is 0.

After switching, the current in L and the voltage across C must have the same value as before switching.

Hence

$$i(0) = E/R . \quad \text{Charge on } C, \; q(0) = 0$$

(b) Circuit differential equation after switching:

$$L \frac{di(t)}{dt} + Ri(t) + \frac{1}{C} \int_0^t i(t)\,dt = E$$

(c) Current transform

Taking Laplace transforms on both sides of the d.e., we have:

$$pL\,\bar{i}(p) - Li(0) + R\bar{i}(p) + \frac{1}{pC}\,\bar{i}(p) = E/p$$

i.e.

$$\bar{i}(p) = \frac{E/p + Li(0)}{pL + R + 1/pC}$$

$$= \frac{E + pLE/R}{p^2 L + pR + 1/C}$$

(d) Current waveform

We wish to manipulate the expression for $\bar{i}(p)$ into a form which can be recognized from the dictionary, and inspection of this suggests the use of forms 11 and 12.

Now

$$p^2 L + pR + \frac{1}{C} = L\left[\left(p + \frac{R}{2L}\right)^2 + \frac{1}{LC} - \frac{R^2}{4L^2}\right]$$

so that if

$$a = \left[\frac{1}{LC} - \frac{R^2}{4L^2}\right]^{\frac{1}{2}} \quad \left(\text{assuming } \frac{1}{LC} > \frac{R^2}{4L^2}\right)$$

$$b = \frac{R}{2L}$$

the denominator is in the correct form.

Also

$$E + pL\frac{E}{R} = E - \frac{E}{2} + \frac{LE}{R}\left(p + \frac{R}{2L}\right) = \frac{E}{2} + \frac{LE}{R}(p + b)$$

so that we may write:

$$\bar{\imath}(p) = \frac{E}{2La} \cdot \frac{a}{(p+b)^2 + a^2} + \frac{LE}{RL} \cdot \frac{p+b}{(p+b)^2 + a^2}$$

from which it follows at once by using nos. 11 and 12 of the dictionary that:

$$i(t) = \frac{E}{2La} \cdot e^{-bt} \cdot \sin at + \frac{E}{R} \cdot e^{-bt} \cdot \cos at$$

a, b being defined above.

Example 6 (see Fig. A.7)

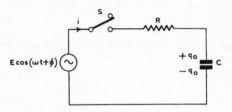

Fig. A.7. Circuit for Example 6.

Switching with an a.c. generator in the circuit
    Switch closes at $t = 0$: initial charge on capacitor C is q_0
(*a*) Initial conditions: charge on $C = q_0$
(*b*) Circuit differential equation after switching:

$$Ri(t) + \frac{1}{C} \int_0^t i(t)\mathrm{d}t = E\cos(\omega t + \phi) - q_0/C$$

In place of the real waveform $E\cos(\omega t + \phi)$, we will use $E\exp(j\omega t + \phi)$. The answer to the problem postulated will then be the real part of the current $i(t)$ given by:

$$Ri(t) + \frac{1}{C} \int_0^t i(t) \cdot \mathrm{d}t = E\exp j(\omega t + \phi) - q_0/C$$

(*c*) Current transform

$$R\bar{\imath}(p) + \frac{1}{pC}\bar{\imath}(p) = \frac{E\exp(j\phi)}{p - j\omega} - \frac{q_0}{pC}$$

i.e.

$$\bar{\imath}(p) = \frac{E\exp(j\phi)}{(p - j\omega)(R + 1/pC)} - \frac{q_0}{pC(R + 1/pC)}$$

(d) Current waveform

The transform is shown as the sum of two parts. The second is a standard form, for:

$$\frac{q_0}{pC(R + 1/pC)} = \frac{q_0}{RC(p + 1/RC)}$$

giving the time waveform $q_0/RC \exp(-t/RC)$.

The first part is most easily handled by expanding it into partial fractions:

$$\frac{1}{(p - j\omega)(R + 1/pC)} = \frac{p}{R(p - j\omega)(p + \alpha)} \quad \text{where } \alpha = \frac{1}{RC}$$

$$= \frac{1}{R}\left[\frac{j\omega}{p - j\omega} + \frac{\alpha}{p + \alpha}\right] \times \frac{1}{\alpha + j\omega}$$

which is now in the form of a sum of two terms of type no. 4 in the dictionary.

Hence, collecting results, we have:

$$i(t) = \frac{E \exp(j\phi)}{R(\alpha + j\omega)}[j\omega \exp(j\omega t) + \alpha \exp(-\alpha t)] - \alpha q_0 \exp(-\alpha t)$$

and

$$\mathrm{Re}.\, i(t) = \frac{E}{R}\left[\frac{\omega}{\alpha^2 + \omega^2}\{\omega \cos(\omega t + \phi) - \alpha \sin(\omega t + \phi)\} + \right.$$

$$\left.\frac{\alpha}{\alpha^2 + \omega^2}(\alpha \cos\phi + \omega \sin\phi) \exp(-\alpha t)\right] - \alpha q_0 \exp(-\alpha t)$$

The solution consists of the steady-state term oscillating at angular frequency ω and a transient part with decay constant, $1/RC$.

A.7.6. HEAVISIDE EXPANSION THEOREM

In Example 6, we found it useful to expand the transform in partial fractions: this is a very common method of attack and justifies its examination in more general terms. In a large number of problems, the transform appears in the form:

$$\bar{f}(p) = \frac{A(p)}{B(p)} \tag{A.74}$$

where $A(p)$ and $B(p)$ are polynomials in p, the degree of $A(p)$ being less than that of $B(p)$ unless the time function involves delta functions, when the degrees are the same.

Suppose $B(p)$ has degree n: it may then be written

$$B(p) = C \prod_{r=1}^{n} (p - p_r) \tag{A.75}$$

in which C is the coefficient of the term in p^n, the symbol $\prod_{r=1}^{n}$ indicates the multiplication of the n factors

$$(p - p_1), \quad (p - p_2), \ldots, (p - p_n)$$

and $p_1, p_2, \ldots, p_n$ are the roots of the equation $B(p) = 0$.

When the degree of $A(p)$ is less than that of $B(p)$, we may write:

$$\frac{A(p)}{B(p)} = \sum_{r=1}^{n} \frac{c_r}{p - p_r} \tag{A.76}$$

thus expanding the transform in partial fractions. This form also assumes that the roots, $p_1, p_2, \ldots, p_n$ are all different. We may find the unknown constants c_r as follows: Consider the root p_r and write:

$$B(p) = (p - p_r) C(p) \tag{A.77}$$

Now consider:

$$\bar{f}(p) - \frac{c_r}{p - p_r} = \frac{A(p)}{(p - p_r) C(p)} - \frac{c_r}{p - p_r} = \frac{A(p) - c_r C(p)}{(p - p_r) C(p)} \tag{A.78}$$

If c_r is the correct coefficient in the partial fraction expansion, the right-hand side of this equation must not involve the factor $(p - p_r)$ in the denominator. This can happen only if the numerator vanishes for p equal to p_r, showing that it too has a factor $(p - p_r)$ to cancel that in the denominator. We must therefore have:

$$A(p_r) = c_r C(p_r) \tag{A.79}$$

thus expressing c_r in terms of known quantities. We now differentiate equation (A.77) to give:

$$\frac{dB(p)}{dp} = B'(p) = C(p) + (p - p_r) C'(p) \tag{A.80}$$

and let p equal p_r, so that:

$$B'(p_r) = C(p_r) \tag{A.81}$$

We now have, from equations (A.79) and (A.81),

$$c_r = A(p_r)/B'(p_r) \tag{A.82}$$

with similar expressions for $c_1, c_2, \ldots, c_n$, giving:

$$\bar{f}(p) = \frac{A(p)}{B(p)} = \sum_{r=1}^{n} \frac{A(p_r)}{(p - p_r) B'(p_r)} \tag{A.83}$$

Each of the terms in the sum is of the form of no. 4 in the dictionary, so that:

$$f(t) = \sum_{r=1}^{n} \frac{A(p_r)}{B'(p_r)} \exp(p_r t) \tag{A.84}$$

Heaviside's theorem thus enables us to invert any transform of the type given by equation (A.74) obtaining the above time function as the answer.

We have imposed two restrictions on the form of $\bar{f}(p)$ in that the degree of the polynomial $A(p)$ must be less than that of $B(p)$ and that $B(p)$ has no repeated roots. These will now be considered.

(a) A(p) and B(p) of the same degree

If $A(p)$ and $B(p)$ have the same degree, the limit of $\bar{f}(p)$ as p tends to infinity is the constant given by the ratio of the coefficients of the terms p^n. [If the degree of $A(p)$ is less than that of $B(p)$, $\underset{p \to \infty}{\text{Lim.}}\ \bar{f}(p) = 0$].

Suppose

$$A(p) = a_n p^n + A_{n-1}(p) \tag{A.85}$$

$$B(p) = b_n p^n + B_{n-1}(p) \tag{A.86}$$

where $A_{n-1}(p)$, $B_{n-1}(p)$ are polynomials of degree $(n-1)$

Then

$$\underset{p \to \infty}{\text{Lim.}}\ \bar{f}(p) = \frac{a_n}{b_n} \tag{A.87}$$

Write

$$\bar{f}(p) = \frac{a_n}{b_n} + \bar{f}_1(p) \tag{A.88}$$

so that

$$\bar{f}_1(p) = \frac{a_n p^n + A_{n-1}(p)}{b_n p^n + B_{n-1}(p)} - \frac{a_n}{b_n} = \frac{b_n A_{n-1}(p) - a_n B_{n-1}(p)}{b_n [b_n p^n + B_{n-1}(p)]} \tag{A.89}$$

in which the numerator now has a lower degree than the denominator. The transform $\bar{f}_1(p)$ can thus be inverted as above, while the constant term a_n/b_n in equation (A.88) leads to $(a_n/b_n)\,\delta(t)$ in the time waveform.

(b) The occurrence of equal roots

General expressions can be derived for roots of any multiplicity and are available in the references. It will suffice here for the sake of illustration to consider the case of one repeated root, when the denominator polynomial may be written as:

$$B(p) = (p - p_0)^2 C(p) \tag{A.90}$$

$C(p)$ being a polynomial with no repeated roots. The effect of the repeated root on the partial fraction expansion is to cause the appearance of a term of the type $d/(p - p_0)^2$. Let us consider then:

$$\bar{f}_1(p) = \bar{f}(p) - \frac{d}{(p - p_0)^2} = \frac{A(p)}{(p - p_0)^2 C(p)} - \frac{d}{(p - p_0)^2}$$

$$= \frac{A(p) - dC(p)}{(p - p_0)^2 C(p)} \tag{A.91}$$

Now, select d so that:

$$A(p_0) = d \cdot C(p_0) \tag{A.92}$$

This means that the numerator also vanishes for p equal to p_0 and a factor $(p - p_0)$ can be cancelled from numerator and denominator, leaving $\bar{f}_1(p)$ as a transform with no repeated roots of the denominator. We now have

$$\bar{f}(p) = \frac{A(p_0)}{(p - p_0)^2 C(p_0)} + \bar{f}_1(p) \tag{A.93}$$

where $\bar{f}_1(p)$ can be handled by the original statement of the expansion theorem. The first term on the right gives by no. 5 of the dictionary the contribution:

$$A(p_0)t \exp(p_0 t)/C(p_0) \tag{A.94}$$

to the time waveform.

The only remaining step is to consider the expression of $C(p_0)$ in terms of the denominator function $B(p)$. We have:

$$B(p) = (p - p_0)^2 C(p)$$

$$B'(p) = 2(p - p_0) C(p) + (p - p_0)^2 C'(p)$$

$$B''(p) = 2C(p) + 4(p - p_0) C'(p) + (p - p_0)^2 C''(p)$$

so that $B''(p_0) = 2C(p_0)$

i.e. $C(p_0) = \tfrac{1}{2} B''(p_0)$ \hfill (A.95)

Example 7
Use of the expansion theorem

The expansion theorem is most useful when the degree of the transform denominator is large. This occurs when multimesh circuits are examined. Consider for example the circuit in Fig. A.8.

The conditions before the switch is closed are $i_1 = i_2 = 0$ and the charge on the capacitor is zero. The switch closes at $t = 0$. We wish to determine $i_2(t)$. The circuit constants satisfy the relations:

$$M = L/\sqrt{2} : L/C = 2R^2 : R/L = \alpha$$

(a) Initial conditions: $i_1(0) = i_2(0) = q = 0$.

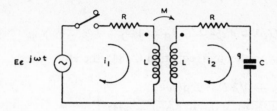

Fig. A.8. Circuit for Example 7.

(b) Circuit differential equations after switching:

$$L \frac{di_1(t)}{dt} + Ri_1(t) - M \frac{di_2(t)}{dt} = E \exp(j\omega t)$$

$$L \frac{di_2(t)}{dt} + Ri_2(t) + \frac{1}{C} \int_0^t i_2(t)\,dt - M \frac{di_1(t)}{dt} = 0$$

(c) Transforms

$$pL\bar{i}_1(p) + R\bar{i}_1(p) - pM\bar{i}_2(p) = E/(p - j\omega)$$

$$pL\bar{i}_2(p) + R\bar{i}_2(p) + \frac{1}{pC}\,\bar{i}_2(p) - pM\bar{i}_1(p) = 0$$

Eliminate $\bar{i}_1(p)$ giving:

$$(pL + R)\left(pL + R + \frac{1}{pC}\right) \bar{i}_2(p)/pM - pM\bar{i}_2(p) = E/(p - j\omega)$$

i.e.

$$\bar{i}_2(p) = \frac{pME}{\left[(pL + R)\left(pL + R + \dfrac{1}{pC}\right) - p^2 M^2\right](p - j\omega)}$$

Substitution of the relations between the circuit constants gives:

$$\bar{i}_2(p) = \frac{\sqrt{2}\, p^2 E}{L(p - j\omega)(p^3 + 4\alpha p^2 + 6\alpha^2 p + 4\alpha^3)}$$

(d) Current waveform

To invert the above transform we use the expansion theorem. We have:

$$A(p) = \sqrt{2}\, p^2 E$$

$$B(p) = L(p - j\omega)(p^3 + 4\alpha p^2 + 6\alpha^2 p + 4\alpha^3)$$

The roots of $B(p) = 0$ are:

$$p_1 = j\omega : p_2 = -2\alpha : p_3 = -\alpha + j\alpha : p_4 = -\alpha - j\alpha$$

Also

$$B'(p) = L(p^3 + 4\alpha p^2 + 6\alpha^2 p + 4\alpha^3) + L(p - j\omega)(3p^2 + 8\alpha p + 6\alpha^2)$$

The expansion theorem, equation (A.84), therefore gives

$$
i_2(t) = \frac{-\sqrt{2}\, E\omega^2 \exp(j\omega t)}{L(-j\omega^3 - 4\alpha\omega^2 + 6j\omega\alpha^2 + 4\alpha^3)} +
$$

$$
+ \frac{\sqrt{2}\, E(2\alpha)^2 \exp(-2\alpha t)}{L(-2\alpha - j\omega)(12\alpha^2 - 16\alpha^2 + 6\alpha^2)} +
$$

$$
+ \frac{\sqrt{2}\, E(-\alpha + j\alpha)^2 \exp(-\alpha t + j\alpha t)}{L(-\alpha + j\alpha - j\omega)[3(-\alpha + j\alpha)^2 - 8\alpha^2(1 - j) + 6\alpha^2]}
$$

$$
+ \frac{\sqrt{2}\, E(\alpha + j\alpha)^2 \exp(-\alpha t - j\alpha t)}{L(-\alpha - j\alpha - j\omega)[3\alpha^2(1 + j)^2 - 8\alpha^2(1 + j) + 6\alpha^2]}
$$

$$
= \frac{\sqrt{2}\, E}{L}\Bigg[\frac{\omega^2 \exp(j\omega t)}{(j\omega + 2\alpha)(\omega^2 - 2j\alpha\omega - 2\alpha^2)} - \frac{2\exp(-2\alpha t)}{(j\omega + 2\alpha)} -
$$

$$
- \frac{j\exp(-\alpha t + j\alpha t)}{(\alpha - j\alpha + j\omega)(1 - j)} + \frac{j\exp(-\alpha t - j\alpha t)}{(\alpha + j\alpha + j\omega)(1 + j)}\Bigg]
$$

If the circuit generator delivers the real waveform $\cos(\omega t)$, the secondary current is given by the real part of the above expression.

Example 8

Two switching operations, as illustrated by Fig. (A.9)

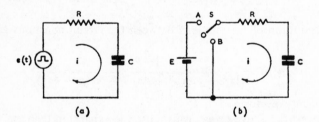

Fig. A.9. Circuit for Example 8.

(*a*) The e.m.f. supplied by the generator is defined by:

$$e(t) = 0 \quad t < 0$$

$$e(t) = E \quad 0 \le t \le T$$

$$e(t) = 0 \quad t > T$$

(b) An equivalent problem in which the switch S is connected to A at $t = 0$ and to B at $t = T$. The capacitor C is assumed to be uncharged for $t < 0$. We may solve this problem in two ways, either treating it as two switching operations as in Fig. A.9(b) or working directly from the transform of $e(t)$.

Solution I

For $0 < t < T$ the differential equation is

$$Ri(t) + \frac{1}{C}\int_0^t i(t)\,dt = E$$

Taking Laplace transforms we have:

$$\left(R + \frac{1}{pC}\right)\bar{i}(p) = E/p$$

$$\bar{i}(p) = E/R(p + 1/RC)$$

so that

$$i(t) = \frac{E}{R}\exp\left(-\frac{t}{RC}\right)$$

At time T

$$i(T) = \frac{E}{R}\exp\left(-\frac{T}{RC}\right)$$

and voltage across C is $v_c(T) = E(1 - \exp[-T/RC])$. At time T, a second switching operation occurs: the voltage across C is unchanged by this and the initial voltage on C is therefore $v_c(T)$.

We now use a new time variable t' equal to $t - T$ which makes the second switching operation occur at $t' = 0$. When S is connected to B, the circuit differential equation is

$$Ri(t') + \frac{1}{C}\int_0^{t'} i(t')\,dt' = -v_c(T)$$

so that:

$$\left(R + \frac{1}{pC}\right)\bar{i}(p) = -v_c(T)/p$$

$$\bar{i}(p) = -v_c(T)/R(p + 1/RC)$$

Inverting the transform:

$$i(t') = v_c(T)\exp\left(-\frac{t'}{RC}\right)\bigg/R$$

To complete the solution, we substitute for $v_c(T)$ and replace t' by $t - T$. Finally, we have:

$$i(t) = \frac{E}{R} \exp\left(-\frac{t}{RC}\right) \quad 0 < t < T$$

$$i(t) = -\frac{E}{R}\left[\exp\left(\frac{T}{RC}\right) - 1\right]\exp\left(-\frac{t}{RC}\right) \quad t > T$$

Solution II

The differental equation

$$Ri(t) + \frac{1}{C}\int_0^t i(t)\,\mathrm{d}t = e(t)$$

is valid for all $t > 0$ if $e(t)$ is defined as

$$e(t) = 0 \quad t < 0$$
$$= E \quad 0 \leq t \leq T$$
$$= 0 \quad t > T$$

Take Laplace transforms of this equation, giving:

$$\left(R + \frac{1}{pC}\right)\bar{\imath}(p) = \bar{e}(p)$$

where

$$\bar{e}(p) = \int_0^T E\exp(-pt)\,\mathrm{d}t = E(1 - e^{-pT})/p$$

Hence

$$\bar{\imath}(p) = \frac{E(1 - e^{-pT})}{R(p + 1/RC)}$$

This transform can be regarded as the sum of two terms, the first $E/R(p + 1/RC)$ giving the time waveform $(E/R)\exp(-t/RC)$, and the second being $-E\exp(-pT)/R(p + 1/RC)$. Comparison with the solution obtained by the direct method above shows that the first contribution gives the total current for $0 \leqslant t \leqslant T$. The second contribution therefore can have no effect until $t > T$: examining the table of general results we see that (g) is of the appropriate form.
We have

$$\mathscr{L}\{g(t)\} = e^{-pT}\mathscr{L}\{f(t)\}$$

so that if

$$\mathscr{L}\{g(t)\} = -Ee^{-pT}/R(p + 1/RC)$$
$$\mathscr{L}\{f(t)\} = -E/R(p + 1/RC)$$
$$f(t) = -(E/R)\exp(-t/RC)$$
$$g(t) = 0 : t < T$$
$$= -(E/R)\exp(-\overline{t-T}/RC) : t > T$$

The solution for the current is therefore:

$$i(t) = (E/R)\exp(-t/RC) : 0 < t < T$$

$$i(t) = (E/R)[1 - \exp(T/RC)]\exp(-t/RC) : t > T$$

agreeing with that found above.

A.7.7. PROOF OF THE SHIFT THEOREM

The general result used above is known as the shift theorem and it is sufficiently important to merit a proof. We show that if

$$g(t) = 0 : t < T \tag{A.96}$$
$$= f(t - T) : t > T$$

then

$$\mathcal{L}\{g(t)\} = e^{-pT}\mathcal{L}\{f(t)\} \tag{A.97}$$

Proof

$$\mathcal{L}\{g(t)\} = \int_0^\infty g(t)\exp(-pt)\,dt$$

$$= \int_T^\infty f(t - T)\exp(-pt)\,dt \tag{A.98}$$

Let

$$t' = t - T$$

Then

$$\mathcal{L}\{g(t)\} = \int_0^\infty f(t')\exp(-pt' - pT)\,dt'$$

$$= e^{-pT}\int_0^\infty f(t')\exp(-pt')\,dt'$$

$$= e^{-pT}\mathcal{L}\{f(t)\} \tag{A.99}$$

Example 9

The occurrence of a very short pulse

We consider again Fig. 9(a), with

$$e(t) = E_1\delta(t)$$

Then, as in solution II for the previous example, we have:

$$\left(R + \frac{1}{pC}\right)\bar{\imath}(p) = \bar{e}(p)$$

where $\bar{e}(p) = E_1$ by no. 13 of the dictionary.
Hence,

$$i(p) = \frac{pE_1}{R(p + 1/RC)}$$

This is a simple example of a transform in which the degrees of the numerator and the denominator are equal. We therefore write the transform as:

$$i(p) = \frac{E_1}{R}\left[1 - \frac{(1/RC)}{p + (1/RC)}\right]$$

and using nos. 4 and 13 of the dictionary we obtain:

$$i(t) = \frac{E_1}{R}\left[\delta(t) - \frac{1}{RC}\exp(-t/RC)\right]$$

We may confirm this result by considering the limiting case of Example 8 in which the pulse length T tends to zero and the pulse amplitude E tends to infinity in such a way that the product ET tends to E_1. This means that $e(t)$ from Example 8 tends to $E_1\delta(t)$ which is the e.m.f. for the present solution. Now consider the current: if T tends to zero and E tends to infinity, the term $E/R \exp(-t/RC)$ which lasts for $0 < t < T$, tends to $(E_1/R)\delta(t)$. In the other term we have,

$$i(t) = \underset{\substack{T\to 0 \\ E\to\infty \\ TE\to E_1}}{\text{Lim.}} \left(\frac{E}{R}\right)\left[1 - \exp\left(\frac{T}{RC}\right)\right]\exp(-t/RC)$$

$$= \text{Lim.} \left(\frac{E}{R}\right)\left[1 - 1 - \frac{T}{RC}\right]\exp(-t/RC)$$

$$= -\frac{E_1}{R}\cdot\frac{1}{RC}\exp(-t/RC)$$

so that the complete solution is

$$i(t) = \frac{E_1}{R}\left[\delta(t) - \frac{1}{RC}\exp\left(-\frac{t}{RC}\right)\right]$$

agreeing with that above.

Example 10
Kelvin's arrival curve, illustrating the application of Laplace transforms to partial differential equations.
We consider a telephone line of length l, having resistance R and capacitance C per unit loop length. The line is uncharged for $t < 0$ and at $t = 0$ switch S (Fig. A.10) is closed. We wish to find the current in the short-

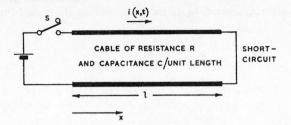

Fig. A.10. Circuit for Example 10.

circuited receiving end. Let x be the distance measured from the sending end. Let $v(x,t)$, $i(x,t)$ be the voltage and current at distance x and time t.

(a) Initial conditions: since the line is uncharged before switching, the voltage $v(x,t) = 0$ for $t < 0$. The voltage cannot change instantaneously, since this would imply an instantaneous change in the stored energy in the capacitance. Hence, $v(x,0) = 0$.

(b) Partial differential equations:

$$\frac{\partial v(x,t)}{\partial x} = -Ri(x,t) : \frac{\partial i(x,t)}{\partial x} = -\frac{C\partial v(x,t)}{\partial t}$$

(c) Transforms: Let

$$\bar{v}(x,p) = \int_0^\infty v(x,t)e^{-pT}\,dt$$

Then

$$\frac{\partial \bar{v}(x,p)}{\partial x} = \int_0^\infty \frac{\partial v(x,t)}{\partial x}\,e^{-pt}\,dt$$

since the operations of differentiation and integration can be reversed.

The current transform $\bar{i}(x,p)$ is defined similarly. Applying these to the equations in (b) and using the condition $v(x,0) = 0$, we get:

$$\frac{\partial \bar{v}(x,p)}{\partial x} = -R\bar{i}(x,p) : \frac{\partial \bar{i}(x,p)}{\partial x} = -pC\bar{v}(x,p)$$

a pair of ordinary differential equations in the variable x. Eliminate $\bar{i}(x,p)$, giving,

$$\frac{\partial^2 \bar{v}(x,p)}{\partial x^2} + pCR\bar{v}(x,p) = 0$$

which has the general solution $\bar{v}(x,p) = K\exp(-\lambda x) + L\exp(\lambda x)$ where $\lambda = (pCR)^{\frac{1}{2}}$ and K, L are amplitude constants, which will be functions of p.

The corresponding general solution for $\bar{i}(x,p)$ is:

$$\bar{i}(x,p) = (\lambda/R)[K\exp(-\lambda x) - L\exp(\lambda x)]$$

The constants K, L are determined by examining the conditions at $x = 0$ and $x = l$.

For $x = 0$, $v(x,t) = E : t > 0$ so that $\bar{v}(0,p) = E/p$

For $x = l$, $\bar{v}(x,t) = 0$ so that $\bar{v}(l,p) = 0$.

Hence,

$$K + L = \frac{E}{p}$$

$$K \exp(-\lambda l) + L \exp(\lambda l) = 0$$

Solving for K, L and substituting in the current transform, we obtain:

$$\bar{i}(x,p) = -\frac{\lambda E}{pR} \cdot \frac{[\exp(2\lambda l - \lambda x) + \exp(\lambda x)]}{1 - \exp(+2\lambda l)}$$

We are interested in the current at the receiving end, where $x = l$, and for this:

$$\bar{i}(l,p) = -\frac{\lambda E}{pR} \cdot \frac{2 \exp(\lambda l)}{[1 - \exp(+2\lambda l)]} = \frac{\lambda E}{pR \sinh(\lambda l)}$$

(d) Current waveform

The transform above is obviously not one of the standard types we have considered. We may, however, invert it by the expansion theorem, the only complication being that the denominator has an infinite number of roots. In the notation used for the expansion theorem we have:

$$A(p) = \lambda E$$

$$B(p) = pR \sinh(\lambda l)$$

The roots of $B(p) = 0$ are $p = 0$ and the roots of $\sinh(\lambda l) = 0$, i.e. $\lambda = 0, \pm j\pi/l, \pm 2j\pi/l, \ldots \pm jr\pi/l, \ldots$

The negative imaginary roots are not relevant since λ is defined as $(pCR)^{\frac{1}{2}}$ and we work with the branch of the square root which is positive imaginary.

When

$$\lambda = \frac{jr\pi}{l}$$

we have

$$p = -\frac{\pi^2 r^2}{l^2 CR}$$

Let

$$u = \frac{\pi^2}{l^2 CR}$$

Then, the roots of $B(p) = 0$ are $p = 0$ and $p = -r^2u$, $r = 1,2,\ldots$
The root $p = 0$ has to be handled very carefully since, when $p = 0$, λ is
also zero. In the transform the numerator has a single zero and the
denominator a double zero, the net effect being as if the denominator
had a single zero. We handle this by the same method as used in the
proof of the expansion theorem, namely by splitting off the term D/p
in the partial fraction expansion corresponding to this root. D is a constant to be evaluated.

From the expansion theorem we therefore have:

$$\bar{i}(l,t) = \sum_{r=1}^{\infty} \frac{A(-r^2u)}{B'(-r^2u)} \cdot \exp(-r^2ut) + D$$

Now,

$$A(-r^2u) = jr\pi E/l : r = 1, 2, \ldots$$

$$B'(p) = R \sinh(\lambda l) + pR \frac{d}{dp} \cdot \sinh(\lambda l)$$

$$= R \sinh(\lambda l) + pRl \cosh \lambda l \cdot d\lambda/dp$$

$$= R \sinh(\lambda l) + \frac{\lambda^2 l}{C} \cosh \lambda l \cdot \frac{CR}{2\lambda}$$

$$B'(-r^2u) = \left[R \sinh(\lambda l) + \frac{\lambda l R}{2} \cosh \lambda l \right]_{\lambda = jr\pi/2}$$

$$= \frac{jr\pi R}{2} (-)^r \qquad r = 1, 2, \ldots$$

so that

$$\frac{A(-r^2u)}{B'(-r^2u)} = \frac{2E}{Rl} (-)^r \qquad r = 1, 2, \ldots$$

To find the constant D, we consider:

$$\frac{E\lambda}{pR \sinh(\lambda l)} - \frac{D}{p} = \frac{[E\lambda - RD \sinh(\lambda l)]}{pR \sinh(\lambda l)}$$

We must select D so that the numerator vanishes for $p = 0$, thus ensuring that the root at $p = 0$ is covered by the term D/p. D is therefore given
by

$$E\lambda = RD \sinh(\lambda l) \quad \text{if} \quad p = \lambda = 0$$

i.e.

$$D = \lim_{\lambda \to 0} \frac{E\lambda}{R \sinh(\lambda l)} = \frac{E}{Rl}$$

Collecting results, we now have the Kelvin arrival curve expression:

$$i(l,t) = \frac{E}{Rl}\left[1 + 2\sum_{r=1}^{\infty}(-)^r \exp(-r^2 ut)\right]$$

When t tends to infinity, $i(l,t)$ tends to E/Rl, a result which is immediately obvious from the form of the circuit. This confirms the validity of the calculation of the constant, D.

Example 11

Pulse-forming network, illustrating the use of the shift theorem (Fig. A.11)

The line of length l has inductance L and capacitance C per unit length. It is charged to a constant voltage V_0 before $t = 0$ and at $t = 0$, S closes. We wish to find the voltage across R.

(*a*) The voltage cannot change on switching, so $v(x,0) = V_0$: also this implies $i(x,0) = 0$

(*b*) Partial differential equations:

$$\frac{\partial v(x,t)}{\partial x} = -L\frac{\partial i(x,t)}{\partial t} : \frac{\partial i(x,t)}{\partial x} = -\frac{C\partial v(x,t)}{\partial t}$$

(*c*) Transforms:

$$\frac{\partial \bar{v}(x,p)}{\partial x} = -pL\bar{i}(x,p) : \frac{\partial \bar{i}(x,t)}{\partial x} = -pC\bar{v}(x,p) + Cv(x,0)$$

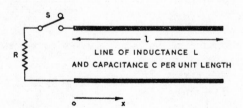

Fig. A.11. Circuit for Example 11.

Eliminate $\bar{i}(x,p)$

$$\frac{\partial^2 \bar{v}(x,p)}{\partial x^2} - p^2 LC\bar{v}(x,p) = -pLCV_0$$

The general solution is:

$$\bar{v}(x,p) = E\exp(\lambda x) + F\exp(-\lambda x) + \frac{V_0}{p}$$

where

$$\lambda = p\sqrt{LC}$$

The corresponding solution for the current transform is:

$$\bar{\imath}(x,p) = -\sqrt{\frac{C}{L}}\,[E\exp(\lambda x) - F\exp(-\lambda x)]$$

The boundary conditions on the line for $t > 0$ are:

$$i(l,t) = 0 : i(0,t) = -v(0,t)/R$$

so that

$$\bar{\imath}(l,p) = 0 : \bar{\imath}(0,p) = -\bar{v}(0,p)/R$$

Hence

$$E\exp(\lambda l) = F\exp(-\lambda l),$$

$$(C/L)^{\frac{1}{2}}(E - F) = \left(E + F + \frac{V_0}{p}\right)\Big/R$$

Solving for E, F and substituting in the voltage transform we have:

$$\bar{v}(x,p) = \frac{V_0}{p}\left[1 + \frac{Z_0(e^{\lambda x} + e^{2\lambda l - \lambda x})}{R - Z_0 - (R + Z_0)e^{2\lambda l}}\right]$$

where

$$Z_0 = \sqrt{L/C}$$

For the voltage across R,

$$\bar{v}(0,p) = \frac{V_0}{p}\cdot\frac{R(1 - e^{2\lambda l})}{(R - Z_0) - (R + Z_0)e^{2\lambda l}}$$

(d) Voltage waveform

(i) Consider first the special case of the resistor matched to the line impedance so that $R = Z_0$.
Then,

$$\bar{v}(0,p) = \frac{V_0}{2p}(1 - e^{-2\lambda l}) = \frac{V_0}{2p}(1 - e^{-pT})$$

if $T = 2l\sqrt{(LC)}$, i.e. the group delay for a line of length $2l$. This transform is essentially the same as that which arose in Solution II for Example 8 and we find:

$$v(0,t) = \tfrac{1}{2}V_0 : 0 < t < T$$

$$= 0 \qquad t > T$$

This waveform is plotted in Fig. A.12.

(ii) In the general case, the transform can be written:

$$\bar{v}(0,p) = \frac{V_0 R}{p(R + Z_0)}\cdot\frac{(1 - e^{-pT})}{(1 - \varrho e^{-pT})}$$

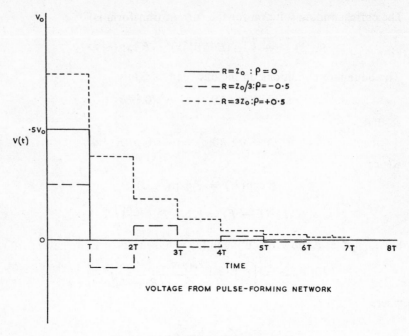

$$R = Z_0 : \rho = 0$$
$$R = Z_0/3 : \rho = -0.5$$
$$R = 3Z_0 : \rho = +0.5$$

VOLTAGE FROM PULSE-FORMING NETWORK

Fig. A.12. Waveforms for Example 11.

where $\varrho = (R - Z_0)/(R + Z_0)$ is the voltage reflection coefficient at the resistor.

Since $|\varrho| < 1$, the magnitude of ϱe^{-pT} is also less than 1 for any positive value of p. We can therefore use the binomial theorem to expand $(1 - \varrho e^{-pT})^{-1}$, obtaining:

$$\bar{v}(0,p) = \frac{V_0(1 + \varrho)}{2p} \cdot (1 - e^{-pT})$$
$$(1 + \varrho e^{-pT} + \varrho^2 e^{-2pT} + \varrho^3 e^{-3pT} + \ldots)$$
$$= \frac{V_0(1 + \varrho)}{2p} [1 - (1 - \varrho)e^{-pT} - (1 - \varrho)\varrho e^{-2pT}$$
$$- (1 - \varrho)\varrho^2 e^{-3pT} - \ldots - (1 - \varrho)\varrho^{r-1}e^{-rpT} - \ldots]$$

in which the transform is now expressed as an infinite series of terms of the type which can be handled by the shift theorem. For two special cases, the series teminates: firstly, if $\varrho = 0$, when we have R matched to Z_0 as considered above, and, secondly, if $\varrho = 1$. The latter implies that R tends to infinity so that the switching operation has no effect and $v(0,t)$ equals V_0.

The general term in the series for $\bar{v}(0,p)$ is:

$$\frac{- V_0(1 + \varrho)(1 - \varrho)\varrho^{r-1}e^{-rpT}}{2p}$$

for whic! 'he waveform is:

$$v(0,t) = 0 : t < rT$$

$$= \frac{V_0}{2}(1 + \varrho)(1 - \varrho)\varrho^{r-1} : t > rT$$

For times between $(r - 1)T$ and rT therefore, we need consider only the first r terms of the series and we find:

$$v(0,t) = \tfrac{1}{2}V_0(1 + \varrho) : 0 < t < T$$

$$= \tfrac{1}{2}V_0(1 + \varrho)[1 - (1 - \varrho)] = \tfrac{1}{2}V_0\varrho(1 + \varrho) : T < t < 2T$$

$$\cdots\cdots\cdots\cdots\cdots\cdots\cdots\cdots\cdots\cdots$$

$$= \tfrac{1}{2}V_0(1 + \varrho)[1 - (1 - \varrho) - (1 - \varrho)\varrho - (1 - \varrho)\varrho^2 - \cdots$$
$$- (1 - \varrho)\varrho^{r-2}]$$

$$= \tfrac{1}{2}V_0(1 + \varrho)\varrho^{r-1} : (r - 1)T < t < rT$$

$$\cdots\cdots\cdots\cdots\cdots\cdots\cdots\cdots\cdots\cdots$$

The waveform is shown in Fig. A.12 for $\varrho = 0\cdot5(R = 3Z_0)$ and $\varrho = -0\cdot5 (R = Z_0/3)$.

A.8. Hilbert Transforms

There is a wide range of transforms by which a function $f(x)$ can be converted into a related function $g(y)$. Most involve an operation of the type:

$$g(y) = \int K(x,y)f(x)\,\mathrm{d}x$$

the function $K(x,y)$ being called the kernel and its nature determining the type of transform obtained. For example, the Fourier transform arises if

$$K(x,y) = \exp(-jxy)$$

and the Laplace transform if

$$K(x,y) = \exp(-xy)$$

One other type of transform, the Hilbert has important applications in communications work, particularly in connection with the relation between the amplitude and phase characteristics of a transfer function. An outline of the way in which the Hilbert transform arises will be given in this section and a simple example will show its usefulness.

A.8.1. RESTRICTION ON THE FREQUENCY DEPENDENCE OF A PASSIVE IMPEDANCE

Suppose a passive two-terminal circuit element has an impedance defined by the complex function $Z(\omega)$ for all real values of the frequency ω. This function can be split into real and imaginary parts so that:

$$Z(\omega) = R(\omega) + jX(\omega) \tag{A.100}$$

where $R(\omega)$, $X(\omega)$ are the resistance and reactance functions respectively. We will now show that there is an intimate relation between these two functions and that if one is known, the other can be calculated from it. To do this, we consider the effect of the application of a unit current impulse at time zero, so that the current through the impedance is

$$i(t) = \delta(t) \tag{A.101}$$

This current will cause a voltage $v(t)$ to appear across the terminals of the impedance. Clearly, $v(t)$ must be zero before the application of the current pulse, since the impedance is passive and will, in general, consist of an impulse $v_0\delta(t)$ occurring simultaneously with the current pulse and a transient $v_1(t)$ persisting after the current pulse terminates. Hence:

$$v(t) = 0 \qquad\qquad : t < 0$$

$$v(t) = v_0\,\delta(t) + v_1(t) : t > 0 \tag{A.102}$$

The transient, $v_1(t)$ is a real function, which is zero for t less than zero and decays to zero as t tends to infinity.

We may also express the current and voltage in terms of Fourier transforms. For the current, we have, from Section A.3.4:

$$i(t) = \frac{1}{2\pi} \int_{-\infty}^{\infty} \exp(j\omega t)\,d\omega \tag{A.103}$$

and the voltage, obtained by multiplying each frequency component by the appropriate impedance, is therefore:

$$v(t) = \frac{1}{2\pi} \int_{-\infty}^{\infty} Z(\omega) \exp(j\omega t)\,d\omega \tag{A.104}$$

The inverse transform gives:

$$Z(\omega) = \int_{-\infty}^{\infty} v(t)\,\exp(-j\omega t)\,dt$$

and on substitution from equation (A.102) for $v(t)$, we have:

$$Z(\omega) = \int_{0}^{\infty} [v_0\delta(t) + v_1(t)] \exp(-j\omega t)\,dt \tag{A.105}$$

The behaviour of $v_1(t)$ as t tends to infinity ensures that the integral can be evaluated. We now split equation (A.105) into its real and imaginary parts to give

$$
\left.
\begin{aligned}
R(\omega) &= v_0 + \int_0^\infty v_1(t)\cos(\omega t)\mathrm{d}t \\
X(\omega) &= - \int_0^\infty v_1(t)\sin(\omega t)\mathrm{d}t
\end{aligned}
\right\}
\qquad \text{(A.106)}
$$

In this pair of equations, $R(\omega)$ and $X(\omega)$ are both expressed in terms of the real function $v_1(t)$ and clearly cannot be independent. In principle at least, the function $v_1(t)$ can be eliminated to obtain a single relation between $R(\omega)$ and $X(\omega)$. The term v_0 in the first equation gives the instantaneous response to the current impulse. This impulse may be thought of as a half-cycle of a wave of infinitely large frequency, and so the term v_0 can be regarded as arising from a signal of infinite frequency. Since the current impulse has unit amplitude, the term v_0 can therefore be equated to R_∞ the resistance of the network at infinite frequency. This interpretation is confirmed by showing that the integral vanishes as ω tends to infinity, provided that $v_1(t)$ remains finite and continuous for any value of t. We may therefore write the first equation in (A.106) as:

$$
R(\omega) - R_\infty = \int_0^\infty v_1(t)\cos(\omega t)\mathrm{d}t
$$

and we will now solve this equation for $v_1(t)$. Multiply both sides of the equation by $\cos(\omega t')\exp(-k\omega)$ and integrate with respect to ω from zero to infinity. Then:

$$
\int_0^\infty [R(\omega) - R_\infty]\cos(\omega t')\exp(-k\omega)\mathrm{d}\omega
$$

$$
= \int_0^\infty \int_0^\infty v_1(t)\cos(\omega t)\cos(\omega t')\exp(-k\omega)\mathrm{d}t\,\mathrm{d}\omega
$$

$$
= \int_0^\infty v_1(t)\left[\int_0^\infty \cos(\omega t)\cos(\omega t')\exp(-k\omega)\mathrm{d}\omega\right]\mathrm{d}t
$$

We now allow the quantity k to tend to zero. In the first integral, k can be placed equal to zero without any difficulty, but the last one has to be evaluated by a limiting process as seen below. Hence:

$$
\int_0^\infty [R(\omega) - R_\infty]\cos(\omega t')\mathrm{d}\omega = \int_0^\infty v_1(t)I\mathrm{d}t
$$

where

$$
I = \operatorname*{Lim.}_{k\to 0}\left[\int_0^\infty \cos(\omega t)\cos(\omega t')\exp(-k\omega)\mathrm{d}\omega\right]
$$

We find by straightforward integration:

$$\int_0^\infty \cos(\omega t) \cos(\omega t') \exp(-k\omega) d\omega$$

$$= \tfrac{1}{2}\left[\frac{k}{k^2 + (t-t')^2} + \frac{k}{k^2 + (t+t')^2}\right]$$

Consider

$$D(x) = \operatorname*{Lim.}_{k \to 0}\left[\frac{k}{k^2 + x^2}\right]$$

clearly

$$D(x) = 0 \quad \text{unless } x \text{ is zero}$$

and

$$D(x) \to \infty \quad \text{if } x \text{ is zero}$$

so that

$$D(x) = c\delta(x)$$

by comparison with the impulse function defined in Section A.3.4. The constant c is obtained by using the property

$$\int_{-\infty}^\infty \delta(x)dx = 1$$

i.e.

$$c = \int_{-\infty}^\infty D(x)dx = \int_{-\infty}^\infty \operatorname*{Lim.}_{k \to 0}\left(\frac{k}{k^2 + x^2}\right)dx$$

$$= \operatorname*{Lim.}_{k \to 0}\left[\int_{-\infty}^\infty \frac{k}{k^2 + x^2}dx\right] = \operatorname*{Lim.}_{k \to 0}\left[\tan^{-1}\left(\frac{x}{k}\right)\Big|_{-\infty}^\infty\right]$$

$$= \pi$$

Hence,

$$D(x) = \pi\delta(x)$$

and

$$I = \tfrac{1}{2}\operatorname*{Lim.}_{k \to 0}\left[\frac{k}{k^2 + (t-t')^2} + \frac{k}{k^2 + (t+t')^2}\right]$$

$$= \frac{\pi}{2}[\delta(t - t') + \delta(t + t')]$$

We now have

$$\int_0^\infty [R(\omega) - R_\infty] \cos(\omega t')d\omega = \frac{\pi}{2}\int_0^\infty v_1(t)[\delta(t - t') + \delta(t + t')]dt$$

The variable t' can be restricted to positive values and since $v_1(t)$ is zero for negative t, the contribution from $\delta(t + t')$ vanishes, leaving:

$$\int_0^\infty [R(\omega) - R_\infty] \cos(\omega t')d\omega = \frac{\pi}{2} v_1(t')$$

Changes in the variables give the equivalent result:

$$v_1(t) = \frac{2}{\pi} \int_0^\infty [R(\omega') - R_\infty] \cos(\omega' t) d\omega'$$

This expression for $v_1(t)$ can now be substituted in the second equation of (A.106) to give:

$$X(\omega) = -\frac{2}{\pi} \int_0^\infty \int_0^\infty \sin(\omega t) \cos(\omega' t) [R(\omega') - R_\infty] d\omega' dt$$

$$= -\frac{2}{\pi} \int_0^\infty [R(\omega') - R_\infty] J d\omega'$$

where

$$J = \int_0^\infty \sin(\omega t) \cos(\omega' t) dt$$

The integral J is difficult to integrate because of the conditions as t tends to infinity, and is most easily evaluated by another limiting process.

$$J = \operatorname*{Lim.}_{\alpha \to 0} \int_0^\infty \sin(\omega t) \cos(\omega' t) \exp(-\alpha t) dt$$

$$= \operatorname*{Lim.}_{\alpha \to 0} \tfrac{1}{2} \left[+ \frac{(\omega + \omega')}{(\omega + \omega')^2 + \alpha^2} + \frac{(\omega - \omega')}{(\omega - \omega')^2 + \alpha^2} \right]$$

$$= \tfrac{1}{2} \left[\frac{1}{\omega + \omega'} + \frac{1}{\omega - \omega'} \right] = \frac{\omega}{\omega^2 - \omega'^2}$$

The final expression for $X(\omega)$ in terms of $R(\omega)$ is therefore:

$$X(\omega) = +\frac{2}{\pi} \int_0^\infty \frac{\omega[R(\omega') - R_\infty]}{\omega'^2 - \omega^2} d\omega' \qquad \text{(A.107)}$$

A corresponding equation, giving $R(\omega)$ in terms of $X(\omega)$, can be found by repeating the above analysis, but first expressing $v_1(t)$ in terms of $X(\omega)$. This equation is:

$$R(\omega) - R_\infty = \frac{2}{\pi} \int_0^\infty \frac{\omega' . X(\omega')}{\omega^2 - \omega'^2} d\omega' \qquad \text{(A.108)}$$

These two equations (A.107) and (A.108), form a pair of Hilbert transforms. $R(\omega)$ can be deduced from $X(\omega)$, and conversely.

The evaluation of the integrals requires some care because of the infinite value of the integrand when ω' equals ω. A more detailed analysis shows that the integrals should be interpreted as limiting values, a small section near ω' equal to ω being exluded, i.e.

$$X(\omega) = +\frac{2}{\pi} \operatorname*{Lim.}_{\delta \to 0} \left[\int_0^{\omega-\delta} + \int_{\omega+\delta}^\infty \right] \frac{\omega[R(\omega') - R_\infty]}{\omega'^2 - \omega^2} d\omega' \quad \text{(A.109)}$$

with a similar change in equation (A.108).

It is easily checked from these equations that $R(\omega)$ is an even function of ω and $X(\omega)$ is an odd function of ω, *i.e.*

$$R(-\omega) = R(\omega) : X(-\omega) = -X(\omega) \qquad \text{(A.110)}$$

This result permits an alternative formulation of the equations.

$$
\begin{aligned}
X(\omega) &= +\frac{2}{\pi} \int_0^\infty \frac{\omega[R(\omega') - R_\infty]}{\omega'^2 - \omega^2} \, d\omega' \\
&= +\frac{1}{\pi} \int_0^\infty \left[\frac{1}{\omega' - \omega} - \frac{1}{\omega' + \omega} \right] [R(\omega') - R_\infty] d\omega' \\
&= +\frac{1}{\pi} \int_0^\infty \frac{[R(\omega') - R_\infty]}{\omega' - \omega} \, d\omega' - \frac{1}{\pi} \int_0^\infty \frac{[R(\omega') - R_\infty]}{\omega' + \omega} \, d\omega'
\end{aligned}
$$

Change the integration variable in the second integral from ω' to $-\omega'$. Then,

$$
\begin{aligned}
X(\omega) &= +\frac{1}{\pi} \int_0^\infty \frac{[R(\omega') - R_\infty]d\omega'}{\omega' - \omega} - \frac{1}{\pi} \int_{-\infty}^0 \frac{[R(\omega') - R_\infty]}{-\omega' + \omega} \, d\omega' \\
&= +\frac{1}{\pi} \int_{-\infty}^\infty \frac{[R(\omega') - R_\infty]}{\omega' - \omega} \, d\omega' \qquad \text{(A.111)}
\end{aligned}
$$

By a similar manipulation of equation (A.108), we obtain:

$$R(\omega) - R_\infty = -\frac{1}{\pi} \int_{-\infty}^\infty \frac{X(\omega')d\omega'}{\omega' - \omega} \qquad \text{(A.112)}$$

Apart from a change in sign equations (A.111) and (A.112) are identical

A.8.2. APPLICATION TO TRANSFER FUNCTIONS

Similar relations to equations (A.111) and (A.112) exist between the real and imaginary parts of many other complex functions, including transfer functions. In dealing with transfer functions, we are, however, more interested in relations between the amplitude and phase characteristics, and some modification to the treatment is required. Let $A(\omega)$ be a transfer function, having a magnitude $|A(\omega)|$ and a phase shift $\phi(\omega)$, so that

$$A(\omega) = |A(\omega)| \exp[-j\phi(\omega)] \qquad \text{(A.113)}$$

Taking logarithms to the base e, we have:

$$\log_e A(\omega) = \log_e |A(\omega)| - j\phi(\omega)$$

Let

$$\alpha(\omega) = -\log_e |A(\omega)| \qquad \text{(A.114)}$$

$\alpha(\omega)$ being therefore, the attenuation of the network in nepers. Then,

$$\log_e A(\omega) = -\alpha(\omega) - j\phi(\omega) \qquad \text{(A.115)}$$

so that $\alpha(\omega)$, $\phi(\omega)$ are the real and imaginary parts of the complex quantity $-\log_e A(\omega)$. We may therefore anticipate that $\alpha(\omega)$, $\phi(\omega)$ will also satisfy the Hilbert transform relations. A full discussion of this requires the use of complex number theory, as given in references 5 and

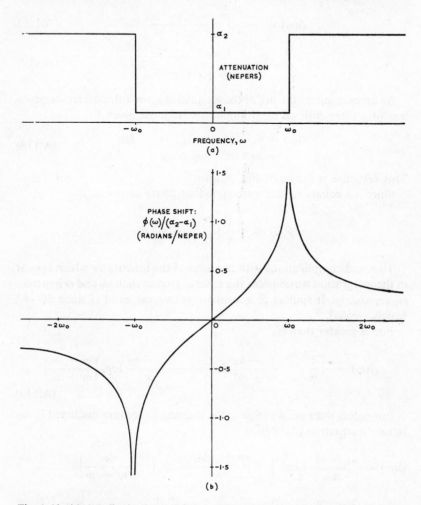

Fig. A.13. (a) Amplitude characteristic and (b) phase characteristic of a low pass network.

6, and it transpires that the anticipated result is true for a class of networks known as minimum phase networks. The implication of the term minimum phase is that the phase variation with frequency is the least possible for a network having the prescribed attenuation function.

Other networks with the same $\alpha(\omega)$ are possible, but involve greater

phase changes. Most networks encountered in communications work are of the minimum phase type.

For minimum phase networks, the attenuation $\alpha(\omega)$ in nepers and the phase change $\phi(\omega)$ in radians, are related by the Hilbert transform pair:

$$\phi(\omega) = +\frac{1}{\pi} \int_{-\infty}^{\infty} \frac{\alpha(\omega') - \alpha_{\infty}}{\omega' - \omega} \, d\omega' \qquad \text{(A.116)}$$

$$\alpha(\omega) = -\frac{1}{\pi} \int_{-\infty}^{\infty} \frac{\phi(\omega')}{\omega' - \omega} \, d\omega' \qquad \text{(A.117)}$$

As an example of the use of these equations, we will consider the problem of a filter with an ideal amplitude response given by

$$\begin{aligned} \alpha(\omega) &= \alpha_1 \quad \text{for} \quad -\omega_0 \leqslant \omega \leqslant \omega_0 \\ &= \alpha_2 \quad \text{for} \quad |\omega| > \omega_0 \end{aligned} \right\} \qquad \text{(A.118)}$$

This response is shown in Fig. A.13(a).

Since α_{∞} equals α_2, the corresponding phase response is

$$\phi(\omega) = +\frac{1}{\pi} \int_{-\omega_0}^{\omega_0} \frac{\alpha_1 - \alpha_2}{\omega' - \omega} \, d\omega'$$

To avoid complications with the signs of the logarithms which appear in the integration we consider the cases ω greater than ω_0 and ω less than ω_0 separately. It suffices to consider positive values of ω since $\phi(-\omega)$ equals $-\phi(\omega)$.

For ω greater than ω_0

$$\phi(\omega) = -\frac{1}{\pi} \int_{-\omega_0}^{\omega_0} \frac{\alpha_1 - \alpha_2}{\omega - \omega'} \, d\omega' = +\frac{\alpha_1 - \alpha_2}{\pi} \log_e \left[\frac{\omega - \omega_0}{\omega + \omega_0} \right]$$

$$\text{(A.119)}$$

For ω less than ω_0, we require the limiting procedure discussed in relation to equation (A.109).

$$\begin{aligned} \phi(\omega) &= \frac{\alpha_1 - \alpha_2}{\pi} \lim_{\delta \to 0} \left[-\int_{-\omega_0}^{\omega - \delta} \frac{d\omega'}{\omega - \omega'} + \int_{\omega + \delta}^{\omega_0} \frac{d\omega'}{\omega' - \omega} \right] \\ &= \frac{\alpha_1 - \alpha_2}{\pi} \lim_{\delta \to 0} \left[+ \log_e (\omega - \omega') \Big|_{-\omega_0}^{\omega - \delta} + \log_e (\omega' - \omega) \Big|_{\omega + \delta}^{\omega_0} \right] \\ &= \frac{\alpha_1 - \alpha_2}{\pi} \lim_{\delta \to 0} \left[\log_e \delta - \log_e (\omega + \omega_0) + \log_e (\omega_0 - \omega) \right. \\ &\qquad\qquad\qquad\qquad\qquad\qquad\qquad\qquad\qquad \left. - \log_e \delta \right] \\ &= \frac{\alpha_2 - \alpha_1}{\pi} \log_e \left(\frac{\omega + \omega_0}{\omega_0 - \omega} \right) \end{aligned}$$

$$\text{(A.120)}$$

The expressions for the phase shifts in equations (A.119) and (A.120) are directly proportional to the difference $(\alpha_2 - \alpha_1)$ between the attenuations in the pass and stopbands. An ideal channel, of the type discussed in Chapter 5, has infinite attenuation in the stopband and so the phase shift is also infinite. Numerical values of $\phi(\omega)/(\alpha_2 - \alpha_1)$ are plotted in Fig. A.13(b).

The group delay for the network can be calculated from the derivative of $\phi(\omega)$, the appropriate frequency for the evaluation being ω equal to zero, at the centre of the passband. Then,

$$\text{Group delay} = \left[\frac{d\phi}{d\omega}\right]_{\omega = o} = \frac{2(\alpha_2 - \alpha_1)}{\pi\omega_o} \qquad (A.121)$$

The group delay becomes infinite if the attenuation in the stop band tends to infinity.

REFERENCES

[1] DAY, W. D., *Introduction to Laplace Transforms for Radio and Electronic Engineers* Iliffe, 1960.
[2] TITCHMARCH, E. C., *The Fourier Integral*, Oxford University Press.
[3] CHURCHILL, R. V., *Operational Mathematics*, McGraw-Hill, 1958.
[4] GUILLEMIN, E. A., *Introductory Circuit Theory*, Wiley & Sons, 1953.
[5] GUILLEMIN, E. A., *The Mathematics of Circuit Analysis*, Wiley & Sons, 1949.
[6] GUILLEMIN, E. A., *Synthesis of Pasive Networks*, Wiley & Sons, 1957.
[7] LEY, B. J., LUTZ, S. G. AND REHBERG, C. F., *Linear Circuit Analysis*, McGraw-Hill, 1959.
[8] BODE, H. W., *Network Analysis and Feedback Amplifier Design*, p. 28, D. van Nostrand, New York 1945.

TABLE A2

Standard Results on Fourier transforms

1. Definition: $\qquad g(\omega) = \displaystyle\int_{-\infty}^{\infty} f(t)\exp(-j\omega t)\,dt$

2. Inversion: $\qquad f(t) = \dfrac{1}{2\pi}\displaystyle\int_{-\infty}^{\infty} g(\omega)\exp(j\omega t)\,d\omega$

3. *General Results*

   (a) *Addition of transforms*

   If $g_1(\omega)$, $g_2(\omega)$ are respectively the transforms of $f_1(t), f_2(t)$, then $g_1(\omega) + g_2(\omega)$ is the transform of $f_1(t) + f_2(t)$.

   (b) *Multiplication by a constant, a*
   If $g(\omega)$ is the transform of $f(t)$, then $ag(\omega)$ is the transform of $af(t)$

   (c) *Frequency shift*
   If $g(\omega)$ is the transform of $f(t)$, $g(\omega - \omega_0)$ is the transform of $f(t)\exp(j\omega_0 t)$

   (d) *Time shift*
   If $g(\omega)$ is the transform of $f(t)$, $g(\omega)\exp(-j\omega\tau)$ is the transform of $f(t - \tau)$

(e) *Convolution theorem*

If $g_1(\omega)$, $g_2(\omega)$ are respectively the transforms of $f_1(t)$, $f_2(t)$, then:

(i) $g_1(\omega) g_2(\omega)$ is the transform of

$$\int_{-\infty}^{\infty} f_1(t - x) f_2(x) \, \mathrm{d}x$$

and (ii) $\dfrac{1}{2\pi} \displaystyle\int_{-\infty}^{\infty} g_1(\omega - u) g_2(u) \, \mathrm{d}u$ is the transform of $f_1(t) f_2(t)$.

4. *Dictionary*

	$f(t)$	$g(\omega)$
1	$0 : t < 0$	$1/(\alpha + j\omega)$
	$\exp[-\alpha t] : t > 0$	$\mathrm{Re.}\,\alpha > 0$
2	1	$2\pi\delta(\omega)$
3	$\exp(j\omega_0 t)$	$2\pi\delta(\omega - \omega_0)$
4	$\cos(\omega_0 t)$	$\pi\delta(\omega - \omega_0) + \pi\delta(\omega + \omega_0)$
5	$\sin(\omega_0 t)$	$-j\pi\delta(\omega - \omega_0) + j\pi\delta(\omega + \omega_0)$
6	$\delta(t)$	1
7	$\delta(t - \tau)$	$\exp(-j\omega\tau)$
8	$1 : \lvert t \rvert > \frac{1}{2}\tau$	$\tau \sin(\frac{1}{2}\omega\tau)$
	$0 : \lvert t \rvert > \frac{1}{2}\tau$	$(\frac{1}{2}\omega\tau)$
9	$\exp(-kt^2)$	$(\pi/k)^{\frac{1}{2}} \exp(-\omega^2/4k)$

In this table $\delta(x)$ is the Dirac delta function, representing an impulse of negligible duration and infinite amplitude. Its properties are:

$$\delta(x) = 0 \text{ unless } x = 0 : \delta(x) \to \infty \quad \text{if} \quad x = 0$$

$$\int_{-\infty}^{\infty} \delta(x) \, \mathrm{d}x = 1 : \int_{-\infty}^{\infty} \delta(x) f(x) \, \mathrm{d}x = f(0)$$

$$\int_{a}^{b} \delta(x - x_0) f(x) \, \mathrm{d}x = f(x_0) \quad \text{provided } a < x_0 < b.$$

TABLE A.3

Standard Results on Laplace Transforms

---

1. **Definition** $\mathscr{L}\{f(t)\} = \bar{f}(p) = \displaystyle\int_{0}^{\infty} f(t) \exp(-pt) \, \mathrm{d}t$

2. **General Results**

(a) $\mathscr{L}\{af(t)\} = a\mathscr{L}f\{(t)\}$ if a is a constant.

(b) $\mathscr{L}\{f_1(t) + f_2(t)\} = \mathscr{L}\{f_1(t)\} + \mathscr{L}\{f_2(t)\}$

(c) $\mathscr{L}\{f'(t)\} = p\mathscr{L}\{f(t)\} - f(0)$

(d) $\mathscr{L}\{f^n(t)\} = p^n \mathscr{L}\{f(t)\} - [p^{n-1}f_0 + p^{n-2}f'(0) + \ldots p^{n-1-r}f^r(0) + \ldots + f^{n-1}(0)]$

(e) $\mathscr{L}\left\{\displaystyle\int_{0}^{t} f(t)\mathrm{d}t\right\} = \dfrac{1}{p}\,\mathscr{L}\{f(t)\}$

(f) $\mathscr{L}\{\exp(-at)f(t)\} = \bar{f}(p + a)$: a is a complex constant.

(g) $\quad \mathscr{L}\{g(t)\} = \exp(-pT)\mathscr{L}\{f(t)\}$ if $g(t) = 0 : t < T$
$$= f(t-T) : t > T$$

(h) $\quad \mathscr{L}\left\{\int_0^t f_1(x)f_2(t-x)\mathrm{d}x\right\} = \bar{f}_1(p)\bar{f}_2(p)$

3. Heaviside Expansion Theorem

If
$$\bar{f}(p) = \frac{A(p)}{B(p)}$$

where $B(p) = 0$ has the roots $p = p_1, p_2, p_3 \ldots p_r \ldots p_n$, then

$$f(t) = \sum_{r=1}^{n} \frac{A(p_r)}{B'(p_r)} \exp(p_r t)$$

where $B(p) = 0$ has the roots $p = p_1, p_2, p_3 \ldots p_r \ldots p_n$, then

This result applies if the roots of $B(p) = 0$ are all different and if the degree of the polynomial $B(p)$ is less than that of $A(p)$. Extensions are discussed in the main text.

4. Dictionary

	$f(t)$	$\mathscr{L}\{f(t)\} = \bar{f}(p)$
1	1	$1/p$
2	t	$1/p^2$
3	t^n	$n!/p^{n+1}$
4	$\exp(at)$	$1/(p-a)$ $[p > \mathrm{Re}\,.\,a]$
5	$t.\exp(at)$	$1/(p-a)^2$
6	$t^n.\exp(at)$	$n!/(p-a)^{n+1}$
7	$\sin at$	$a/(p^2+a^2)$
8	$\cos at$	$p/(p^2+a^2)$
9	$t\sin at$	$2ap/(p^2+a^2)^2$
10	$t\cos at$	$(p^2-a^2)/(p^2+a^2)^2$
11	$\exp(-bt)\sin at$	$a/[(p+b)^2+a^2]$
12	$\exp(-bt)\cos at$	$(p+b)/[(p+b)^2+a^2]$
13	$\delta(t)$	1
14	$\delta(t-T)$	$\exp(-pT)$
15	$\begin{cases} at : 0 \leqslant t \leqslant T \\ aT : t \geqslant T \end{cases}$	$(a/p^2)[1-\exp(-pT)]+$ $+ (aT/p)\exp(-pT)$

Bessel Functions

B.1. Definitions

In discussing the properties of angle modulation, we are faced with the problem of determining the Fourier expansion of an expression such as $\cos(x \sin \theta)$. The coefficients of such expansions involve Bessel functions, and for the present purposes we may define these functions as the coefficients of a suitable Fourier series. It is convenient to work with the exponential series and the function which we expand is taken as exp $(jx \sin \theta)$: this function can be split into a real part $\cos(x \sin \theta)$ and an imaginary part $\sin(x \sin \theta)$. The function $\exp(jx \sin \theta)$ is periodic in θ with period 2π and can therefore be written as a Fourier series:

$$\exp(jx \sin \theta) = \sum_{n=-\infty}^{\infty} c_n \exp(jn\theta) \qquad (B.1)$$

The coefficients c_n obviously depend on x and we define the nth coefficient to be the nth order Bessel function of argument x, $i.e.$ we write:

$$c_n = J_n(x) \qquad (B.2)$$

(Bessel functions are frequently introduced as the solutions of a particular type of differential equation. The approach used here is more suited to our purpose since it defines the functions in relation to the problem of interest. We will show later that the functions so defined do satisfy the appropriate differential equation.)

The coefficients can be determined by the usual rules of Fourier analysis, and so we have:

$$c_n = J_n(x) = \frac{1}{2\pi} \int_{-\pi}^{\pi} \exp(jx \sin \theta - jn\theta) \mathrm{d}\theta \qquad (B.3)$$

The integral on the right-hand side of the above equation can be evaluated numerically and so tables of $J_n(x)$ for different values of n and x can be constructed. We will find, however, that many useful properties of these functions can be deduced without carrying out the numerical calculations. The first such property is that $J_n(x)$ is always real if x is real, and this is proved by a simple manipulation of the integral. We split the integration range into the regions 0 to π and $-\pi$ to 0, and in the second region we replace θ by $-\theta$, thus obtaining:

$$J_n(x) = \frac{1}{2\pi} \int_0^\pi \exp(jx \sin \theta - jn\theta) \, d\theta$$

$$+ \frac{1}{2\pi} \int_0^\pi \exp(-jx \sin \theta + jn\theta) \, d\theta$$

$$= \frac{1}{\pi} \int_0^\pi \cos(x \sin \theta - n\theta) \, d\theta. \tag{B.4}$$

The integrand in the final integral is purely real and so $J_n(x)$ is real for any real values of n and x.

We next obtain a relation between $J_n(x)$ and $J_{-n}(x)$ by replacing $\pi - \theta$ for θ and using the result that $\sin(\pi - \theta) = \sin \theta$. Then

$$J_{-n}(x) = \frac{1}{\pi} \int_0^\pi \cos(x \sin \theta - n\pi + n\theta) \, d\theta$$

$$= \frac{1}{\pi} \int_0^\pi [\cos(n\pi) \cos(x \sin \theta + n\theta)$$

$$+ \sin(n\pi) \sin(x \sin \theta + n\theta)] \, d\theta \tag{B.5}$$

We are only interested in integral values of n so that

$$\cos(n\pi) = (-)^n : \sin(n\pi) = 0 \tag{B.6}$$

Also, from equation (B.4) we have:

$$J_{-n}(x) = \frac{1}{\pi} \int_0^\pi \cos(x \sin \theta + n\theta) \, d\theta \tag{B.7}$$

so that equation (B.5) becomes:

$$J_n(x) = (-)^n J_{-n}(x) \tag{B.8}$$

This equation, coupled with the result that $J_n(x)$ is always real, enables us to deduce several other Fourier series involving Bessel functions.

B.2. Fourier Series involving Bessel Functions

Equation (B.1) can be written as:

$$\exp(jx \sin \theta) = J_0(x) + \sum_{n=1}^\infty [J_n(x) \exp(jn\theta) + J_{-n}(x) \exp(-jn\theta)] \tag{B.9}$$

Replace $J_{-n}(x)$ by $(-)^n J_n(x)$

Then:

$$\exp(jx \sin \theta) = J_0(x) + \sum_{n=1}^{\infty} \{J_n(x)[\exp(jn\theta) + (-)^n \exp(-jn\theta)]\} \tag{B.10}$$

Now

$$\exp(jn\theta) + (-)^n \exp(-jn\theta) = 2 \cos(n\theta) \qquad \text{if } n \text{ is even}$$
$$= 2j \sin(n\theta) \qquad \text{if } n \text{ is odd} \tag{B.11}$$

Hence

$$\exp(jx \sin \theta) = J_0(x) + 2 \sum_{n=2,4,\ldots} J_n(x) \cos(n\theta)$$
$$+ 2j \sum_{n=1,3,\ldots} J_n(x) \sin(n\theta) \tag{B.12}$$

Since $J_n(x)$ is always real, this equation can easily be split into real and imaginary parts:

$$\cos(x \sin \theta) = J_0(x) + 2 \sum_{n=2,4,\ldots} J_n(x) \cos(n\theta)$$
$$= J_0(x) + 2J_2(x) \cos 2\theta + 2J_4(x) \cos(4\theta) + \cdots \tag{B.13}$$

$$\sin(x \sin \theta) = 2 \sum_{n=1,3,\ldots} J_n(x) \sin(n\theta)$$
$$= 2J_1(x) \sin \theta + 2J_3(x) \sin 3\theta + 2J_5(x) \sin(5\theta) + \cdots \tag{B.14}$$

Two further series can be obtained by replacing θ by $\pi/2 - \theta$

$$\cos(x \cos \theta) = J_0(x) - 2J_2(x) \cos(2\theta) + 2J_4(x) \cos(4\theta)$$
$$- 2J_6(x) \cos(6\theta) + \cdots \tag{B.15}$$

$$\sin(x \cos \theta) = 2J_1(x) \cos \theta - 2J_3(x) \cos 3\theta + 2J_5(x) \cos(5\theta) - \cdots \tag{B.16}$$

B.3. Bessel's Differential Equation

Differentiating equation (B.3) with respect to x gives:

$$\frac{\mathrm{d}J_n(x)}{\mathrm{d}x} = \frac{j}{2\pi} \int_{-\pi}^{\pi} \sin \theta \exp(jx \sin \theta - jn\theta) \mathrm{d}\theta \tag{B.17}$$

$$\frac{\mathrm{d}^2 J_n(x)}{\mathrm{d}x^2} = -\frac{1}{2\pi} \int_{-\pi}^{\pi} \sin^2\theta \exp(jx \sin \theta - jn\theta) \mathrm{d}\theta \tag{B.18}$$

Also, if integration by parts is used, the following alternative expressions for $J_n(x)$ can be obtained:

$$J_n(x) = \frac{1}{2\pi} \int_{-\pi}^{\pi} \exp(jx \sin \theta) \frac{d[\exp(-jn\theta)]}{-jn}$$

$$= \frac{1}{2\pi} \left[\frac{\exp(jx \sin \theta - jn\theta)}{-jn} \right]_{-\pi}^{\pi}$$

$$+ \frac{1}{2\pi jn} \int_{-\pi}^{\pi} jx \cos \theta \exp(jx \sin \theta - jn\theta) d\theta$$

$$= \frac{x}{2\pi n} \int_{-\pi}^{\pi} \cos \theta \exp(jx \sin \theta) \frac{d[\exp(-jn\theta)]}{-jn}$$

$$= \frac{x}{2\pi n} \left[\frac{\cos \theta \exp(jx \sin \theta - jn\theta)}{-jn} \right]_{-\pi}^{\pi} +$$

$$+ \frac{x}{2\pi jn^2} \int_{-\pi}^{\pi} (-\sin \theta + jx \cos^2\theta) \times [\exp(jx \sin \theta - jn\theta)] d\theta$$

$$= \frac{x}{2\pi n^2} \int_{-\pi}^{\pi} [x - x \sin^2\theta + j \sin \theta] \exp(jx \sin \theta - jn) d\theta \quad \text{(B.19)}$$

The values at the limits cancel because of the periodicity of all the functions involved. The right-hand side of equation (B.19) can be expressed in terms of $J_n(x)$ and its derivatives by using equations (B.17) and (B.18), giving:

$$J_n(x) = \frac{x^2}{n^2} J_n(x) + \frac{x^2}{n^2} \frac{d^2 J_n(x)}{dx^2} + \frac{x}{n^2} \frac{d J_n(x)}{dx} \qquad \text{·(B.20)}$$

i.e.

$$\frac{d^2 J_n(x)}{dx^2} + \frac{1}{x} \frac{d J_n(x)}{dx} + \left(1 - \frac{n^2}{x^2}\right) J_n(x) = 0 \qquad \text{(B.21)}$$

Since this equation is of the second order, it must have two independent solutions. $J_{-n}(x)$ is also a solution, but it is simply related to $J_n(x)$ and is not independent of it. The second solution, usually denoted by $Y_n(x)$ and called the Bessel function of the second kind, is not required for a discussion of frequency modulation but does occur in certain electromagnetic field problems.

B.4. Power Series for $J_n(x)$

An expression for $J_n(x)$ as a series of powers of x can be obtained either by expanding $\exp(jx \sin \theta)$ as a power series and integrating the result term by term according to the integral definition, or by finding a series solution for the differential equation. Either method results in straightforward algebra and the series expansion so obtained is:

$$J_n(x) = \frac{1}{n!}\left(\frac{x}{2}\right)^n - \frac{1}{(n+1)!}\left(\frac{x}{2}\right)^{n+1} + \frac{1}{2!(n+2)!}\left(\frac{x}{2}\right)^{n+2} + \dots$$

$$+ \frac{(-)^r}{r!(n+r)!}\left(\frac{x}{2}\right)^{n+r} + \dots \qquad \text{(B.22)}$$

The most significant feature of this expression is that $J_n(0)$ is zero for all values of n except $n = 0$ and that $J_0(0)$ equals unity. These results can be verified directly from the integral in equation (B.3). The leading term of $J_n(x)$ is proportional to x^n and this means that as n increases, the curve of $J_n(x)$ against x rises more and more slowly for x near zero.

B.5. Behaviour of $J_n(x)$ for Large Values of x

An approximate solution to the differential equation can be found for large values of x. We begin by a change in the equation to eliminate the first derivative.
Let

$$J_n(x) = F_n(x)/\sqrt{x} \qquad \text{(B.23)}$$

Then

$$J_n'(x) = \frac{F_n'(x)}{\sqrt{x}} - \frac{F_n(x)}{2x^{3/2}} \qquad \text{(B.24)}$$

$$J_n''(x) = \frac{F_n''(x)}{\sqrt{x}} - \frac{F_n'(x)}{x^{3/2}} + \frac{3F_n(x)}{4x^{5/2}} \qquad \text{(B.25)}$$

where primes are used to denote derivatives. Substitution of these expressions in equation (B.21) and simplification of the resulting differential equation gives

$$F_n''(x) + \left[1 - \frac{n^2 - \frac{1}{4}}{x^2}\right]F_n(x) = 0 \qquad \text{(B.26)}$$

If x^2 is much larger than $n^2 - \frac{1}{4}$, the term $(n^2 - \frac{1}{4})/x^2$ can be neglected in comparison with unity and we obtain the approximation:

$$F_n''(x) + F_n(x) = 0. \qquad \text{(B.27)}$$

which has the general solution:

$$F_n(x) = A_n \cos(x + \phi_n) \qquad \text{(B.28)}$$

A_n, ϕ_n are integration constants whose value will depend on the particular Bessel function which is being considered. We therefore see that for sufficiently large values of x (*i.e.* $x \gg n$), the following approximation holds:

$$J_n(x) \doteqdot \frac{A_n}{\sqrt{x}} \cos(x + \phi_n) \qquad (B.29)$$

This result shows that for large values of x, $J_n(x)$ behaves like a sine wave with progressively decreasing amplitude. It is possible to show by methods beyond the scope of the present treatment that:

$$A_n = \left(\frac{2}{\pi}\right)^{\frac{1}{2}} \qquad (B.30)$$

$$\phi_n = \frac{-(2n+1)\pi}{4} \qquad (B.31)$$

B.6. Expansions for Frequency Modulated Waves

(*i*) SINUSOIDAL MODULATION

The expression for the frequency modulated carrier is (equation (3.48))

$$\begin{aligned} f(t) &= A_c \cos[\omega_c t + m_p \sin(\omega_m t)] \\ &= A_c \mathrm{Re}.\{\exp j[\omega_c t + m_p \sin(\omega_m t)]\} \\ &= A_c \mathrm{Re}.[\exp j\omega_c t][\exp\{jm_p \sin(\omega_m t)\}] \qquad (B.32) \end{aligned}$$

From equations (B.1) and (B.2):

$$\exp\{jm_p \sin(\omega_m t)\} = \sum_{n=-\infty}^{\infty} J_n(m_p) \exp(jn\omega_m t) \qquad (B.33)$$

so that:

$$f(t) = A_c \mathrm{Re}.\left[\sum_{n=-\infty}^{\infty} J_n(m_p) \exp(j\omega_c t + jn\omega_m t)\right] \qquad (B.34)$$

Since we have shown that $J_n(m_p)$ is a real quantity, we can take real parts without difficulty and obtain:

$$f(t) = A_c \sum_{n=-\infty}^{\infty} J_n(m_p) \cos(\omega_c t + n\omega_m t) \qquad (B.35)$$

the result quoted as equation (3.49).

(*ii*) MODULATING SIGNAL CONTAINING TWO SINUSOIDAL TERMS

Suppose the modulated carrier is:

$$f(t) = A_c \cos[\omega_c t + m_{p1} \sin(\omega_{m1} t) + m_{p2} \sin(\omega_{m2} t)] \qquad (B.36)$$

Then

$$f(t) = A_c \text{Re.} [\exp(j\omega_c t) \exp\{jm_{p1} \sin(\omega_{m1} t)\} \exp\{jm_{p2} \sin(\omega_{m2} t)\}]$$

$$= A_c \text{Re.} \left[\exp(j\omega_c t) \left\{ \sum_{n=-\infty}^{\infty} J_n(m_{p1}) \exp(j\omega_{m1} nt) \right\} \right.$$

$$\left. \times \left\{ \sum_{m=-\infty}^{\infty} J_m(m_{p2}) \exp(j\omega_{m2} mt) \right\} \right]$$

$$= A_c \text{Re.} \left[\sum_{n=-\infty}^{\infty} \sum_{m=-\infty}^{\infty} J_n(m_{p1}) J_m(m_{p2}) \exp[j\omega_c t + jn\omega_{m1} t \right.$$

$$\left. + jm\omega_{m2} t] \right]$$

$$= A_c \left[\sum_{n=-\infty}^{\infty} \sum_{m=-\infty}^{\infty} J_n(m_{p1}) J_m(m_{p2}) \cos(\omega_c t + n\omega_{m1} t + m\omega_{m2} t) \right]$$

$$\text{(B.37)}$$

This shows that the modulated carrier contains sidebands of frequencies $\omega_c + n\omega_{m1} + m\omega_{m2}$ where n, m may have any positive or negative integral value. It should be noted that there is no simple relation between the results in equations (B.35) and (B.37).

FM Threshold

C.1. Estimate of Threshold Level

The analysis of the noise performance of an FM discriminator given in Section 6.6 is only valid when the input signal-to-noise ratio exceeds a certain threshold level as indicated in Fig. 6.16. An estimate of this threshold level will now be obtained.

We begin by postulating a form of demodulator which could provide the desired output $k\omega_d$, where ω_d is the frequency deviation. Suppose this comprises a zero-crossing detector producing an impulse each time the input waveform crosses from a positive to a negative value and an averaging circuit with a time constant much larger than the period of the carrier but smaller than the period of the highest modulation frequency. The output is thus proportional to the number of zero crossings per second and since only the positive to negative crossings are included this number equals the instantaneous frequency, $f_c + f_d$. The impulse strength and integration time constant can be selected to make this output equal to $2\pi k \times$ (instantaneous frequency) and after removal of the d.c. term corresponding to the carrier frequency, f_c, we are left with $k\omega_d$ as desired.

We shall later require to know what contribution is made to the output by a single pulse. Suppose the integration time is T. Then the output $2\pi k f$, where f is instantaneous frequency, may be regarded as arising from the sum of fT pulses each of duration T. A single impulse thus gives an output of amplitude $2\pi k/T$ and duration T. Since T is short in relation to modulation frequencies, we may thus deduce that an extra impulse from the zero-crossing detector produces an impulse of strength $2\pi k$ (amplitude−duration product), i.e. $2\pi k\delta(t-\tau)$ where τ is the time at which this impulse occurs at the demodulator output.

We now examine the response of this ideal demodulator to the carrier waveform plus noise, i.e. to the input signal defined by:

$$v(t) = [E_c + x_1(t)]\cos\omega_c t + y_1(t)\sin\omega_c t$$

$$= A(t)\cos[\omega_c t + \theta(t)] \tag{C.1}$$

where

$$\theta(t) = \tan^{-1}\left[\frac{y(t)}{E_c + x_1(t)}\right] \tag{C.2}$$

If $x_1(t)$ and $y_1(t)$ are small relative to E_c, the value of $\theta(t)$ will never be appreciably different from zero. The effect of $\theta(t)$ on the instants,

τ_r, at which $v(t)$ changes from a positive to a negative value, will be to introduce a random departure from the values when the noise is absent, *i.e.*

$$\tau_r = (4r+1)\pi/2\omega_c + \varepsilon_r : r = 0, 1, 2, \ldots \qquad (C.3)$$

where

$$\varepsilon_r = -\theta(\tau_r) \qquad (C.4)$$

so that ε_r is a random variable with statistical properties which may be deduced from those of $x_1(t)$ and $y_1(t)$. Continuing with this line of argument we find that the output from the ideal detector, described above, contains noise terms identical to those derived in Section 6.6.

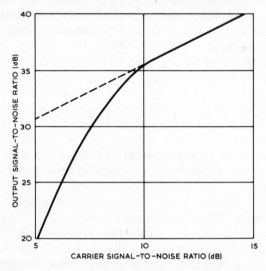

C.1. FM noise performance, showing threshold. —— from equation (C.10). ......
from equation (6.75).

The argument in the above paragraph remains valid so long as the angle $\theta(t)$ may be represented by a noise-like waveform with a limited excursion about its zero value. A new factor arises, however, if the excursion of $\theta(t)$ is such that it swings through a total angle of 2π in some interval, t_1. If this swing occurs in a positive sense, i.e. $\theta(t)$ increases by 2π, then an extra zero crossing, additional to those predicted by equation (C.3), will occur during the time interval, t_1, and an additional impulse is therefore generated by the zero-crossing detector. If the demodulator output is fed to a loudspeaker, this extra impulse would be audible as a click. A corresponding result is obtained if $\theta(t)$ decreases by 2π, one impulse now being lost and again a click is heard. We will now estimate the extent of this click noise, which is additional to the noise calculated in Section 6.6, by finding an expression for the number of clicks per second.

We first note that $x_1(t)$ and $y_1(t)$ are baseband signals with frequencies in the range $-\frac{1}{2}B$ to $\frac{1}{2}B$, where B is the receiver bandwidth. The sampling theorem shows that $x_1(t)$, $y_1(t)$ and hence $\theta(t)$ may be specified in terms of B independent samples per second. Suppose $\theta(t)$ changes by approximately 2π in a time interval of $2/B$. Examination of equation (C.2) shows that this can only happen if $E_c + x_1(t)$ is successively positive, negative and positive at three sampling instants t, $t + 1/B$ and $t + 2/B$. A necessary condition that a click be heard is therefore that $E_c + x_1(t) < 0$ at one of the sampling instants. This argument covers both types of click since the condition does not specify whether $\theta(t)$ increases or decreases by 2π. If further we assume that clicks occur relatively rarely, then we may ignore the possibility that $E_c + x_1(t) < 0$ for two successive sampling instants. We now have:

Number of clicks per second

$$= B \times (\text{probability that } E_c + x_1(t) < 0) \qquad \text{(C.5)}$$

$$= \frac{B}{\sigma\sqrt{2\pi}} \int_{-\infty}^{-E_c} \exp(-x^2/2\sigma^2) dx$$

$$= \frac{B\sigma}{E_c\sqrt{2\pi}} \exp(-E_c^2/2\sigma^2)$$

$$= (B/2\sqrt{\pi S_{\text{in}}}) \exp(-S_{\text{in}}) \qquad \text{(C.6)}$$

where σ, S_{in} have been defined in Section 6.6.

Each click corresponds to an impulse at the demodulator output. As argued earlier such additional impulses may be approximated by delta functions of strength $2\pi k$ and the succession of these impulses constitutes additional noise to be added to that already calculated in Section 6.6. The noise spectrum of this additional click noise is obtained as in the discussion of shot noise in Section 6.2.3. The energy per impulse within the audio band of width b is therefore $2b \times (2\pi k)^2$ and the noise power is:

N_c = energy per impulse $\times$ number of impulses per second

$$= 8\pi^2 k^2 b \times (B/2\sqrt{\pi S_{\text{in}}}) \exp(-S_{\text{in}}) \qquad \text{(C.7)}$$

The noise given by equation (6.74) is

$$N = 4\pi^2 k^2 b^3 / 3BS_{\text{in}} \qquad \text{(C.8)}$$

showing that the total noise contribution is

$$N + N_c = \frac{4\pi^2 k^2 b^2}{3BS_{\text{in}}} \left[1 + \frac{3B^2}{\sqrt{\pi}b^2} \sqrt{S_{\text{in}}} \exp(-S_{\text{in}}) \right] \qquad \text{(C.9)}$$

We now consider the limiting case of modulation with a frequency deviation, $f_d = B/2$, the maximum value permitted by the receiver bandwidth. Then, from equation (6.67), output signal power

$$= 2\pi^2 k^2 (B/2)^2$$

and so

$$S_{out} = \frac{3S_{in}(B/2b)^3}{1 + \dfrac{3B^2}{\sqrt{\pi}b^2} \sqrt{S_{in}} \exp(-S_{in})} \tag{C.10}$$

Fig. C.1. has been plotted from this equation for the broadcasting case, $B = 150$ kHz, $b = 15$ kHz. The threshold can be regarded as being about 10 dB, the value for S_{in} at which the solid curve corresponding to equation (C.10) begins to fall below the linear approximation obtained by ignoring click noise. It should be noted that this estimate of 10 dB is a little below the value of 12 dB deduced from practical observations, as mentioned in Section 6.6.

C.2. Low-threshold FM Receiver

Multi-channel telephony signals are often transmitted over radio links by frequency modulation. The individual channels are assembled on a frequency-division-multiplex basis to provide a base band which extends for example from 60 kHz to 1.025 MHz for a 240 channel system. The base band signal is then used to frequency modulate the carrier which will usually be in the microwave region and thus have a frequency of several GHz. In studies of such systems a reasonable estimate of the modulating signal can be made by assuming that it consists of white noise – this is realistic since the individual signals in each of the channels are independent. The results given in Section 3.4.1 for the bandwidth, B, of the modulated carrier require modification and a good approximation is

$$B = 2b + 8f_{dr} \tag{C.11}$$

where b is the highest modulating frequency used and f_{dr} is r.m.s. value of the frequency deviation. This result is consistent with those in Table 1 of Section 3.4.1, if the peak frequency deviation which occurs is $4f_{dr}$. In practice for the 240 channel system, B is taken as 24 MHz giving a maximum value of 2.75 MHz for the r.m.s. frequency deviation.

The analysis given in the previous section remains applicable provided allowance is made for the expression of the frequency deviation as an r.m.s. value and provided that the final answer provides the output signal-to-noise ratio in an individual telephone channel. The result for the channel centred on the base-band frequency, f_i, is

$$\text{Output signal-to-noise ratio} = \frac{2Bf_{dr}^{2} S_{in}}{bf_i^{2}}$$

above the F.M. threshold. This expression applies when the base band signal is used to modulate the carrier directly and depends on the centre frequency, f_i, since the noise power in the output increases by 6 dB per octave as noted earlier. Pre-emphasis of the modulating signal can be used to provide equal signal-to-noise ratios in all the output channels.

The threshold for such an arrangement can be estimated as discussed in the previous section. F.m. f.d.m. transmission of this kind is used in satellite communications and it is then essential to achieve the maximum efficiency. Although there is no technique by which the signal-to-noise ratio given by the expression above can be improved it is possible to reduce the threshold at which click noise becomes important. The way in which this can be done is best seen by examining the block diagram in Fig. C.2 of a receiver in which the local oscillator is varied in sympathy with the incoming carrier.

Suppose that the carrier frequency is f_c and that the instantaneous frequency deviation is f_d. Then the input to the mixer has instantaneous frequency $f_c + f_d$. Further, suppose that the instantaneous local oscillator frequency is $f_{LO} + kf_d$, where f_{LO} is a constant value such that $f_c - f_{LO} = f_I$, the mid-band frequency of the I.F. amplifier. The time-dependent contribution to the local-oscillator frequency can be provided by using a voltage-controlled oscillator, the output frequency of which is proportional to a voltage derived from the discriminator connected to the I.F. amplifier. Under these conditions, the instantaneous I.F. frequency is $(f_c + f_d) - (f_{LO} + kf_d) = f_I + (1-k)f_d$. The frequency deviation in the I.F. amplifier is therefore $(1-k)f_d$ and is therefore reduced by the factor $(1-k)$ compared with the deviation in a conventional receiver with a fixed local oscillator. This argument is equally applicable whether f_d applies to a wanted signal or to noise and it is plausible to expect that the signal-to-noise ratio is not affected by the introduction of feedback to the local oscillator. This is a valid argument in respect of the noise which can be regarded as imposing frequency modulation on the carrier but it does not hold for the click noise contribution. The reason for this is that click noise is produced by noise peaks which exceed the carrier and the occurrence of such peaks depends on the total noise present. In the feedback arrangement of Fig. C.2, the frequency deviation is reduced by the factor $(1-k)$ and so the I.F. amplifier bandwidth can also be reduced by a factor which will be considered below. The total signal-to-noise ratio in the I.F. amplifier can therefore be increased relative to its value at the mixer input. Since the threshold depends on total signal-to-noise ratio the possibility of lowering the threshold exists.

The feedback loop controlling the local oscillator must be capable of

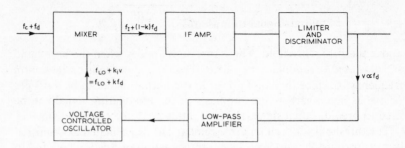

Fig. C.2. F.M. feedback receiver.

responding to changes in the instantaneous frequency deviation and must therefore have a minimum bandwidth corresponding to the modulation frequencies which are present. This leads to the conclusion that the low-frequency amplifier driving the voltage-controlled oscillator must have a bandwidth at least as great as the base bandwidth b and that the minimum I.F. bandwidth will be $2b$. Confirmation of this argument is provided by eqn. (C.11): if this equation is applied to the I.F. amplifier, which handles an FM signal containing modulating frequencies up to b and having an r.m.s. frequency deviation of $(1-k)f_\mathrm{d}$, the required I.F. bandwidth B_I is:

$$B_\mathrm{I} = 2b + 8(1-k)f_\mathrm{dr} \tag{C.12}$$

which tends to $2b$ if k tends to unity. If the I.F. bandwidth is restricted too much, distortion arises as a result of intermodulation. A detailed examination of this distortion leads to the conclusion that a value for B_I of approximately $6b$, i.e. three times the minimum possible value, is acceptable. For the 240 channel system, with b and f_dr equal to 1 MHz and 2.75 MHz, the value of $1-k$ corresponding to a 6 MHz I.F. bandwidth calculated from eqn. (C.12) is $1/5.5$ implying that the feedback reduces the frequency deviation by 5.5 times.

The reduction from the r.f. bandwidth of B to the I.F. bandwidth of B_I implies that the total signal-to-noise ratio in the I.F. amplifier exceeds the total signal-to-noise ratio at the mixer input by $10 \log_{10}(B/B_1)$ db. If the signal-to-noise ratio at the threshold point has the same value in the I.F. amplifier as it would have in a conventional receiver, it follows that the threshold signal-to-noise ratio has been reduced by $10 \log_{10}(B/B_1)$ db. For the 240 channel system this works out to be $10 \log_{10}(24/6)$, i.e. 6 db. This estimate is an optimistic one and the practical improvement lies between 4 and 5 db.

An alternative technique for threshold reduction is indicated in Fig. C.3. A tracking filter is inserted before the mixer and a feedback loop is

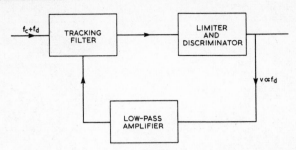

Fig. C.3. Low-threshold F.M. receiver using a tracking filter. The centre frequency of the tracking filter is controlled by v and is maintained at f_c+f_d.

used to adjust the centre frequency of this filter to the value of the instantaneous frequency of the input. The filter bandwidth may be restricted to a value comparable to that used in the I.F. amplifier of the feedback receiver with a corresponding increase in total signal-to-noise ratio. A detailed analysis shows that the reduction in the threshold value is similar to that obtained with the feedback receiver.

Bibliography

This bibliography provides references from which readers may continue their studies of the topics in the present book or enlarge their knowledge of communications practice. The items listed have been selected as of appropriate level and no attempt has been made to provide a comprehensive guide to the very wide range of literature covering the field of communications.

SCHWARTZ, M. 'Information Transmission, Modulation and Noise' McGraw-Hill, New York, second edition, 1972.
A more mathematical treatment of much of the material in this volume and a useful introduction to more advanced study.

SCHWARTZ, M., BENNET W. R. and STEIN, S. 'Communications Systems and techniques' McGraw-Hill, New York, 1966.

HILLS, M.T. and EVANS, B. G. 'Telecommunications Systems Design. Vol. 1. Transmission Systems' Allen and Unwin, London, 1973.
The above two referemces cover the technology of telecommunication systems.

TERMAN, F. E. 'Electronic and Radio Engineering' McGraw-Hill, New York, fourth edition, 1955.
A classic text-book first written in 1932 but frequently updated. Contains a wealth of detail on practical aspects of communications.

CATTERMOLE, K. W. 'Principles of Pulse Code Modulation' Iliffe, London, 1969.
A comprehensive account of the principles of PCM with some detail on circuitry.

ANGELAKOS, D. J. and EVERHART, T. E. 'Microwave Communications' McGraw-Hill, New York, 1968.
Describes the technology and principles of typical microwave systems including radar.

SKOLNIK, M. I. 'Introduction to Radar Systems' McGraw-Hill, New York, 1962.
Covers the technology of modern radar.

MATHEWS, P. A. 'Radio Wave Propagation' Chapman and Hall, London, 1965.

BURROWS, W. G. 'VHF Radio Wave Propagation in the Troposphere' Intertext, London, 1968.

KING, R. A. 'Electrical Noise' Chapman and Hall, London, 1966.

Developments in telecommunications occur with increasing rapidity. Excellent sources of review articles covering such developments are IEE Reviews (special articles in the *Proceedings of the Institution of Electrical Engineers*) and *Spectrum* (published monthly by the Institute of Electrical and Electronic Engineers, New York). Examples of such articles are:

REDMOND, J. 'Television broadcasting, 1960-70: BBC 625-line services and the introduction of colour' *Proc. I.E.E.*, vol. 117, p. 1469, 1970.

DAVIES, D. H., FIELDING, C. C. and GIRLING, F.E.J., 'Radar' *Proc. I.E.E.*, vol. 118, p. 1071, 1971.

BACK, R. E. G., WILKINSON, D. and WITHERS, D. J. 'Commercial satellite communication' *Proc. I.E.E.*, vol. 119, p. 929, 1972.

BROWN, W. M. and PORCELLO, L. J. 'An introduction to synthetic aperture radar' *Spectrum*, p. 52, Sept. 1969.

FORNEY, G. D. 'Coding and its application in the space communication' *Spectrum*, p. 47, June 1970.

SCHINDLER, H. R. 'Delta modulation' *Spectrum*, p. 69, Oct. 1970.

HERSCH, P. 'Data communications' *Spectrum*, p. 47, Feb. 1971.

McKENZIE, A. A. 'What's up in satellites' *Spectrum*, p. 16, May 1972.

Index